Studies in History and Philosophy of Science

Volume 47

Series Editor
Stephen Gaukroger, University of Sydney, Australia

Advisory Board
Rachel Ankeny, University of Adelaide, Australia
Peter Anstey, University of Sydney, Sydney, Australia
Steven French, University of Leeds, UK
Ofer Gal, University of Sydney, Australia
Clemency Montelle, University of Canterbury, New Zealand
Nicholas Rasmussen, University of New South Wales, Australia
John Schuster, University of Sydney/Campion College, Australia
Koen Vermeir, Centre National de la Recherche Scientifique, Paris, France
Richard Yeo, Griffith University, Australia

More information about this series at http://www.springer.com/series/5671

Julián Simón Calero
Editor

Jean Le Rond D'Alembert: A New Theory of the Resistance of Fluids

Springer

Editor
Julián Simón Calero
Department of Logic, Philosophy
and History of Science
UNED
Madrid, Spain

Translators
Verónica H. A. Watson
INTA
Madrid, Spain

Larrie D. Ferreiro
George Mason University
Fairfax, VA, USA

ISSN 0929-6425   ISSN 2215-1958   (electronic)
Studies in History and Philosophy of Science
ISBN 978-3-319-88529-2   ISBN 978-3-319-68000-2   (eBook)
https://doi.org/10.1007/978-3-319-68000-2

© Springer International Publishing AG 2018
Softcover re-print of the Hardcover 1st edition 2018
This work is subject to copyright. All rights are reserved by the Publisher, whether the whole or part of the material is concerned, specifically the rights of translation, reprinting, reuse of illustrations, recitation, broadcasting, reproduction on microfilms or in any other physical way, and transmission or information storage and retrieval, electronic adaptation, computer software, or by similar or dissimilar methodology now known or hereafter developed.
The use of general descriptive names, registered names, trademarks, service marks, etc. in this publication does not imply, even in the absence of a specific statement, that such names are exempt from the relevant protective laws and regulations and therefore free for general use.
The publisher, the authors and the editors are safe to assume that the advice and information in this book are believed to be true and accurate at the date of publication. Neither the publisher nor the authors or the editors give a warranty, express or implied, with respect to the material contained herein or for any errors or omissions that may have been made. The publisher remains neutral with regard to jurisdictional claims in published maps and institutional affiliations.

Printed on acid-free paper

This Springer imprint is published by Springer Nature
The registered company is Springer International Publishing AG
The registered company address is: Gewerbestrasse 11, 6330 Cham, Switzerland

Abstract

It is recognised that the *New Theory of the Resistance of Fluids* was a turning point in the genesis of the Fluid Mechanics, for the first time the motion of a fluid was reduced to two differential equations in partial derivatives relating the velocities and positions. Therefore, the physical problem of the fluid motion was converted into a mathematical one.

However, the rapid evolution of the discipline, jointly with the difficulties inherent in the work overshadowed d'Alembert's contributions, so that the studies have been carried out up to now are only partial ones and in some cases his findings have been undervalued. We think that the book deserves an in-depth study in order to bring both merits and defects to light.

Firstly, we present a translation into English of d'Alembert's book, in which we have tried to follow his original words, even when sometimes it results a bit literal and unidiomatic.

The subsequent commentaries are divided in two parts. One is dedicated to the circumstances in which this book was written and the state of the Fluid Dynamics at that time. D'Alembert's ideas on basic concepts, such as forces, pressures, fluids and dynamics are also included. Finally, a brief analysis of the *New Theory* is made in order to make the next stage easier.

The second part is dedicated to the *New Theory*. In order to highlight the genuine contribution of d'Alembert to Fluid Mechanics and to extract his core contribution, we have preferred to analyse the book by subjects, giving preference to those related with the resistance over those other with little, if any, relation with it. The reorganization of the articles and the use of his previous Latin *Memoire*, that preceded the *New Theory*, contribute to mitigate the negative opinions about the lengthiness and tortuosity that sometimes have been attributed to it. We have made an effort to show how he manages to find the differential equations ruling the velocities around the body and the streamlines pattern. Also how he contrives to join the pressure forces over the body with its loss of momentum, which turns out to be the resistance.

Contents

Part I A New Theory of the Resistance of Fluid

1 Principles of Dynamics and Hydrodynamics Necessary for the Understanding of the Subsequent Propositions.......... 19

2 General Principles of the Equilibrium of Fluids............... 25

3 General Principles of the Pressure Fluids, in Motion or at Rest... 29

4 On the Pressure That a Fluid Exerts on a Body at Rest and Immersed in It...................................... 39
    4.1 Observations Necessary for the Understanding of the Subsequent Propositions......................... 39
    4.2 On the Fluid Pressure at the First Instant of the Impulse...... 52
    4.3 Method for Determining the Fluid Velocity at Any Point..... 53
    4.4 On the Pressure of the Fluid at Each Moment............. 57
    4.5 The Resistance of a Plane Figure........................ 61
    4.6 Notes on Our Solution to the Problem of Fluid Pressure...... 61
    4.7 Reflections on the Experiments That Have Been Made or That Can Be Made to Determine the Pressure of the Fluids........ 62

5 On the Resistance of Fluids to the Bodies Moving Therein........ 67
    5.1 General Observations on the Various Classes of Fluids....... 67
    5.2 The Resistance of Non-elastic and Indefinite Fluids.......... 68
    5.3 On the Use of Pendulum Experiments to Determine the Resistance of Fluids Whose Velocity Is Very Small....... 77
    5.4 Examination of an Hypothesis Which Would Lead to Strange Paradoxes on the Resistance of Fluids.................... 80
    5.5 About the Resistance of Non-elastic and Finite Fluids........ 83

	5.6 On the Resistance of Elastic Fluids	85
	5.7 Principles Necessary for Determining the Pressure of an Elastic Fluid	88
6	**Oscillations of a Body Floating in a Fluid**	**91**
	6.1 Rectilinear Oscillations	91
	6.2 Curvilinear Oscillations	92
	6.3 Oscillations of a Body of Irregular Shape	97
7	**On the Action of a Fluid Stream That Exits from a Vessel and Strikes a Plane**	**103**
8	**Application of the Principles Outlined in This Essay in the Research of the Motion of a Fluid in a Vessel**	**113**
9	**Application of the Same Principles to Some Research on Streams in Rivers**	**117**
10	**Appendix**	**105**
	10.1 Reflections on the Laws of the Equilibrium of Fluid	121

Part II Introduction to the *Essay*

11	**General Considerations**	**135**
	11.1 The Manuscript and the Essay	135
	11.2 Fluid Mechanics in the Eighteenth Century	139
12	**D'Alembert's Dynamic Conceptions**	**147**
13	**Forces and Fluids in the *Essay***	**153**
	13.1 Forces and Pressures	153
	13.2 Conception of *Fluids*	156
	13.3 Experience and Experiments	158
14	**Brief Analysis of the Contents of the *Essay***	**161**
	14.1 The Preliminaries	162
	14.2 Body in a Fluid Stream	163
	14.3 Fluid Impulsive Velocity	164
	14.4 Body Moving in Fluid at Rest	165

Part III Analysis of the *Essay*

15	**The *Essay's* Introduction**	**169**
16	**The Preliminaries**	**173**
	16.1 Principles of Dynamics	173
	16.2 General Principles of the Fluid Equilibrium	176
	16.3 Pressure on Submerged Bodies	179
	16.4 Motion in Tubes	183

17	**Resistance of a Body Moving in a Fluid**	187
	17.1 Fluid in Motion	187
	17.2 Impulsive Motion	208
	17.3 Body in Motion	216
18	**Other Resistances and Fluids**	221
	18.1 Friction and Viscosity	221
	18.2 Resistance in Non-elastic and Finite Fluids	223
	18.3 Resistance of Elastic Fluids	226
19	**Experiments and Theories**	231
	19.1 Reflections on Experiments	231
	19.2 Experiment with Pendulums	237
20	**Other Motions**	243
	20.1 Motion in a Vessel	243
	20.2 Streams in Rivers	246
	20.3 Jet Stream Against a Plate	247
	20.4 Hypothesis of the Plane Sections	252
21	**The Oscillation of Floating Bodies**	257
	21.1 Rectilinear Oscillations	258
	21.2 Curvilinear Oscillations	259
	21.3 Irregular Bodies	261
22	**Reflections on Fluid Equilibrium**	265
	Annexes	273
	Notes on the Translation and Manuscript	283
	Bibliography	287
	Index	291

Part I
A New Theory of the Resistance of Fluid

Translated by
Verónica H. A. Watson, Larrie D. Ferreiro

Julián Simón Calero

By M. D'ALEMBERT, of the Académie Royale de Sciences of Paris, of the Prussia, and of the Royal Society of London.

IN PARIS
Chez DAVID the elder, Libraire, rue S. Jacques, à la Plume d'or.
MDCCLII
WITH THE KING APPROBATION AND PRIVILEGE

To Monseigner the Marquis D'Argenson Minister of State

Sir,

The famous scholars and writers who approach you in such great numbers would appreciate the honour that I pay you. The respect that they show is all the more sincere and all the more justifiable than you would ever have demanded. You are owed such true and flattering feelings because of the ease and humility with which you welcome talent and this ease alone makes the society of the Great and the men of Letters each worthy of the other. Your activity is both useful and pleasant by the extent of knowledge which gives you the approval of the most enlightened sections of our nation and is forever a continuous example, to all who surround you, of modesty, honesty, public spiritedness and all the other virtues beloved and valued in our times. Essentially recognised as a philosopher through your feelings and actions, this seldom-found conduct is joined by an even rarer one, which encompasses so many more, that of wearing your virtues with a total lack of ostentation.

May your example, your Eminence and that of your illustrious House teach the majority of our patrons, greater in number now the better to promote the good and the glory of literature, that the true way to honour the worth of the candidate is to honour oneself by the manner in which you single him out and honour him.

I am Sir, your most humble and most obedient servant, d'Alembert.[1]

---

[1] We thank Christina Demitriou for the translation of this letter.

Mr Nicole and Mr le Monnier who have been named to examine a work of Mr d'Alembert, which is titled *Essay of a new Theory of the resistance of Fluids*, having made their report, the Academy has judged this Work worthy of publishing. In faith of which I have signed the present Certificate. At Paris the 22nd of December.

Grand-Jean de Fouchy, Life Secretary of the Royal Academy of Sciences.

Introduction

[I][2]

Although the physics of the Ancients was neither as unreasonable nor as limited as some modern philosophers think or say, it seems that they were not versed in the so called physical-mathematical sciences, which consist in the application of the calculus to natural phenomena. The matter that I undertake to treat in this book is one of those which seems to have been the least studied from this point of view. I say from this point of view, because the knowledge of the resistance of fluids is an absolute necessity for the construction of ships that the Ancients may have pushed further than us, they could not have been lacking in knowledge up to a point: it is more than likely that experience had provided early on a few rules to determine the shock and pressure of the water. But these rules, limited without doubt regarding their practice, and, so to speak, purely traditional, have not come down to us.

With regards to the theory of this resistance, it is not surprising that the Ancients had ignored it. If we may so express it we must take into account even their ignorance for not to having wanted to achieve what was impossible for them to know and not to have tried to make us believe that they had succeeded. The most subtle geometry is permitted in order to try out this theory; the geometry of the Ancients, while also very deep and very wise, could not go as far. It is rather the semblance of what they felt, as their method of philosophizing was wiser than we commonly imagine. In this respect the modern geometricians knew how to provide themselves with more help, not because they were superior to the Ancients, but because they came later. The invention of differential and integral calculus has put us in position to follow by any means the movement of bodies even in their elements or last particles.

It is only with the help of these calculi that we are allowed to penetrate the fluids and to discover the play of their parts, the action that these innumerable atoms of which a fluid is composed exert on each other, and which at the same time appears

---

[2]The Introduction has been divided in seven sections to make easier to read.

to be both united and divided, dependent and independent of each other. Also the internal mechanism of the fluids, so dissimilar to that of solid bodies that we touch, and subject to completely different laws, ought to be a particular object of admiration for philosophers, if the study of nature, of the most simple phenomena, and even of the very elements of matter, had not accustomed them to be unsurprised at anything, or rather to be equally amazed at everything. As unenlightened as the common people were about the first principles of all things, they neither had, nor could have had, no more advantage than in the combination that they make of these principles and the consequences they draw from them, and it is in this kind of analysis that mathematics are useful to them . However, even with this help, the resistance of fluids still enclosed such considerable difficulties, that the efforts of the greatest men have been limited up to here to giving us a slight sketch.

After reflecting for a long time on this important matter with all the attention I am capable of, it seemed to me that the slight progress made so far comes from the fact that the true principles have not yet been grasped, and the matter should be treated according to these principles. So I thought it my duty to apply myself to seek them and how to apply the calculus to them, if this was possible. As these two objects must not be confused and maybe modern geometricians have not paid enough attention to this point. Often the desire to make use of the calculus is what determines the choice of the principles; instead they should first examine the principles in themselves, without thinking ahead to bend them by force to calculus. Geometry, which must obey Physics only when it meets with it, sometimes commands it. Should it occur that the question which one wishes to examine is too complicated for all the elements to enter into the analytical comparison that one wants to make, the most inconvenient are separated and substituted by others less annoying, but also less real, and, despite hard work, it is surprising to reach to a result contradictory to nature; as if after having disguised, truncated or altered it, a purely mechanical combination could give it to us.

I have proposed to myself to avoid this problem in the work I offer today. I have searched the principles of the fluid resistance, as if the analysis should not enter therein for anything, and I have tried to apply the analysis to these principles once found. However, before I expound my work and the degree to which I have pushed it, it will not come amiss to explain what has been done on this issue so far.

[II]

Newton, to whom the Physics and Geometry are so indebted, is the first, that I know of, who has undertaken to determine by the principles of Mechanics the resistance that a body moved in a fluid undergoes and to confirm his theory by experiments.[3] In order to reach the solution of such a thorny issue more easily and perhaps to present it in a more general way, this great philosopher envisages the fluid from two

---

[3] Cf. Book 2, Section VII, "The motion of fluid, and the resistance made to projected bodies".

different points of view. At first, he looks at it as a cluster of elastic particles, which tend to move away from each other by a centrifugal or repulsive force, and that are placed freely at equal distances. He also assumes that this cluster of particles composing the resisting medium has very little density respect to the body, so that the parts of the fluid pushed by the body can move freely without communicating the motion they have received to the neighboring parts. According to this hypothesis, Newton finds and proves the laws of the resistance for such a fluid; laws fairly well known, so that we may be dispensed from reporting them here. From these laws it follows that the resistance of a cylinder in such a fluid, namely the force that retards its motion at every moment, is equal to the weight of a cylinder of fluid with the same base, and whose height would be twice that from which a heavy body should fall in order to acquire the same velocity as that at which the cylinder moves. Newton makes it clear that, in this same assumption, the resistance of a globe would be half that undergone by the circumscribed cylinder.

In his book titled *Discourse on the laws of the communication of the Motion*, the famous Jean Bernoulli determined the resistance of fluids from the same hypothesis and he represents this resistance by a fairly simple formula. In one of my works I gave the demonstration that Bernoulli had removed;[4] a demonstration in which I apply any figure and any arrangement that can be supposed in the parts of the fluid over the generality.

But it must be confessed that this formula is insufficient to determine the resistance we seek. In all fluids which are known to us the particles are immediately adjacent by some of their points, or at least they act upon each other almost as if they were. Thus any body moved in a fluid necessarily pushes at one and the same instant a large number of particles placed in the same row, and where each one receives a velocity and a different direction according to its position. It is therefore extremely difficult to determine the motion communicated to all these particles and consequently the motion that the body loses at every instant.

These reflections had not escaped M. Newton. He recognized that his theory of the resistance of a fluid composed of elastic scattered globules, if it may be expressed thus, cannot be applied either to dense and continuous fluids where the particles immediately touch themselves, such as water, oil, quicksilver; or to fluids whose elasticity comes from a cause other than the centrifugal force of their parts, for example for the compression and the expansion of these parts, as the air we breathe appears to be. He also recognizes that in the same case wherein the fluid was such as he has imagined, it should be also assumed that the velocity of the moving body is large enough for the centrifugal forces of the parts of the fluid to have no time to act, and thus to alter by this action the resistance coming from the single force of inertia. From this it follows that this first part of Newton's theory, and that of M. Bernoulli is but a commentary of the former, are rather a research due to pure curiosity as they are not applicable to nature.

---

[4] The *Traité de l'équilibre et du mouvement des fluides.*

Also the illustrious English Philosopher has not thought it necessary to maintain that. He considers the fluids in the compression state where they really are, as being composed of particles adjacent to each other, and this is the second point of view from which he envisages them. The method he employs in this new hypothesis for solving the problem is to find first the velocity of a fluid stream that escapes from a cylindrical vessel through a horizontal hole made at the bottom of the vessel, and the pressure that a circular flat surface, exposed to the action of this current, would suffer. For determining this pressure M. Newton employs a kind of approximation and testing that would be difficult to give our readers the idea of here. We will be satisfied by observing that this pressure depends on the height of the fluid, or what amounts to the same thing, on the velocity with which it escapes, the diameter of the hole, and that of the circular flat surface. By next increasing the capacity of the vessel to infinity, but keeping the diameter of the hole, and substituting the movement of the circular surface for that of the fluid, M. Newton discovers that the resistance undergone by this surface is equal to the weight of a cylinder which has the surface as its base and for its height half of one from where a heavy body should fall to acquire an equal velocity to the actual velocity of the circular surface. According to M. Newton the weight of such cylinder can then represent the resistance that a solid cylinder of any length undergoes at each moment, having as base the previous circular area in question; because whatever length of the cylinder is, the base is the only part exposed to the shock of the fluid. Finally, using an argument which will be discussed below, M. Newton equals the resistance of a globe to that of the circumscribed cylinder. He also reaches the conclusion that a dense, continuous and compressed fluid, such as it really is in nature, produces a resistance four times lower than a cylindrical body and all things being equal, twice less than a spherical body, as in the case of the fluid with elastic globules of the first hypothesis.

But this second theory of M. Newton, although more in agreement with the nature of the fluids, is still subject to many difficulties. In first place, it has as basis the method by which this great geometrician determines the motion of a fluid escaping from a cylindrical vessel; certainly a very ingenious method, but insufficient and faulty. The cataract that M. Newton assumed to be formed by the falling of the fluid cannot exist, as M. Jean Bernoulli has made clear in his *Hydraulica*,[5] because the fluid, that is supposed to flow into the cataract and to fall with all the force of its weight without exerting any lateral pressure, cannot resist the pressure of stagnant fluid surrounding it.[6]

In second place, if we refer several experiments made by the skillful physicists, the pressure of a fluid in motion upon a circular surface is equal to the weight of a

---

[5]Cf. §LX, Scholium V. In *Opera Omnia*, Vol. 4, p. 483.

[6]The text is not clear. Following the *Mss*.2, "nullam in cataractae parietibus esse pressionem, qua susteneri possit pressio fluidi extra cataractam stagnantis; proinde cataractam ob fluidi ambientis pressionem omnino debere":. "There is not any pressure upon the cataract walls which can support the pressure of the stagnant fluid outside of the cataract. Therefore, the cataract must be destroyed completely due the ambient fluid pressure."

cylinder whose height would be equal to what a heavy body should fall from in order to acquire the same velocity of the fluid; from which it follows that this pressure is double that determined by M. Newton using calculus.

In third place, by this new theory of the pressure in continuous fluids Newton finds that the resistance undergone by a globe is equal to what a circumscribed cylinder would undergo; while by his theory of the resistance of non-continuous fluid he finds that the resistance of the globe is only half that of the cylinder. Let us discover here what Newton relies on to establish the equality of resistance between the globe and the cylinder in the second case. According to him, if a cylinder, a sphere and a spheroid whose widths or bases are equal, are placed in the middle of a cylindrical channel so that the axes of these body coincide with that of the channel, these bodies will oppose an equal obstacle to the movement of the water in the channel, because the spaces through which the fluid flows between the cylindrical channel and each body are equal among them, and the fluid must move in the same way in equal spaces. This is the only proof that M. Newton gives of this fundamental proposition, a proof which does not seem very strong

Because the space between the cylinder and each of the three bodies is only the same in the plane where the greatest width or common base of the bodies is located; in any other plane parallel to this the space between the cylinder and each of the bodies is different, and therefore the fluid would be unable to move there in the same way.

Moreover, even when the fluid moved with the same velocity in these different spaces, it does not follow that these bodies underwent an equal pressure. Because for example the water flowing between the channel and the cylinder presses the cylinder so that it acts on its sides in lines perpendicular to the axis of the cylinder; whereby it follows that the [pressure on the] cylinder walls destroy each other, and that the true pressure supported by the cylinder comes only from the action of the fluid that strikes the frontward base, and which is not widespread through the empty space between the cylinder and the channel. Instead, the fluid flowing between the channel walls and the surface of the sphere acts on the sphere surface along lines perpendicular to its surface and therefore located obliquely respect to the axis of the sphere, whereby is easy to conclude that the forces acting on each side of the axis are not destroyed completely as in the case of the cylinder, but are partially destroyed and partially contribute to form a single and unique pressure. This is much greater as the direction of the primitive forces forms a more acute angle with the axis of the sphere. Nothing is then less proved than this alleged equality of resistance of the globe and of the circumscribed cylinder.

Finally, M. Newton supposes that the parts of the fluid, which by their oblique and unnecessary motions can delay the movement of the fluid in the channel, must be regarded as frozen and at rest, and as adhering to the front and rear surface of the body; a hypothesis that is probably true to some extent, but it presented in so vague way, that it seems intended to rather circumvent the difficulty of the problem more than overcome it.

Despite these observations, we have nevertheless to admire the efforts and the wisdom of this great philosopher, who, after finding the truth so happily in a large number of other issues, dared to pave the first road to solve a problem that no one before him had ever attempted. Thus this solution, though not very exact, shines throughout with inventive genius, this mind fertile in resources that nobody has possessed in a higher degree than him.

[III]

Helped by the assistance that geometry and mechanics provide us today in greater abundance, is it any wonder that we take a few more steps in the difficult and extensive race that he has opened to us? Even the errors of great men are instructive, not only by the views that they provide for ordinary persons, but for the useless steps they save us. The methods that have misled them, seductive enough to dazzle them, would have deceived us like them; it was necessary that they be tempted by them, so that we could know the pitfalls. The difficulty is to imagine another method, but often this difficulty consists more in choosing what we will follow, than following once it is well chosen. Among the different routes that lead to a truth, some have an easy entry, these are the ones we throw ourselves upon at first; and if no obstacles are encountered until a certain path has been traversed, then, as one only admits with pain to have made endeavored in vain, then a way to avoid these obstacles is sought when it not possible to overcome them. Other roads, on the contrary, do not present obstacles to their entry; their access may be painful, but once these obstacles are overcome, the rest of the way is easy to traverse.

It must be admitted to the rest, that the majority of the geometrician who have attacked Newton concerning the resistance of fluids have not been more fortunate than him; almost all have given us a lot of calculations instead of true principles. However M. Daniel Bernoulli, who has united much light and philosophical spirit to great sagacity in geometry, should be exempted. As he is the one who has gone most deeply into this matter, he also seems to be the one to have the best knowledge about the difficulties that it entails. In the second volume of the Memoirs of Petersburg (year 1727) he proposes a formula for the resistance of fluids, whose principles are different from those of M. Newton, but he does not seem to have been highly satisfied, because he admits that this formula gives the resistance as being four times that resulting from the experiments. The illustrious author then seeks by ordinary methods the ratio of resistance of a fluid formed by any spheroids, and after using these methods he establishes that the resistance of the globe is half that of the cylinder; proposal he fought later in his *Hydrodynamica*. Indeed, the hypothesis upon which this rests is not very accurate, because it is necessary to assume that the parts of the fluid, when they strike the cylinder or globe, are either annihilated or at least rebound in such a way that they do not encounter any other particle. This hypothesis and some others, whose insufficiency is easy to appreciate, are the expressed or implied basis of almost all works published so far on the resistance of fluids, and consequently leave much to be desired in these works.

In 1741, in Volume VIII of the same Memoirs of Petersburg, the great mathematician of whom we are speaking gave a very ingenious and much more direct method to determine the pressure that a fluid stream escaping from a vessel exerts against a plate. Although it is supported by experiments the formula he proposes for this does not seem to beyond doubt, as we hope to show in one of the chapters of this book. The details of this review are too geometrical for us to give the idea in this Introduction.

Anyway, M. Daniel Bernoulli accepts that this theory of the pressure of a fluid stream against a plate would not be very useful for determining the pressure of a plate fully submersed in a fluid, because the motion of the fluid particles is very different in the two cases. Indeed, in the case where the stream strikes the plane, as soon as the particles of the fluid arrive at the plane they change direction, so that they move parallel to the plane, and they slide along the plate following the latest [plane] direction; which does not occur when the plane is fully immersed in a deep fluid. As soon as the fluid particles leave the front surface of the plane on which they have slipped, they are pushed and returned to the rear surface by the moving fluid which surrounded them at right and left; so that their direction, from being parallel to the plate, becomes perpendicular, or at least makes a very large acute angle with this plate, as everyday experience shows. Now then this reflux of particles and the pressure that may result on the posterior surface must alter the pressure that the front surface undergoes.

[IV]

It follows from all we have said so far, that the theory of fluid resistance, although handled by so many great mathematicians, is still very imperfect in its actual elements. These reasons have led me to address this matter in a completely new way, without borrowing anything from those who have preceded me in the same work. The theory that I present in this book, or rather I will give the principles, has, it seems to me, the advantage of not resting on any arbitrary assumption; I assume only what no one can deny, that a fluid is a body composed of very small particles, separate and able to move freely.

The resistance that a body undergoes when it impacts with another is just, strictly speaking, the quantity of motion it loses. When the motion of a body is altered, we can consider this motion as being composed of the one the body would have in the following instant and of another which is destroyed. From this it is not difficult to conclude that all the laws of the communication of motion among the bodies are reduced to the laws of equilibrium. It is also to this principle to which I have reduced the solution of all problems of dynamics in the first book I published in 1743.[7] I have had frequent occasions to show its fertility and simplicity in the various treaties that I have published since, and perhaps it would be useful to enlighten us to some extent on the very obscure metaphysics of the percussion of

---

[7] *Traité de dynamique.*

bodies and on the laws to which the resistance is subjected. Whatever it is, this principle applies naturally to the resistance of a body in a fluid; and also to the laws of equilibrium between the fluid and the body, to which I reduce the investigation of this resistance. But one must not imagine that this research, although facilitated by this means, is as simple as that of the communication of motion between two solids bodies. Indeed let suppose that we had the advantage, of which we are deprived, of knowing the figure and the mutual arrangement of the particles that compose the fluid: the laws of their resistance and their action will surely be reduced to the known laws of motion; because the investigation of the motion communicated from a body to any number of surrounding corpuscles is only a dynamic problem, for whose solution we have all the mechanical principles that we could wish. However, the greater the number of particles is, the more difficult becomes to apply the calculus to the particles[8] in a simple and convenient way; therefore such a method would be scarcely practicable in the search for the resistance of fluids. But we are rather far from having all the *data* needed to be able to use this method. We ignore not only the figure and the arrangement of the parts of the fluids; we even ignore how these parts are pushed by the body and how these parts move among themselves. Anyway there is such a large difference between a fluid and an aggregate of solid corpuscles, that the laws of pressure and equilibrium of the fluids are very different from the laws of pressure and equilibrium of the solid. Experience alone has taught us in detail the laws of hydrostatics, what the most subtle theory never have made us suspect; and nowadays although experience has unveiled these laws, no one has yet been able to find a satisfactory hypothesis for explaining and reducing them to the known principles of the statics.

However, this ignorance did not prevent great progress made in the hydrostatics. Since the philosophers cannot deduct the laws of their equilibrium immediately and directly from the nature of the fluids at least they have reduced them to a single principle of experiment, *the equality of the pressure in all directions*; principle that has been taken (lacking a better one) as the fundamental property of fluids, and as the one to which it was necessary to refer all others. Indeed, condemned as we are to ignore the first properties and internal contexture of the bodies, the only resource left to our shrewdness is to try at least to capture in each subject the analogy of the phenomena and to recall all them in a small number of primitive and basic facts. Thus, Newton, without assigning the cause of the universal gravitation, did not leave to prove that the system of the world is only rested on the laws of this gravitation. Nature is an immense machine whose main springs are hidden to us; we do not see this machine except through a veil which conceals from us the interplay of the most delicate parts. Among the most striking, and maybe if we dare say it, the coarsest parts that this veil allows us to glimpse or discover, there are several that the same spring sets in motion, and that's mainly what we must seek to unravel.

---

[8] In the original says "aux principes"; we think it is a misprint for "aux particules".

Therefore we cannot flatter ourselves even to deduce the nature of the theory of fluid resistance and action, let us confine ourselves to deduce it, if possible, from the hydrostatic laws that now are well founded, and on which several leading geometricians, whom I mentioned in my *Traité des Fluides*, have worked on successfully. The purely experimental knowledge of these laws compensates the figure and the arrangement of the parts of the fluids, and perhaps makes the problem easier to solve than if we were confined to this former knowledge.

[V]

Therefore I commence this book by showing how the laws of the resistance of fluids depend on the laws of their equilibrium, from where quite general and, it seems to me, new and useful theorems result on the motion of a system of bodies or corpuscles which act one upon another. Next I expound in relatively few words the already known theory of the fluid equilibrium, and I make several comments on this theory which could be considered of some importance.

From the above the laws of the pressure of a fluid either in motion or at rest are derived in a rather simple way.

This research led me to that of the pressure of a fluid that strikes a body at rest. I see first that the question is reduced to finding out the pressure of the thread of the fluid that slides immediately over the surface of the body. For this it is necessary to know the velocity of the particles of this thread, which I determine by two different methods, which perhaps the geometricians may not find unworthy of their attention. Once this velocity is found, the fluid pressure is necessarily deduced, but the formula of this pressure requires a very complicated analysis of which I show the principles.

Next I come to the laws of the resistance of a fluid when the body is moved and the fluid is at rest; and by a new and unique method I demonstrate that the pressure of a fluid moving with a variable velocity against a body at rest is equal to the resistance that this body would undergo in the fluid at rest when moved with similar velocity. A proposition hitherto assumed as true by all hydrodynamic authors, but whose rigorous proof is however quite difficult, as I flatter myself that my readers will be convinced.

To render my theory more general, I give formulas for the resistance of the fluid taking into account the weight, the friction and the viscosity of the particles. I specially seek the laws of resistance in the case when there is a void between the fluid and the rear part of the body, which as I prove can occur, even when the fluid is not elastic. But I must admit that here the calculation throws very little true light, and that perhaps it may be very difficult to submit the case in question to the experiment itself.

Having thus developed my principles, I examine a hypothesis which several hydrodynamic authors have used so far and I wish to make clear that if such a hypothesis was followed in order to determine the resistance of a fluid, the resistance would be nil, which is contrary to all experience.

I deal next with the action of a fluid stream coming out from a vessel and that strikes a plane, and I find that the pressure found is a little lower than the weight of a cylinder that would have for its base the width of the stream and for the height twice that of the fluid in the vessel; a result that agrees perfectly with the exact and numerous experiments made by the Academy of Petersburg. Finally, I join to all these researches thoughts on the resistance of elastic fluids, a matter which had been barely touched until now, and on which I try to give some principles; but according to all appearances, the resistance will be never well known by the theory alone.

These are the main objects of this book. To make my principles even more worthy of the attention of physicists and geometricians, I thought it would be appropriate to pay attention to how they can be applied to different issues that have a more or less immediate relation to the matter I am dealing with; such as the movement of a fluid flowing, be it in a vessel or in any channel, the oscillations of a body floating in a fluid when the center of gravity of the submerged and non-submerged parts is not in the same vertical line, and other problems of this kind.

[VI]

Moreover, having proposed to myself to demonstrate everything in this book rigorously, I found in the proof itself of the simplest propositions more difficulties than could have been naturally suspected, .and it was not without difficulty that concerning this matter I managed to demonstrate the most generally accepted truths and it was not without pain that I managed to prove in this subject the truths more generally accepted and the least precisely proven so far. But after having sacrificed the easiness of calculation to the security of the principles, I should have naturally expected that the application of the calculation to these principles would be most painful, and this is also what happened to me. It seems to me even very likely that at least in some cases the solution will entirely resist analysis. It is up to savants to decide about this point; I thought I had worked very usefully if I had succeeded in such a difficult matter, either to set myself or to make others discover how far the theory can reach, and the limits where it should stop.

When I speak here of the limits to which the theory must be prescribed, I only envisage those that can be available with the current supports, not with those which could be obtained in the future and that are still to be found. Because in any matter whatsoever, one should not be too hasty in raising a wall of separation between nature and the human mind. For having learned to be wary of our industry, let us prevent us from distrusting in excess. In the impotence that we feel every day in overcoming as many obstacles as appear to us, we would be without doubt too content if we could at least judge at first glance until where our efforts can reach. But such is all the strength and weakness of our mind; it is often as dangerous to pronounce on what it cannot do, as on what it can. How many modern discoveries exist of which the Ancients had not even any idea? How many lost discoveries there are which challenged us too lightly? And how many others are reserved for our posterity which we deemed impossible?

I would have liked to compare my theory of the resistance of fluids with the experiments that several famous physicists have made in order ☐☐to determine it. But after reviewing these experiments, I found so little agreement among them, that there is not, it seems to me, any fact still perfectly found on this point. It does not take more to show how delicate these experiments are. Also some persons very skilled in experimental physics having undertaken recently to restart the experiments, have almost abandoned the project due to the difficulties in the execution. The multitude of forces, either active or passive, here becomes complicated to such a degree that it is somehow impossible to determine separately the effect of each one; and for example, to distinguish which comes from the force of inertia from that resulting from the viscosity, and these with the effect that can produce the weight and the friction of the particles. Moreover, when the effects of each of these forces and the law that they follow were unraveled in one single case, we would be well justified to conclude that in a case where the particles had acted quite differently, as much by their number as by their direction, their positions and velocities, would not the law of the effects be totally different? This subject could be in the number of those where the experiments made in small scale have almost no analogy with experiments made in large scale, and are sometimes even contradictory them; where each particular case requests, so to speak, an isolated experiment, and where therefore the overall results are still very faulty and very imperfect.

Finally, even when the experiment gives us the most exact and sharpest formulas on the resistance of fluids, it will still be very difficult to compare these formulas with those given by the theory. Because the calculation of the latter is extremely complicated if it is not supported on any arbitrary and vague hypothesis. But either this drawback to the analysis itself must be rejected, or it should be attributed to difficulties that other will surmount more happily than I, it seems to me that at least that no one can form any doubt on the truth of my principles. I even think to ensure that if after determining the formula of resistance by the long and painful method that these principles have forced me to use, this formula will be contradicted by the experiment; such a contradiction would came only, it seems to me, from some purely analytical assumptions that the application of geometry to physics necessarily entails. In this case it would be necessary, it seems to me, to renounce fully to any theory of fluid resistance and to consider it as one of those questions on which the calculation can have no hold.

[VII]

For the rest, the difficulties of the calculations which I have just spoken of did not seem so striking to the famous Royal Academy of Sciences and Belles Letters of Prussia, and this consideration alone will be sufficient to induce me to avoid a decisive tone, which is not in any way convenient to me. Having proposed for the price of 1750 the Theory of Fluid Resistance, this wise Company has judged it convenient to postpone this award and to exhort the authors to make clear with supplements the agreement of their calculations with the experiment; a condition about which, however, it had not mentioned at all in its program of 1748. It was natural to believe that then it asked simply for the true principles hitherto unknown, and whose research seemed to be the object of a satisfactory work. I thought to have discovered these principles and therefore I could compete for the awards. The piece that I sent to Berlin on this object in the month of December 1749 is, with a few additions, the book I give today. I contented myself in this piece in making seen the agreement of my principles with the best known facts of the fluid resistance: such are the ratio of the resistance with the square of speed; the changes that the fluid viscosity causes in this ratio, especially when speed is very small; the pressure of a fluid stream coming out from a vessel and that strikes a plate, the pressure determined, as I have said, by accurate experiments; and some other similar phenomena. The Academy, not having considered these researches as sufficient, demands nowadays the formulas of the resistance all calculated and which are agreement with experiments to be done. But feeling myself neither with wisdom enough, nor strength enough, nor courage enough to finish in a short time a work so delicate, so long and so painful, I thought I should refrain from competing again; other reasons, in whose detail it is useless to enter, confirmed me in this resolution. However, as it seemed to me that this *Essay* could be useful, I thought, for me to assure the possession of what it contains, that I have to bring it to light before the publication of the judgment of the Academy. I wish for the interest I take in the advancement of the Science, that the Judges appointed by this illustrious Company, which without any doubt have not proposed this question without assuring themselves if that solution was possible, would find something to entirely be satisfy themselves in the works that will be sent to them for the contest.

For me, who has felt that the difficulty of calculations would make perhaps impossible for me the comparison of theory and experiment that others could do with more success, I have confined myself, as I have said, to show the agreement of my principles with the more certain and best known facts; in all the rest I leave much to be done to those who work in the future on the same matter. Maybe my sincerity will be found very far from this pageantry of which one never renounces when exposing their works; but only to my work should be given the place it can have. I do not flatter myself for having pushed to perfection a theory that so many great men have barely begun. The title of essay that I give to this book responds exactly to the idea that I have of it, but I think to be at least on the true road; and without daring to prepare the way I can have made, I will applaud with pleasure the efforts of those who will be able to go further than me, because in the search for truth, the first duty is to be fair. I still think I should give to those who afterwards will go into this matter more deeply an advice notice which I start profiting from myself: not to erect too lightly the formulas of the algebra as truths or physical propositions. The spirit of calculus, that has hunted the spirit of system, rules perhaps quite too much in its turn. Because in every century there is a dominant flavor of philosophy; this flavor almost always drifts to some prejudices and the best philosophy is that which has the least of them in its train. It would be better, no doubt, that it was never subject to any particular tone. The different knowledge acquired and collected by scholars would be easier to join together and form a whole. But every science seems somehow to receive and shake successively the law from those who are most honored or most neglected, and philosophy takes, so to speak, the coloring of the minds where it finds itself. In a metaphysician it is usually all systematic, in a geometrician it is often all calculus; the method of the latter, generally speaking, is probably the safest, but must not abuse of it and believe that everything will be there reduced to it; otherwise we would not progress in transcendent geometry more than to be proportionally more limited by the truths of physics. We look like a man who had the sense of sight contrary to that of touch, or in which one of these senses would be not perfected more that at the expense of the other. As much can be achieved of utility in the application of geometry to physics, the more circumspect one must be in this application. It is to the simplicity of its object that geometry owes its certainty; as the object becomes more complex, confidence darkens and walks away. It is necessary to know when to stop on what is unknown, not to believe that the words *Theorem* and *Corollary* by some secret virtue make the essence of a demonstration and that writing at the end of a proposition what must be demons*trated* will show what it is not.

Chapter 1
Principles of Dynamics and Hydrodynamics Necessary for the Understanding of the Subsequent Propositions

Prop. I. Theorem

1. Let any system be composed of as many bodies as desired, which I designate as A, B, C, D, etc. and let us suppose that these bodies are impelled by any forces $\varphi, \psi, \pi, \upsilon$, etc., being A by the force φ, B by the force ψ, etc., and that at any instant these bodies move with velocities V, U, v, u, etc., being A with velocity V, B with velocity U, etc. It is easy to see that these bodies, if they were not impelled by the forces, φ, ψ, π, etc., and besides there was not any obstacle to their movement, they would retain in the next instant the velocities V, U, v, u with same direction. But because of the impelling forces and the mutual action that these bodies can exert ones upon another, we suppose that in the next instant their velocities are changed to V', U', v', u', etc. It is obvious that each one of the first velocities V, U, v, u can be considered as composed of the velocities V', V''; U', U''; v', v''; u', u''; thus at the start of the second instant, that I call dt, the body A actually tends to move with the velocities V', V'', φdt, the body B with velocities U', U'', ψdt, the body C with velocities v', v'', πdt, the body D with velocities u', u'', ωdt, etc. But (by the hypothesis) of these three velocities, which each body is impelled by, only one remains to each one; namely, the velocity V' for body A, the velocity U' for the body B, the velocity v' for the body C, the velocity u' for the body D. Therefore if the bodies A, B, C, D turned in to move with the single velocities V'', φdt; U'', ψdt; v'', πdt; u'', ωdt; there would not be any motion in the system; or what is the same, the system would be at rest or in equilibrium. At *rest* if the bodies are completely separated and loosed, not acting each upon others; in *equilibrium* if these bodies are linked or adjacent, in such a way they can exercise a mutual action one upon the other.

In the first case, the velocity V'' will be equal and directly contrary to φdt, and similarly U'' will be equal and directed contrary to ψdt, etc. In the second case, it will be sufficient for the balance, and therefore for the rest, that the forces $A \times V''$, $A \times \varphi dt$, $B \times V''$, $B \times \psi dt$, $C \times v''$, $C \times \pi dt$, $D \times u''$, $D \times \omega dt$, etc. destroy each other.

This principle is of a very general use in order to solve all issues of dynamics. It will be seen in this work how useful it is for determining the resistance of fluids.

Corollary I

2. Let the forces $\varphi, \psi, \pi, \omega$, etc. be equal to zero, it is obvious that if the bodies A, B, C, D etc. tended to move only with the velocities V', U'', V''', u'' etc, they would be in equilibrium among them. Hence it follows that the equilibrium would still remain if, retaining the same direction, they tended to move with velocities gV''', gU''', gv'', gu'' etc., where g is a coefficient or any number. Because the powers that are in equilibrium and remain, whatever change that they might undergo, provided that they retain the same direction and the same ratio among them.[1]

Corollary II

3. Having allways excluded the forces φ, Ψ, etc. regarding them as null, we assume that the velocities V, U, v, u, etc., with which the bodies A, B, C, D, move or tend to move in any instant, become by whatsoever cause gV, gU, gv, gu (g expressing any coefficient) and retaining the same direction; I say that the velocities, which would have been in the next instant $V', U'\ v', u'$, will be gV', gU', gv', gu' with the same direction that they would have had the velocities V', U', v', u'. To make the demonstration easier to understand, we take only two bodies A, B (Fig. 1.1) and being Aa, Bb the infinitely small spaces that these two bodies will describe in the instant dt with velocities V, U, and $a\alpha, b\beta$ the infinitely small spaces that they would describe in the next instant with velocities V', U'. Let Aa, Bb be extended until $aa' = Aa$, and $bb' = Bb$ and let the parallelograms $\alpha a', \beta \beta'$ be completed. It is obvious (*art.* 1) that the bodies A, B would be in equilibrium if they tended to traverse the small spaces $a\alpha', b\beta'$ in the instant dt. Indeed, these small spaces $a\alpha'$, $b\beta'$ represent the velocities V''', U''', because the velocities aa' and bb', that is to say V and U are composed of the velocities $a\alpha, \alpha a'$ and $b\beta, \beta\beta'$, and that the velocities V', U' are represented by $a\alpha$ and $b\beta$.

Now let us imagine that the bodies A, B move following Aa and Bb with velocities gV, gU; it is easy to see that in this case they trasverse the spaces Aa, Bb in an instant equal to dt/g; and in next instant dt/g they would tend to move along aa' and bb', that is to say, following $a\alpha, \alpha a', b\beta, \beta\beta'$. Now well (*hyp*) the bodies A, B, as they would tend to move $a\alpha'$ and $b\beta'$ in the next instant dt, are in equilibrium; so they will also be in equilibrium if they tend to describe the same space in the time dt/g (*art.* 2). Therefore the bodies A, B will describe actually in the second time dt/g the spaces $a\alpha, b\beta$, so the velocities V', U' will change to gV', gU' maintaining the same direction. Now what we have demonstrated here for a system of two bodies, obviously will demonstrated in the same way for as many bodies as one wishes. So etc.

---

[1] "Si potentiæ illæ omnes augeantur aut minuantur in ratione quavis data". *Mss.*8. "If all those powers were to increase or decrease in whatever ratio given."

Fig. 1.1

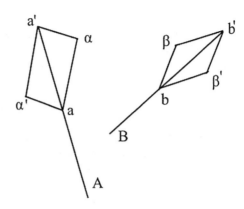

Corollary III

4. The demonstration would be the same if one or many of the velocities V, U, v, u, etc. were nil. Let, for example, the velocity U of the body B be equal to zero, and U' its velocity in the next instant, then it will be $Bb = 0$, $bb' = 0$ and the sides $b\beta$, $b\beta'$ of the parallelogram $\beta\beta'$ will be equal and placed in a straight line. So that the velocity U, which is assumed zero, may be regarded in this case as consisting of the equal and opposite velocities U', U''. That said, the demonstration remains the same, so if the body A tends to move in any instant with the velocituy gV, and the body B is at rest, in the next instant the body A will move with the velocity gV' and the body B with the velocity gU'.

Corollary IV and Fundamental

5. Let any system be with as many bodies as one likes A, B, C, D, etc. which are not impelled by any accelerative force, and which, in the beginning, are supposed at rest. When any velocity along any direction is impressed to only one of these bodies, for example the body A, I say that the bodies A, B, C, D will describe curves, certainly different from one another; but however each one in particular will be always the same, whatever the initial velocity imparted to the body A was, provided it was impressed in the same direction. Let α be the initial velocity impressed to the body A, which by the mutual action of the bodies A, B, C, D is changed in the first instant dt to V, and let U, v, u be the velocities that in the first time take the bodies B, C, D under this action. Suppose then that in the second instant of time dt, these velocities change into V', U', v', u'; it is clear by the previous Corollary that if the velocity impressed to the body A had been $g\alpha$ maintaining the same direction, the actual velocities of bodies A, B, C, D in the first instant dt/g would have been gV, gU, gv, gu, without changing direction. Therefore (*Corol.* 3) in the next instant dt/g these velocities will change to gV' gU' gv' gu' and they will have the same direction that the velocities V', U', v', u' would have had. Therefore, whether the initial velocity imparted to the body A is α, or $g\alpha$, where g is an arbitrary coefficient, the bodies A, B, C, D, etc. will always describe the same curve, however with this difference: that if in the case of the impressed velocity $= \alpha$, any portion of each of

these curves is described during a time t, the same portion will be described during the time t/g in the case of the impressed velocity $= g\alpha$. Therefore, the time that every body take to travel any part of the curve that describes will be inversely proportional to the initial velocity impressed to the body A.

Corollary V
6. Let x be the rectilinear or curvilinear space described by one of the bodies, for example A, in the first case when the impressed velocity is $= \alpha$, and let γ be the velocity of the same body A when it has described this space x.

It is clear (*art. 5*) that if the initial impressed velocity had been $g\alpha$, the velocity at the end of the space x would have been $g\gamma$; because $g\alpha/g\gamma = \alpha/\gamma$. Therefore whatever the initial impressed velocity to the body A is, the same velocity that the body will have at the end of the space x will always be in the same ratio with the initial velocity. Therefore if usually Q is denominated the initial velocity impressed to the body A, q its actual initial velocity, and finally u its velocity at the end of the space x, the fraction Q/u will be proportional to some function of x, and it will be of the same fraction q/u as well. Therefore, naming X this function of x, it will give $q/u = X$, or taking the logarithmic differentials $-du/u = \xi dx$, being ξ also a function of x.

Corollary VI
7. So far we have excluded all accelerating or retarding forces. But if it is assumed that each body is animated by a force proportional to the velocity, in this case all the theorems proved in *Corol.* 2. 3. 4. etc. will be met. This last observation will be useful in the sequel to determine the resistance of fluids taking into consideration friction. Moreover, all previous theorems are, if I am not mistaken, entirely new.

Prop. II. Theorem
8. *Let a solid body be submerged in a restful and non-elastic fluid; and, by excluding of all the accelerative forces acting as much upon the body as the fluid, let suppose that this body is given any impulse; I say:*

- 1st. That whatever the initial velocity impressed to the body is, provided it is impressed in the same direction, the body will always describe in the fluid the same line, either straight or curved; but the time it will employ to traverse any part of this line will be in inverse ratio of its initial velocity. This evident from *article* 5.
- 2nd. That any particle of the fluid will always describe the same curve, whatever the initial velocity impressed to the body is, and that in the moment when the body has finished describing the space x, the velocity of the particle will always be in a given ratio with the velocity of the body with the same instant. This is a continuation of the same *article* 5.

3rd. [2]If the resistance of the fluid is assumed to depend only on the velocity of the moved body, it can not be proportional to functions other than to the square of this velocity. Because, let the initial velocity be g and at the end of the space x the velocity $= u$ or zg, representing z a variable, t the time taken to traverse the space x, and $\varphi(u)$ a function of the velocity to which the resistance is proportional. It will give by the general principle of the accelerative forces that $\varphi(u) dt = -du$, or $= -\frac{udu}{\varphi(u)} = -\frac{zdzg^2}{\varphi(zg)}$. Now let in another case the initial velocity be g', the velocity at the end of the space x will be zg' (art. 3) and it will give $dx = -\frac{zdzg'^2}{\varphi(zg')}$. Therefore comparing these two values of dx, it will give $\frac{g^2}{\varphi(zg)} = \frac{g'^2}{\varphi(zg')}$, equation that must be acomplished in general, irrespective of z, and which can not be met unless $\varphi(zg) = z^2 g^2$. Therefore, $\varphi(u) = u^2$. Q.F.D.

Corollary

9. Now in general let be R the resistance of the fluid, whether it depends on the velocity alone, or on any other amount combined with it; it will give $Rdx = -udu$, and $dx = -udu/R$. [3]Now (art. 6), it gives in general $dx = -du/u\xi$, so $u/R = -1/u\xi$, and so $R = \xi u^2$. Therefore in general the resistance of the fluid is always proportional to the square of the velocity multiplied by any function of the distance traversed by the body.

So since ξ is a function of u/g (art. 6) it follows that the resistance R is as the product u^2 times of a function of u/g.

Scholium I

10. We will demonstrate later that the fluid resistance (excluding weight, friction and elasticity) is actually proportional to the square of the velocity, so that the function ξ of the space traversed is reduced to a constant.[4] This proposal has been regarded hitherto as true by all authors who have dealt with the action of fluids, and many have demonstrated it in their way. But it seems to me that the proofs they give are not very satisfactory. Because each of them is based on this single reason: the more velocity the moving body has, the more it communicates to the fluid particles, and the more it recieves at the same time from the particles of the same fluid. Well now, it seems to me that nobody can deny that this reasoning is rather vague. Others, claiming to treat this matter more accurately, found the resistance proportional to the square of velocity, making all the assumptions we discussed in the Introduction, and whose insufficiency we have demonstrated.

Morever, all these proofs, although not very convincing, all joined toghether in the same conclusion, may make one suspect that certainly this conclusion is true, and that the resistance of the fluids is actually proportional to the square of the

---

[2]In this article the letter g is used as the initial velocity, while in the preceding and followings articles it is a multiplicative factor. Also the velocity relation at any point is here called z with the same meaning as the former X.

[3]Strictly speaking the body mass should be introduced in the formula as, $Rdx = -mudu$.

[4]See Art. 87.

velocity of the bodies which move therein. This is what we will discuss in more depth later.

Scholium II

11. In general, it is obvious from the nature of our proof that in any system of bodies which act on each other (excluding gravity, and all other external forces) the force by which the motion of each body is altered at every moment is proportional to the product of the square of the velocity times any function of the espace traverse.

Besides this, it is obvious from *art.* 5 that a body which moves in a single homogeneous fluid, or which passes from a fluid into another, will always describe the same curve, regardless of the initial velocity, provided it has the same direction. Therefore a globe, for example, which passes obliquely from one fluid into another, must always describe the same curve in its path if its angle of incidence on the lower fluid does not change, whatever its initial velocity may be. This proves, to say in passing, that if the light refraction is atributed to the resistance of the medium, it can not be assumed that the difference of color, that is to say the different refrangibility of the rays, comes from the difference in their velocities. See here my *Treaty of the Equilibrium and Motions of Fluids*, I. III. Ch II.

Scholium III

12. It follows from all the principles establised so far, that the laws of the resistance of fluids depend heavily on their equilibrium laws. We will therefore expound the general laws of the Hydrostatics in the following chapter.

Chapter 2
General Principles of the Equilibrium of Fluids

Propos. III. Theorem
13. Let ABCD (Fig. 2.1) *be a fluid or any portion of fluid in which the particles are impelled by any forces, so that they are equilibrium; I say, if, from any point* P *placed inside that mass of fluid, are drawn the straight lines* PA, PB *to any two points* A, B *of the surface ABCD, the point* P *will be equally pressed along* BP *and along* AB*, or, what comes to be the same, the fluid contained in the channel or siphon* APB *will be in equilibrium. In fact, nobody ignores that when a fluid is in equilibrium, each particle P is pressed equally in all senses.*

Scholium I
14. Even though the principle of the equilibrium of the rectilinear channels is, as one can see, a very natural consequence of fluid pressure in all directions; however, I must admit here that the late Mr. *Maclaurin* was the first who made use this principle, which he applied to the important research on the figure of the Earth. See his *Treatise of fluxions* art. 639, and his Treatise *On the Cause of the ebb and flow of the seas,* Paris 1740.

Corollary I
15. If any point *p* is taken in *BP*, the pressure at *p* along *Bp* will be equal to the one along *Ap*, so that the fluid enclosed within the rectilinear channel *ApB* would be in equilibrium. Now the fluid enclosed in the *APB* channel would be as well; therefore the fluid enclosed in any triangular channel *ApP* will be in equilibrium.

Corollary II
16. Thus the channel or rectangular siphon *APCB* (Fig. 2.2) would also be in equilibrium; because drawing *BP*, we will see that the *APB* channel would be in equilibrium and the *PBC* channel would be also (*art.* 15). So, etc.

Fig. 2.1

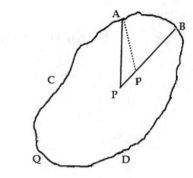

Fig. 2.2

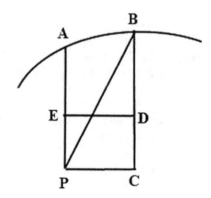

Corollary III
17. If *ED* is drawn parallel to *PC*, it will be seen that the channel *AEDB* will be also in equilibrium, therefore the rectangular channel *EDCP* must be also.

Corollary IV
18. Let *AP* (Fig. 2.3) be any curvilinear channel.[1] I say that the fluid enclosed in the channel will also be in equilibrium, because taking the axis *Pp* infinitely small, as well as the arc *Pp'*, we will see (by the *article* 15) that the channels *APp, App'* are each other in particular in equilibrium. Therefore the channel *APp'* will also be, and it will be proved similarly that the channel *APp'A* will be in equilibrium, as well as the channel *BRQP;* or the rectilinear channel *APRB*; therefore the curvilinear channel *APB* will be also in equilibrium. Thus the principle of the equilibrium of curvilinear channels is but a corollary of the simple principle of balance of rectilinear triangular channels finishing in the fluid surface. Principle due to M. MacLaurin.[2]

---

[1] We think that the definitions of the channels are not clear enough.
[2] In the *Mss*.19, he says that Clairaut had proved that, but with under a less general view.

Fig. 2.3

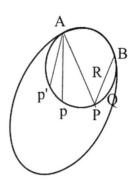

Corollary V[3]

19. Let M, N, O, Q (Fig. 2.4) be four points or particles of the fluid, infinitely close each other, and located in such a way that $MNQO$ is an infinitely small rectangle. Let A be any fixed point inside or outside the fluid and in the $MNOQ$ plane, AP parallel to MO, and APM a right angle. It is assumed that the forces impelling the points M, N, O, Q act in the plane $MNOQ$, or APM. It is obvious that instead of power acting on M, for example two forces, can be assumed, one acting along MO, parallel to AP, the other along MN, parallel to AZ; and similarly for the other points N, O, Q. Let be $AP = x$, $PM = y$, R the force on point M along MO, and Q the force of the same point along MN, $MO = \alpha$, $\beta = MN$. Now imagine that the impelling[4] forces of the points M, N, O, Q are proportional to any function of the distances of these points to the lines AZ and PA. Finally, to make the proportion, let us that the fluid is heterogeneous and the density δ of any particle M is proportional to any other function of the lines AP and PM; in this case the force of the point N is along NQ will be[5] $R + \beta \frac{dR}{dy}$ and the density of the column $NQ = \delta + \beta \frac{dR}{dy}$, thus the force of the column MO along MO being $\alpha \times R\delta$, that of the column NQ along NQ will be $\alpha \left(R + \frac{\beta dR}{dy} \right) \left(\delta + \frac{\beta d\delta}{dy} \right)$. By the same reasoning, it will be found that the force of the column MN, being $\beta Q\delta$, the force of point O along OQ will be $+\alpha \frac{dQ}{dx}$, and the force of the column OQ along $OQ = \beta \left(Q + \frac{\alpha dQ}{dx} \right) \left(\delta + \frac{\alpha d\delta}{dx} \right)$; because (*art.* 17) the rectangle channel $MNQO$ must be in equilibrium. Therefore the force of the columns MN and NQ along MN and NQ must be equal to that of columns MO and OQ along MO and OQ. Therefore (neglecting what should be neglected, that is to say, where the quantities $\alpha\beta\beta$ and $\beta\alpha\alpha$ are found) it

---

[3]The letter Q is used in two senses: as a geometrical point and as a force.

[4]In the *Essay* says "forces accélératrices". We think this is a misprint for "forces sollicitatrices", because in the *Mss.* 20 is "vires sollicitatrices".

[5]In general I refer to $\frac{dR}{dy}, \frac{dR}{dx}, \frac{d\delta}{dy}$ etc as the coefficients that dx, dy, etc would have in the differentiation of the quantities R, δ, which (*hyp.*) are functions of x and y. M *Fontaine* was the first who imagined this expression that is extremely convenient. (*Original note*).

Fig. 2.4

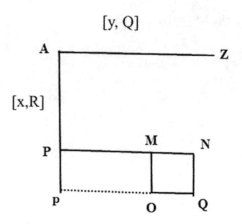

will give $\beta Q\delta + \alpha R\delta + \alpha\beta\frac{\delta dR}{dy} + \alpha\beta\frac{Rd\delta}{dy} = \alpha R\delta + \beta Q\delta + \alpha\beta\frac{\delta dQ}{dx} + \alpha\beta\frac{Qd\delta}{dx}$. Therefore $\frac{Qd\delta}{dx} + \frac{\delta dQ}{dx} = \frac{Rd\delta}{dy} + \frac{\delta dR}{dy}$, or what is the same thing $\frac{d(Q\delta)}{dx} = \frac{d(R\delta)}{dy}$.

Corollary VI

20. Therefore, if the fluid is homogeneous, that is to say, if the density δ is constant, it will give $\frac{dQ}{dx} = \frac{dR}{dy}$; a proportion that was already known, but that nobody, it seems to me, had yet demonstrated by a method as simple as we have just done. This equation will be very useful to us in the following to determine the laws of the resistance of homogeneous fluids, and the equation $\frac{d(Q\delta)}{dx} = \frac{d(R\delta)}{dy}$ to determine those of the elastic fluids.

Chapter 3
General Principles of the Pressure Fluids, in Motion or at Rest

Propos. IV. Problem

21. *Let* **MNGH** *(Fig. 3.1) be a homogeneous fluid without weight, and that either it is of an indefinite extension or enclosed in a reservoir of any size and shape. Place a solid body* **BCDE** *anywhere one wishes inside this fluid, taking around that body a fluid portion limited by the surface* **FOKL** *and assuming that all the particles of either the fluid or the solid contained by the* **FOKL** *surface are impelled by such forces that an equilibrium exists between the fluid and the solid. The pressure that the fluid exerts on any point* D *of the solid body is impelled.*

Let *FB, OD* be any lines that are terminated by the body surface and by the fluid surface; it is obvious that the particles of the *FO* surface are impelled by forces that are either absolutely zero, or at least are perpendicular the *FO* surface. In fact, *FOKL* can be considered as the external surface of a fluid in equilibrium, since the particles of fluid located outside the space *FKPL* are not required (*hyp.*) by any force.[1] Now, the fluid in the channel *FBDO* must be in equilibrium (*art.* 18), therefore the weight of the *OD* channel is equal to the weight of the *FBD* channel. Therefore the pressure in point *D* will be the same as if this point was pressed perpendicular to the *BDC* surface by a force equal to the weight of the *FBD* channel.

Corollary I

22. Let *FB* (Fig. 3.2) be a straight line, the weight of the particle Z along $ZB = \varphi$, $FZ = z$, the pressure in *B* will be equal to which that becomes $\int \varphi dz$ when $z = FB$, which I call *K*. Let be $BD = s$ as well, and the weight of the particle *V* along $VD = \pi$; the weight of the *BD* channel will be $\int \pi ds$. Therefore the pressure that the particle *Dd* along *DG* undergoes, perpendicular to $D\delta$, will be equal to $D\delta \times (K + \int \pi ds)$;

---

[1] In *Mss*.25, it is said: "multis (hyp) viribus agitantur": "they are impelled by many forces".

Fig. 3.1

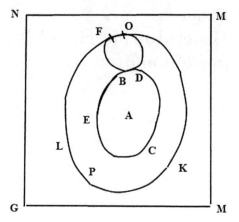

Fig. 3.2

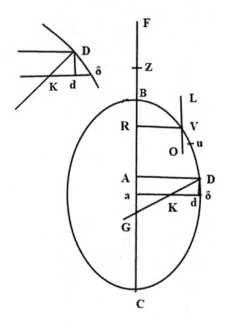

therefore the pressure resulting from the former along Dd, that is to say parallel to BC, will be $D\delta \times (K + \int \pi ds) \times \frac{Dd}{DK} = d\delta \times (K + \int \pi ds)$, (because of the similar triangles DKD, $dD\delta$). Therefore, if the line FB is very small, it can be assumed without noticeable error, that the pressure D parallel to BC is $d\delta \times \int \pi ds$.

3 General Principles of the Pressure Fluids, in Motion or at Rest

Corollary II

23. Now let ψ be the gravitation force in the point V along $RV=y$, $BR=x$ (Fig. 3.3), this will give $\pi = \frac{\psi dx}{ds}$. Therefore $\pi ds = \psi \, dx$ and so $\int \pi ds = \int \psi dx$; so the pressure in V along Vu will be $\int \psi dx$[2]; that is to say, it will be equal to the weight of a straight column VN in which the parties would be impelled by the variable force ψ.[3] In the same way the pressure over the point u along uV will be equal, for the same reason, to the weight that the column Nu would have; therefore the pressure at the point V along VN will be equal to the weight of the column Vu, for which this theorem is deduced.

If the parties V of the fluid contiguous to the $BDCE$ surface are impelled along VO, parallel to the axis BC, by a power ψ, which is different (if desired) for each point V; I say that the weight that the body $BDCE$ undergoes in virtue of all these forces will be directed from C to B and will be equal to the weight that the body would have along BC, if all the parts contained in each ordinate QV were pushed parallel to BC by the same force ψ acting on the corresponding point V.

Fig. 3.3

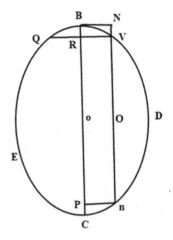

---

[2] In the original "so the pressure in Vu will be...". We think it is a lapse, we have followed the *Mss*.27: "Igitur pression in V secundum Vu erit...": "Therefore, the pressure at V following Vu will be [ ], that is, of the same weight that the column NV, whose parts will be impelled by the same force ψ..."

[3] Probably the term "variable" is a lapse, because the gravitation is obviously constant. Furthermore, in the *Mss*. 27, this word is missed out: "cujus partes sollicitarentur eadem vi ψ...", "which parts are impelled by the same force ψ".

Fig. 3.4

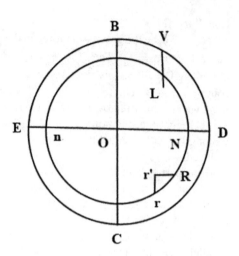

Corollary III[4]

24. Let be *BDCE* (Fig. 3.4)[5] in a channel reentering itself and filled with fluid [whose parts *BD* y *BE* are impelled by any forces, while the parts *DC*, *EC* are not pressed by any force], and that the points *N*, *n* are those which correspond the largest width *Nn* of the channel. Let suppose that these two points *N*, *n* are impelled parallel to *BC* by a force φ [resulting from the forces acting upon *BD* and *BE*]; I say that the pressure which results along *BC* will be $=\varphi \times Nn$. Because the pressure φ that acts on point *N*, it also acts [by the nature of the fluids] on all points *R* of the portion *NRC*; so that the pressure *R*, acting perpendicularly to the channel walls, is $Rr \times \varphi$; from this pressure another results along rr' that will be $= Rr \times \varphi \times \frac{Rr'}{Rr} = \varphi \times Rr'$; therefore all the pressure along $CB = \varphi \times \int Rr' = \varphi \times Nn$.

Corollary IV

25. Let be *BVDNCEB* a channel whose all parts are impelled along *VL* by a constant force $= \psi$, the pressure of this channel along *BC* will be (*art*. 23) $\psi \int ydx$, designating $\int ydx$ as the mass of the body *BDCE*. Let suppose, besides this, that the parties of the channel *BVDN* are impelled by variable forces π, which act along *VD*, so that these forces π end at point *N*, which corresponds to the largest ordinate. The resulting pressure from *B* to *C* will be $\int dy \int \pi ds$ (*art*. 22); now let Δ be the value of $\int \pi ds$ at *N*, it is clear that the pressure at *N* is Δ, and that this pressure (*art*.

---

[4]In order to make this article three sentences taken from the *Mss*.28 are added. "Sit BDCE canaliculus fluido plenus, cujus partes BL, Bl viribus quibuscumque sollicitentur, partes vero LC, lC nulla vi premantur. Sit φ presio in L et in l, orta a viribus in BL et Bl agentibus, dico...". *Mss*.28". "Let *BDCE* a small cannel full of fluid, whose parts *BL*, *Bl* are impelled by whatever forces, while the parts *LC*, lC are not pressed by any force. Let φ the pressure at *L* and *l* due to the forces which act in *BL* and *Bl*, I say..."

[5]In the original the letter *O* is missing and *r* and r' are changed.

24) is the same in all parts of the NC channel; therefore the pressure from C to B coming from the NCE channel will be $\Delta \times b$, b designing the largest ordinate Nn. Therefore if we call G what $\int dy \int \pi ds$ becomes when $y = ON$,[6] the total pressure along BC will be $= \psi \int y dx - \Delta \cdot b + G$.

Note
26. So far we have looked at the body BDCE as a plane figure, or, what is the same thing, as a solid generated by the parallel movement of a plane figure. However if this solid were generated by the revolution of the figure BDCE around the axis BC, then in this case naming the ratio of the circumference of the circle to the radius as 2π, in the above formulas $\int y^2 dx$ should be substituted instead of $\int y dx$, $\frac{\pi b^2}{4}$ instead of b, and $2\pi y dy$ instead of dy.

Propos. V. Problem
27. Let ABCD (Fig. 3.5) *be either a pipe or channel of indefinite length, whose walls* AB, CD, *are extremely close each other, and whose width is always the same in its upper part* FABG, *after that it grows from* A *till* C, *or at least it is variable. Then let us suppose that in this channel a homogeneous fluid flows without weight, so that in the indefinite and cylindrical part* FABG *the fluid velocity is uniform and*

Fig. 3.5

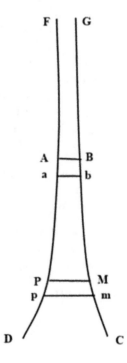

---
[6]In the original $y = NL$, it is a misprint.

always the same. The fluid velocity is impelled at any point P of the channel ABCD, and the pressure at the point P.

1st. It is obvious that all parts of the fluid contained in any slice *PM* always have the same velocity, at least very nearly, both because *PM* is assumed to be very small and because one can imagine a certain viscosity in the particles of the fluid, whereby the particles that are contiguous each other in a slice *PM* adherence to each other, and they have an equal velocity. For the same reason, all parts of the slice *AB* have the same velocity. Therefore while the particles AB come into *ab*, the particles *PM* will come into *pm*, so $PMmp = ABba$ or $PM \times Pp = AB \times Aa$, because *PM* and *AB* can considered as perpendicular to *Pp* and *Aa*. Therefore the velocity at the point *P* is to the velocity at point *A* as *Pp* to *Aa*, that is to say, as *AB* to *PM*. Therefore taking $PM = y$, $AB = \beta$, the constant velocity at *A* is b,[7] and the velocity at *M* or *P* is *u*, this will give $u = \frac{b\beta}{y}$.

2nd.[8] Let $AP = x$ be and *dt* the instant used to traverse *Pp*; it is clear that at the end of the instant *dt* the velocity *u* becomes $u + du$, in such a way that when the *PM* particles pass to *pm*, the velocity with which they tend to move becomes $u + du$ (I use $+du$, although the velocity actually decreases from *P* to *p*, the width of the channel from *A* to *P* is assumed growing in the Figure, but as *du* is negative when *x* increases, it follows that $u + du$ is actually less than *u*). Now then, the velocity *u* is composed of $u + du$ and $-du$; whence it follows (*art.* 1) that if the slice *PM* were impelled by the single infinitely small velocity $-du$, or what is the same thing, by the only accelerating force $\frac{-du}{dt}$, the fluid contained in the *ABCD* channel would be in equilibrium. Therefore the pressure at *P* will be the same if the particles *PM* of each slice were impelled by a force $\frac{-du}{dt}$; well now, in this case it is found that making $Pp = ds$, the pressure at *P* would be $\int Pp \times \frac{-du}{dt} = \int ds \times \frac{-du}{dt}$. Therefore because $ds = udt$, it will give the pressure at *P* as $\int -udu = \frac{b^2-u^2}{2} = b^2 \frac{y^2-\beta^2}{2y^2}$.

Corollary I

28. If (for any reason whatsoever) the velocity of the fluid in the cylindrical portion *ABGF* was not always the same, so that *b* was variable, then putting in the place of *b* any other variable *v*, I we t will have $u = \frac{v\beta}{y}$ and $-du = \beta \times \frac{(-ydv+vdy)}{y^2}$. Therefore the pressure in P would be $\frac{-\beta dv}{dt} \times \int \frac{ds}{y} + \beta v \int \frac{dsdy}{y^2 dt}$, taking *v*, *dv* and *dt* constants, because the pressure which is sought is not the sum of pressures in a time *t*, but the pressure in an instant *dt*. Therefore if in $\frac{dsdy}{y^2 dt}$ one puts in the place of *dt* its value $\frac{ds}{u}$ or $\frac{yds}{\beta v}$, it will give the pressure en $P = \frac{-\beta dv}{dt} \times \int \frac{ds}{y} + \beta^2 v^2 \times \left(\frac{1}{2\beta^2} - \frac{1}{2y^2} \right)$.

---

[7]The letter *b* is used here as velocity.

[8]In the beginning of the paragraph the symbol *x* is used for the vertical distance, but later it is changed to *s*. Also *P* is used for a geometric point and for the pressure.

3 General Principles of the Pressure Fluids, in Motion or at Rest

Fig. 3.6

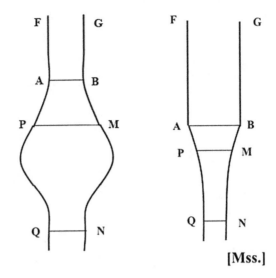

[Mss.]

Corollary II
29. If the fluid is assumed to be heavy, then taking g for the natural gravity, it is clear that the PM particles impelled by the forces $g - \frac{du}{dt}$ will be (*art.* 1) in equilibrium between them. Therefore, 1st, if the velocity v is constant, the pressure will be $\int ds \left(g - \frac{du}{dt} \right) = g \cdot AP + b^2 \frac{(y^2 - \beta^2)}{2y^2}$ to which must be added $g \times FA$. 2nd, if the velocity v is variable, the pressure will be $g \cdot AP - \frac{\beta dv}{dt} \times \int \frac{ds}{y} + \beta^2 v^2 \left(\frac{1}{2\beta^2} - \frac{1}{2y^2} \right) + \left(g - \frac{dv}{dt} \right) \cdot FA$.

Scholium I
30. If the ordinates PM decrease from A to P, then the fluid velocity will increase from A to P, and the pressure will be from P to A. Therefore in this let the case Q (Fig. 3.6[9]) be the site where the channel width is the least, and therefore the fluid velocity the greatest; it will be found that the pressure at P is equal to half of the square of the velocity at Q minus half of the square of the velocity in P. Such that, the pressure is highest in A and null in Q.

But perhaps one can say, how can it be that the pressure is not null at A, and on the contrary it is greater than at another point? Because if there is some pressure at A along AF, it must necessarily have an equal pressure along FA; now then, the fluid (*hyp.*) moves uniformly from F to A, therefore at A there cannot be any pressure along FA. I answer that having supposed the channel $AFBG$ of indefinite length, the pressure at A is supported by the only mass of fluid $AFBG$. In fact, if the fluid

---
[9]That article is somewhat confused. The original Fig. 3.6 (left side) does not correspond with the narrative description given. Rather it agrees with the right one, taken from *Mss*.34 and Fig. 12.

Fig. 3.7

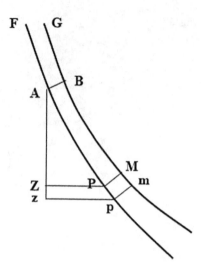

contained in the cylindrical channel *AFBG* was not supposed undefined, then it would be necessary that the velocity decreases at every moment, so that the velocity increases in the narrowed channel *ABMP*; for the same reason that when a body strikes another that moves the same direction, the velocity of the rear body decreases and that of the previous body increases. To make it more noticeable, let *l* be the length of the channel *FABG* assumed to be finite, and imagine that each particle of this channel has received a velocity V that must change to U due to the communication with the part *AQNB*; the velocity in *PM* will be $\frac{v\beta}{y}$, and the pressure at *AB* along *FA* will be equal to the pressure at *AB* along *PA* (*art.* 1); from where $(V-U)l = U \int \frac{ds\beta}{y}$ is obtained, therefore $U = \frac{Vl}{l + \int \frac{\beta ds}{y}}$; so $V-U$ is not zero unless *l* is undefined. In any other case it will give $U < V$.

Corollary III

31. If the pipe is not vertical but inclined, as shown in Fig. 3.7, then drawing the vertical *AZ* and the horizontal *PZ*, it will be necessary to put $g \cdot AZ$ instead of $g \cdot AP$ in the two formulas of the precedent corollary, because the quantity gds changes into $g \times \frac{Zz}{Pp} \times Pp = g \cdot Zz$.

Furthermore, if there is no accelerative and external force acting upon the fluid more other than the natural gravity, we will give in the case of the *art.* 29, $\frac{dv}{dt} = g$ and in the case of the preceding article. $\frac{dv}{dt} = gh$, naming the cosine of the inclination of the pipe *FA* as *h*. Thus in the first case the pressure will be equal to $g \cdot AP - \beta g \int \frac{ds}{y} + \beta^2 v^2 \left(\frac{1}{2\beta^2} - \frac{1}{2y^2} \right)$ and in the second case it will be $g \cdot AZ - \beta gh \int \frac{ds}{y} + \beta^2 v^2 \times \left(\frac{1}{2\beta^2} - \frac{1}{2y^2} \right)$.

Corollary IV

32. Since $\frac{\beta v}{y}$ is the velocity at P or M, in general let $v\rho$ be the velocity at M, and [in consequence] the pressure will be $\frac{-dv}{dt} \times \int \rho ds + \frac{v^2}{2}[1 - \rho^2]$, g being $= 0$. This expression will be of great use in the sequel.

Scholium II

33. In general, whether the fluid heavy or not, it can be assumed that the velocity v is equal to that of a body impelled by the gravity g and falling from the height h.[10] Therefore in the case of *art*. 27 we will have $v^2 = 2gh$, and the pressure $P = gh (\rho^2 - 1)$, thus the pressure P would be the same as that of a column of stagnant fluid of gravity g and height $h (1 - \rho^2)$. By this we see that the formula found here for the amount of pressure can be used for registering and comparing easily known pressures.

Scholium III

34. So far we have assumed the fluid density constant. If it was not, let δ be the fluid density at P or M, and δ' the density at A (Fig. 3.5). I say that the velocity at P will be $\frac{\beta v \delta'}{y \delta}$. Since assuming the mass of $ABba$ equal to that of $PMmp$, we will have $Aa \times \delta' \times Bb = PM \times Pp \times \delta$. Therefore making $\frac{\beta v}{y} = v\rho$, and $\delta = \frac{\delta'}{\sigma}$, we will have the velocity at $M = v\rho\sigma$, from where the pressure will be $-\frac{\delta' dv}{dt} \int \frac{\sigma \rho ds}{v} - \delta' v \int \frac{dsd(\rho\sigma)}{vds}$, that is to say (because $dt = \frac{ds}{v\rho\sigma}$) equal to $\frac{-\delta' dv}{ds} \int \rho ds - \delta' v^2 \int \rho d(\rho\sigma)$, g being zero.

---

[10]Here we have followed the *Mss.* 36, because the *Essay* is not clear. The original text is "…la vitesse v égale à celle qu'acquêterait un corps sollicité par la pesanteur g et tombant de la hauteur h soit variable, soit constante. Donc dans le premier cas on aura $bb = 2gh$…". It is not clear what the words variable or constant refer to. The same is applicable to the mentioned first case. In the *Mss*.36. "Generatim sive fluidum grave, sive non grave supponatur, potest spectari velocitas v tanquam debita altitudini h ex qua corpus gravitate g sollicitatum caderet. Unde in casu art. 31… erit vv = 2gh." *Mss*.36. "In general, assuming a fluid either heavy or not heavy, the velocity v can be considered as the one due to a height h, from which a body falls impelled by the gravity g. From this in the case of art. 31 it would be $vv = 2gh$.". It is worth noting that the *Mss.* 36 ends as "Cæterum, methodus qua in præsenti capite pressionem fluidorum sive quiescentium, sive motorum definivi, ex genuinis mechanic principiis deducta mihi videtur." That is "For the rest, it seems to me that the method that I have defined in this chapter for the fluid pressure, be they at rest or in motion, is deduced from the genuine principles of mechanics."

Chapter 4
On the Pressure That a Fluid Exerts on a Body at Rest and Immersed in It

35. In order to determine the resistance that a fluid, either in motion or at rest, produces on a body that moves therein, we must first to determine the action that a moving fluid exerts against a body at rest. Because we will show in the next chapter that the whole theory of the resistance of fluids depends upon that, therefore we begin by expounding our research on this subject.

4.1 Observations Necessary for the Understanding of the Subsequent Propositions

36. Let *QqGH* (Fig. 4.1) be a homogeneous fluid and without weight, either indefinite or enclosed in a vessel of any size and figure. The fluid moves from *Q* to *H*, and let a solid body *AECD* be submerged in the fluid, which, notwithstanding the action that the fluid exerts upon it, remains at rest for any cause that may be, for example, the resistance of a power pushing the body from *C* to *A*, while the fluid pushes from *A* to *C*. The pressure of the fluid upon the body *ADCE* is impelled.

1st. It is obvious that the particles of the fluid, if the body *ADCE* did not obstruct them, must describe the lines parallel among them *Tf*, *OK*, *PS*, etc., but the presence of the body, when the lines approach at some distance from the body, makes that they have to change their direction gradually at *F*, *K*, *S*, etc. and describe the curves *FM*, *Km*, *Sn*, etc. Those lines will differ more from a straight line as they are nearer to the surface *ADC*, and on the contrary less different from a straight line as they are more distant from that surface. So at a certain distance from the body *ADEC*, eg. *ZY*, these curves will become straight lines; and the fluid contained in the space *ZYHQ* will move uniformly, in the same way as if the solid body *ADCE* was not in the fluid. It is necessary to say the same about the fluid which is at the other side of *AEC*, and if this part *AEC* is equal and similar to *ADC*, the curves which the

Fig. 4.1

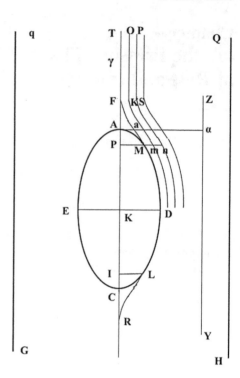

particles of the fluid describe at the side of *AEC* will be completely similar and equal to those which are described at the side *ADC*.

2nd. Besides, since the body *ADCE* is assumed at rest and excluding all accelerative forces that could act upon the fluid, it is obvious that the motion of the fluid must be assumed to be in a permanent state; that is to say that the curves *FD, Km* described in any instant by the particles are always the same; so that the particles which have described, for example the straight line *OK*, will always describe the curved line *Km*.

3rd. Every body in motion that changes direction only makes this change by imperceptible degrees. From the foregoing, it follows that the particles moving in the axis *TF* do not arrive at vertex *A* of the body. Because if they arrived at *A*, then because of the right angle *FAa*, their direction *TA* must change in an instant to another direction that will make with the former *TA* a finite angle. Therefore the particles that move in the axis *TF* will start leaving this direction at least at some small distance from *A*, for example *F*, and they will describe the curve *FM* that touches the *TF* line in *F*, and the surface of body in *M*. Next this curve will coincide and will slide exactly on the surface of the solid body *MDL* until a point *L* where it will leave this surface, to reach and touch the axis *TAC* at *R*. From this it follows that there are spaces *FAM, CLR* in front and behind the solid, where the fluid is necessarily stagnant. The same should be said on the other side *AEC*.

4 On the Pressure That a Fluid Exerts on a Body at Rest and Immersed in It

4th. Let us suppose for convenience that the part *AEC* of the body is perfectly similar and equal to the part *ADC*, in this case the action of the fluid will be exactly the same on both sides, this is why we will only pay attention to the part *ADC*. Now, let α be the velocity of the particles of the fluid in any instant, this velocity becomes α' in the next instant and let us assume that the velocity α is composed of velocities α' and α''; it is obvious (*art.* 1) that the particles of the fluid, if they tend to move with the single velocity α'', would be in equilibrium, and in this case, the fluid pressure would be the same as if it was stagnant and its parts were subjected to the accelerative force α''/dt. Now then, let be α constant, that is to say, $\alpha = \alpha'$ and then the particles are moved in a straight line, it will give $\alpha'' = \alpha' - \alpha = 0$. Therefore the body cannot undergo any pressure other than the fluid particles, whose velocity, or direction, or both are changed by the encounter with the body.

5th. So let α and α' be the velocities of these particles in two consecutive instants (it is not necessary to notice that these quantities α and α' are indeterminate and different for each particle). It is obvious that these particles would be in equilibrium if they were impelled to move by the accelerative force α''/dt. Therefore if γ is the point where the particles that describe the *TF* line starts changing the velocity, the pressure at *D*, for example, will be equal to the pressure that a fluid contained in the γFMD channel, whose parts were driven by the force α''/dt different for each one, would exert. The question, therefore, is reduced to finding both the curvature of the channel γFMD and the forces α''/dt in this channel.

I remark at first that no pressure can result from the particles contained in the portion *FM*, which touches the axis at *F* and the surface at *M*. To prove this, I assume that the particle *a* (Fig. 4.2) of the part *FM*, describes in any instant the small line *ab*, and in the next instant the line *bc*. Let us make *bd* equal and in the straight line with *ab*, it is clear that the particle *a*, when it comes to *b*, would describe in the next instant the line *bd*, if nothing were to prevent it. But as it is forced to describe *bc*, it follows that the velocity *ab* or *bd*, that it had in the previous instant, can be considered (*art.* 1) as composed of the velocity *bc*, that it has in the next instant, and another velocity *cd* which must be destroyed. Therefore putting *bi* parallel to *dc*, and *ie* perpendicular to *bc*, it is clear that the particle *b* impelled by the forces *be, ei* must remain in equilibrium. That said, I say that *be* will be equal to zero; said in general, that the accelerative or retarding force of the particle *b* along *bc* must be null. Because if it was not, let us draw *bm* (Fig. 4.3) perpendicular to *Fb* and *nq* which is infinitely close; therefore the part *bn* of the fluid contained in the channel *bnqm* would have some pressure from *b* to *n* or from *n* to *b*. Therefore since the fluid in the channel *bnqm* must be in equilibrium, it would need also some action, at least upon one of parts *bm, qm, qn* to counteract the action of the part *bn*. But it has been proven that the fluid is stagnant in the space *FAM*; so there is no force that can act on *bm, mq, qn*; so the pressure on the channel *bn* along *bn* or *nb* is null. Therefore the force along *be* (Fig. 4.2) of the particle *b* is equal zero; hence *bi* or *cd* is perpendicular to *bc*, therefore there is not any pressure in the channel *FM*, if this is not coming from the upper part γF (Fig. 4.1) or from the force *ei* (Fig. 4.2). But as the latter is perpendicular to the channel walls, it follows that it does not exert

Fig. 4.2

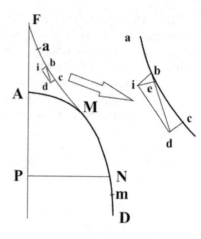

Fig. 4.3

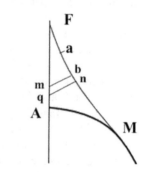

any pressure from F to M; therefore the point M undergoes no pressure other than that which t can come from the part γF (Fig. 4.1).

From above it follows that the velocity in the curve FM is either constant if it is finite or infinitely small if it is variable. Because in the first case, the force along be will be absolutely null; and in the second it will be infinitely small of the second order, and consequently it can be considered as null. We will make clear later that is the second case which takes place here,[1] that is to say that the fluid velocity along FaM must be infinitely small, or at least so small that it can be treated as zero. Whence it follows that the velocity of the fluid, before starting to change the direction at F, begins to change the magnitude in some point γ above the point F; so that, it decreases after γ until F to become very small in F.

Corollary I
37. Therefore the pressure on any point D comes both from the part γF and from the fluid particles that are in the channel MD. Now then as these latter particles move along the body surface; the force α''/dt, destroyed in each one, is composed of

---

[1]This will be in §.52-53.

4 On the Pressure That a Fluid Exerts on a Body at Rest and Immersed in It

another two, one along the surface MD, the other perpendicular to that surface. Let call the first of these forces π, the second π'; we will see easily that the point D is pressed perpendicular to the surface MD: 1st, by the sum of π forces in the curve MD; 2nd, by the force π' which acts upon the single point D. Now then, this latter force, acts only on a single point D, being infinitely small compared to the sum of π forces acting on the infinite number of particles placed in the curve MD, it follows that the pressure upon point D is the sum only of the π forces. Therefore taking in the arc MD any infinitesimal portion (Fig. 4.2) $Nm = ds$, the pressure in D perpendicular to the surface of the body will be equal to $\int \pi ds$, and this quantity $\int \pi ds$ must be taken in such a way that $\int \pi ds = 0$ in M.

Corollary II

38. Therefore to determine the pressure at D, it is necessary to know the force π at any point N (Fig. 4.2). So let u be the velocity of the particle N following Nm in any instant and $u + du$ its velocity in the next instant; it will give (*art.* 1) $\pi = -du/dt$. The question is therefore reduced to finding the velocity u of any point N following Nm. The following propositions are committed to doing this.

Proposition VI. Theorem

39. *Whatever the velocity and density of the moving fluid and mass of the body ADCE are (Fig. 4.1) as long as the body always keeps the same figure and the same volume; I say that each of the curves FaMD, Kmd, which are all different from each other, will be always the same.*

At first I will prove that it can be assumed that each of these curves is always the same; next I will prove that they must be necessarily to be supposed as such.

I

[1st.] Let U be the velocity of any particle m, when the velocity at γ is a. Next let us assume a similar body, with the same figure and the same volume, exposed to another fluid current whose velocity and density are whatever; and finally, let us suppose that in both cases the curves FaM, Km, etc. and the two points γ, F are the same. I am going to prove that this assumption is legitimate. Let ga be the velocity at γ, g being an arbitrary coefficient; I say that the curves can remain the same, provided that the velocity at m is gU, that is to say in general, whenever the velocity at any point is changed at a ratio of g to 1 without changing the direction. Indeed, the ratio of the velocity U at m to the velocity a depends only on the mutual distance of the curves FM, Km at m, because the ratio of velocities U and a depends on the width of the channel contained between the curves FM and Km. Therefore these curves can remain the same, provided that the ratio of velocities U, a, does not change; that is to say, provided that U becomes gU, a becomes ga.[2]

2nd. When the velocity is a at γ and U at m, the force a''/dt represents (*art.* 36) the force that must be destroyed in each particle, so that the parts of the fluid

---

[2] "Eadem autem remanet hæc ratio, quando velocitas in M est. gU, in g vero ga" *Mss*.42.I. "If this same reason is maintained, when the velocity at M is gU, at G it is ga".

impelled by the force α''/dt would be in equilibrium among them. Now then, if the parts of a fluid, whose density is δ, are driven by any forces π are in equilibrium, it is obvious that the equilibrium subsists if the force π becomes πg and density δh, g and h being arbitrary coefficients, provided that the direction of the acting force on each particle remains the same. Therefore the equilibrium of the fluid whose parties are animated by the force α''/dt will not be disturbed if the fluid density is changed at wish and each force α''/dt becomes $g\alpha''/dt$, maintaining the same direction; now then the curves described by the fluid particles always remain the same (*hyp.*). It is obvious that if the velocities U generate[3] the force α''/dt, the velocities gU will generate $g\alpha''/dt$. Therefore the force $g\alpha''/dt$ will be destroyed and therefore one can assume that the curves FM, Km are the same in both cases.

II
Now I say, it follows from that that these curves are necessarily the same. Since the particles of the fluid *can* always describe the same curves in the two cases, as we have just proved. Therefore they really *must* describe them, since both the fluid density and its velocity were given, and the figure and mass of the body, the way that each particle must traverse is necessarily determined and unique. This reasoning is completely analogous to the one which is accepted by all geometers: if a body is thrown in the vacuum, in the case of Newtonian attraction, there is always a conic section that it *can* describe. Therefore it *must* actually describe this section, since the way it should traverse is necessarily unique and determined.

Corollary I
40. Therefore whatever the fluid velocity a, its density and body mass are, U/a will always be constant for the same point m, although different for different points; as a becomes ga, U becomes gU, now then $U/a=gU/ga$. Moreover, the velocities U and gU will have the same direction at m, since the curves described by the point m are the same in both cases.

Corollary II
41. Therefore if in general one assumes $U/a = \rho$, the quantity ρ will neither depend on the fluid density nor the body mass, but only on the figure and volume of the body, and the position of the point m. Therefore making $AP = x$ and $Pm = z$, ρ will be a function of x and z that will vary depending on the body figure $ADCE$.

Corollary III
42. Therefore, since the velocity has always the same direction, if this velocity is split into two others, one parallel to AP, that I call aq, the other perpendicular to AP, that I call ap, q and p will be functions of x and z which will depend neither on the velocity, nor on the fluid density.

---

[3]"donnent" in the original, but in *Mss.*42.I is "exurgat", whose meaning is rather to generate.

Fig. 4.4

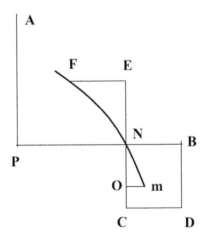

Proposition VII. Theorem

43. *Let us assume that any fluid particle N (Fig. 4.4)*[4] *describes the two contiguous sides and infinitely small FN, Nm of the curve FNm, and let* aq *be the velocity of particle N in N parallel to AP;* ap *its velocity at N perpendicular to AP, q and p being (art. 42) unknown functions of AP (x) and PN (z). Finally, let* dq = Adx + Bdz, *and* dp = A'dx + B'dz, A, B *and* A', B' *are similarly unknown functions of x and z. I say that the force along* NB *perpendicular to AP, which must be destroyed in the particle N, will be* $-(B'p - A'q)a^2$.

Because when the particle N is at N, its force along NB, which should be destroyed, is the excess of the velocity that it has at F along FE over the velocity that it has at N along NB. Now then the velocity at N along NB is equal to ap. Therefore the velocity at F along FE is equal to $ap - a \cdot FE \times \frac{dp}{dz} - a \cdot NE \times \frac{dp}{dx}$, or $a \times (p - FE \times B' - NE \times A')$; now then the velocity at F along FE is to the velocity at N along NB, as FE to mO or apdt, it has $NE = \frac{FE \times q}{p}$. Therefore it will give $ap : ap - a \cdot FE \times B' - \frac{aq \cdot FE \times A'}{p} :: apdt : FE$. Therefore (considering FE as infinitely small, and consequently rejecting of its expression the third order quantities) it will find $FE = apdt\left(1 - aB'dt - \frac{aA'qdt}{p}\right)$. Therefore $E - Om = a^2pdt^2 \times \left(-B' - \frac{A'q}{p}\right)$; therefore the force in N along NB, that is to $\frac{FE-Om}{dt^2} = (-pB' - A'q)a^2$. This Q.F.D.

Scholium

44. By similar reasoning $aq : aq - a \times \frac{NE \times dq}{dx} - \frac{a \cdot NE \times p}{pq} \times \frac{dq}{dz} :: aqdt : NE$ can be found where $\frac{NE-NO}{dt^2}$ is obtained (that is to say, the force which must be destroyed at N along NO) equal to $(-Aq - Bp)a^2$.

It is worth noting that the lines FN, Nm are always in a plane passing through the axis of the body when the body is a solid of revolution. In the following

---

[4] In this figure B, D and C are not used, however NB = FE and NC = NE.

Fig. 4.5

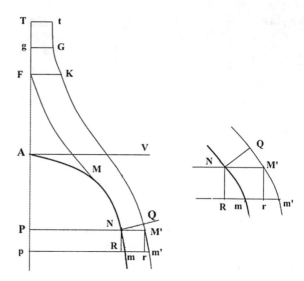

propositions, we only consider the types of solids generated by the revolution of figure ADC (Fig. 4.1) about the axis AC, and we shall pay attention to only one section ADC for the axis, because the calculation must be the same for all the rest.

Proposition VIII. Theorem

45. *Being the same things as in art. 43, I say that* $B' = -a - \frac{p}{z}$ *and* $A' = B$.

Let KQM' (Fig. 4.5) be the curve described by the fluid particles infinitely close to the surface AMN, and let the ordinates PNM', pnm', be drawn infinitely close; NR perpendicular to pm' and NQ to AN or $QM'm'$.

1st. It is obvious that the fluid velocity in N along Nm is in inverse ratio of the conical surface described by the revolution of NQ around FP. Therefore the velocity at N is as $\frac{1}{NQ \times PN}$. Therefore if the velocity along Nm is called U, and $PN = z$ or y is made, it will give $= \frac{\alpha^2 a}{Uz}$, a being the velocity at g and α a constant in order to keep the law of the homogeneity.[5]

2nd. Now, the velocity U along Nm is composed of the velocity along NR that I call aq, and the velocity parallel to Rm, that I call ap, so that $U : qa :: Nm : NR$; now then because of the similar triangles QNM', NRm, we have $Nm : NR :: NM' : NQ$, so $U \times NQ = aq \times NM'$. Therefore as $NQ = \frac{\alpha^2 a}{Uz}$, it follows that $NM' = \frac{\alpha^2}{qz}$.

3rd. Let p and q be functions of AP (x) and $PN(y)$, or in general functions of $AP = x$ and $PM' = z$, that is to say, let be the velocity of any particle a function of the

---

[5] The point N can be seen either as on the body surface, or as belonging in general to a line in any curve FN, whether contiguous to the body or not but described by the fluid particles. The ordinate PN of this curve is generally called z, and when it becomes the ordinate of the body itself, I call it y. In this and the following articles, the FN curve is not regarded as adjacent to the body surface, but away from the body at such distance as liked. It is for not multiplying the figures that we consider it as adjacent to the body in the Fig. 4.5. [*Original note*].

4 On the Pressure That a Fluid Exerts on a Body at Rest and Immersed in It   47

distances to the lines AV and AP, it is clear that aq being the velocity at M' along $M'r$, the velocity at N along NR will be: $aq + a \cdot NM' \cdot \frac{dq}{dz} = aq + \frac{a^2 a}{qz} \times \frac{dq}{dz}$. For the same reason the velocity at M' parallel to rm' will be: $ap + \frac{a^2 a}{qz} \times \frac{dp}{dz}$. Besides that, being $NM' = \frac{a^2}{qz}$, it will give: $mm' = \frac{a^2}{qz} + a^2 \cdot Pp \cdot \frac{d}{dx}\left(\frac{1}{qz}\right) + Rm \cdot a^2 \cdot \frac{d}{dz}\left(\frac{1}{qz}\right) = \frac{a^2}{qz} + a^2 \cdot dx \cdot \frac{d}{dx}\left(\frac{1}{qz}\right) + a^2 \cdot dz \cdot \frac{d}{dz}\left(\frac{1}{qz}\right)$. Therefore: $rm' = Rm + mm' - Rr = dz + a^2 dx \cdot \frac{d}{dx}\left(\frac{1}{qz}\right) + a^2 dz \cdot \frac{d}{dz}\left(\frac{1}{qz}\right)$. Now then the velocity along $M'r$ is to the velocity at M' parallel to rm', as $M'r$ to rm'.

Therefore it will give the next equation: $\dfrac{aq + \frac{a^2 a}{qz}\frac{dq}{dz}}{ap + \frac{a^2 a}{qz}\frac{dp}{dz}} = \dfrac{dx}{dz + a^2 dx \cdot \frac{d}{dx}\left(\frac{1}{qz}\right) + a^2 dz \cdot \frac{d}{dz}\left(\frac{1}{qz}\right)}$. Now then $\frac{dx}{dy} = \frac{q}{p}$ because the velocity at N parallel to dx is equal to aq, and parallel to dz is ap. Therefore we will have (neglecting quantities where a^4 is found and dividing the other for $a^2 a$) the following equation: $\frac{1}{pqz}\frac{dq}{dz} - \frac{1}{ppz}\frac{dp}{dz} = -d\frac{d}{dx}\left(\frac{1}{qz}\right)\frac{q^2}{p^2} - \frac{q}{p}\frac{d}{dz}\left(\frac{1}{qz}\right)$. Then $\frac{pdq - qdp}{zp^2 qdz} = \frac{qz^2 dq}{z^2 q^2 p^2 du} + \frac{zqdq}{z^2 pq^2 du} + \frac{q^2}{pz^2 q^2}$. Therefore $-\frac{dp}{dz} = \frac{dq}{dz} + \frac{p}{z}$. Hence, if $dq = Adx + Bdz$ and $dp = A'dx + B'dz$ are made, it will give[6]: $B' = -A - \frac{p}{z}$.

4th. Now let T be any point above γ. The fluid contained in the channel $TNM't$ and driven by the forces a''/dt, must be in equilibrium (art. 1), that is to say, the pressure of the channel NM' along NM' joined with the force of the channel $TFMN$ along FMN must be equal to the force of the channel $tKQM'$; because in the channel Tt there is not any pressure, since the velocity at T, t is uniform and rectilinear. Now then the pressure on M that comes from channel $TFMN$ is (art. 27) $(a^2 - U^2)/2$ and the pressure in M' coming from the channel $tKQM'$ is for the same reason $(a^2 - U'^2)/2$ (naming U' the velocity at M' along Nm'). Therefore $(U'^2 - U^2)/2$ is equal to the pressure of channel NM' along NM'. But $U^2 = (p^2 + q^2)a^2$. And $U'^2 = (p'^2 + q'^2) a^2 = (p^2 + q^2)a^2 + a^2 \cdot NM' \cdot \frac{d(p^2 + q^2)}{dz} = (p^2 + q^2)a^2 + \frac{a^2 \cdot a^2}{qz} \cdot \frac{d(p^2 + q^2)}{dz}$. Therefore: $\frac{U'^2 - U^2}{2} = -\left(\frac{pdp}{dz} + \frac{qdq}{dz}\right)\frac{a^2 a^2}{qz} = \frac{a^2 a^2}{z}\left(-\frac{B'p}{q} - B\right)$. Now then the force of channel NM' along NM' is (art. 43): $NM' \cdot p \cdot \left(-B' - \frac{Aq'}{p^2}\right) = \frac{a^2 a}{qz} p\left(-B' - \frac{A'q}{p}\right)$. Then we will have: $\frac{a^2 a}{z}\left(-\frac{B'p}{q} - B\right) = \frac{a^2 a}{qz} p\left(-B' - \frac{A'q}{p}\right)$. Therefore $B = A'$. Q.F.D.

From those the following theorem results. Let qa be the velocity of the fluid particles parallel to AP, ap its velocity parallel to AV, and let be $dq = Adx + Bdz$, A and B being functions of x and z, it will give $dp = Bdx - Adz - pdz/z$ or $d(pz) = zBdx - Azdz$.

Corollary I
46. So $Adx + Bdz$ and $zBdx - Azdz$ must be exact differentials. We will see later how A and B can be determined by these conditions, or what is the same, q and p.

---

[6]This sentence is unnecessary and confusing. If only to remind us that $A = dq/dz$ and $B' = dp/dz$. There is also a misprint dy instead dp. This does not appear in the Mss.47-3rd.

Corollary II

47. I do not need to announce that the same law that the quantities q and p follow has so less place for the upper part FM and adjacent curves as for the part MD in contact with the body surface and adjacent curves. It should be only noted that as the curves FM, MD do not belong to the same equation, the values of q and p in the curve FM will be determined for an equation other than that for the curve MD, although in one and in the other $dq = Adx + Bdz$ and $d(pz) = zBdx - zAdz$ must be accomplished.

Another Proof of Proposition VIII

48. The equations $dq = Adx + Bdz$ and $Bdx - Adz - pdz/z$ can still be found by another slightly more general method than the previous one. I will explain this method here very willingly, since it will be very useful to us later in the sequel of these researches to determine the resistance which a fluid at rest produces to body that moves therein.

First of all let N, C, D, B (Fig. 4.6) be four fluid particles, infinitely close each other, separated from the body as desired, and placed so that $NCDB$ is rectangular parallelogram; N', C', D', B' four other fluid particles, forming a rectangle, so that the axis AP is the common section of the two planes $NCBD$, $N'C'B'D'$, and NN', BB' infinitely small arcs described from the center P.

Now let us imagine that the particles N, C, D, B reach (Fig. 4.7) n, c, d, b. I say that $ncdb$ can be taken without error for a rectangular parallelogram. Because having drawn Mnb', $b'bd'$, nc, Gco, the parallelogram rectangle $nb'd'c$ is formed; it is obvious that the triangles $nc'c$, bdo' are infinitely small of the third order (since the line nc' is infinitely small of the first one, and the difference of lines Gc, Mn is infinitely small of the second order). Therefore the difference of triangles $nc'c$, $bd'o$ is infinitely small of the fourth order. Likewiseilt must be said of the difference between the triangles $nb'b$, cod; therefore the area of the figure $nbdc$ can be estimated equal to that of figure $nb'd'c$. Hence the small parallelepiped whose bases (Fig. 4.6) are $NN'B'B$, $CC'D'D$ will be changed into another one.

Fig. 4.6

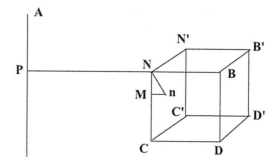

Fig. 4.7

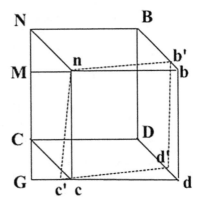

Now having made (Fig. 4.7) $NM = aqdt$, $NC = \alpha$, $NB = \beta$, it will give: $CG = aqdt + \alpha dt\frac{adq}{dx}$ and nc' or $nc = NC + CG - NM = \alpha + aqdt + \alpha dt\frac{adq}{dx} - aqdt = \alpha + \alpha dt\frac{adq}{dx}$. For the same reason it will find: $nb' = \beta + \beta dt\frac{adp}{dz}$. And if $NN' = k$ (Fig. 4.6) is done, then it is obvious that N goes to n, the quantity k will become $k\left(\frac{PN+Mn}{pN}\right) = k + \frac{kapdt}{z}$. Now then, as the particles $N, C, D, B, N', C', D', B'$ become (Fig. 4.7) $n, c, d, b, n', c', d', b'$, etc., it must be that the infinitely small portion of fluid enclosed in the first parallelepiped has to be equal to that which will fill the second parallelepiped. Therefore: $nc' \times nb' \times \left(k + \frac{kapdt}{z}\right) = NC \times NB \times Nn$. So $\alpha\beta k + k\beta\alpha dt\frac{adp}{dz} + k\beta\alpha dt\frac{adq}{dx} + \frac{k\beta apdt}{z} = \alpha\beta k$. So $\frac{dp}{dz} + \frac{dq}{dx} + \frac{p}{z} = 0$. So $B' = -A - \frac{p}{z}$, as the 3rd of the *art*.45.

Now, the force at n along nb is (*art.* 43) $a^2p\left(-B' - \frac{A'q}{p}\right)$; and the force at n along nc' or $nc = a^2q\left(-A - \frac{Bp}{q}\right) = a^2(-qA - Bp)$ (*art.* 44); now then these forces must be destroyed (*art.* 1), and it will give (*art.* 20): $\frac{d(qA+Bp)}{dz}a^2 = \frac{d(qA'+B'p)}{dx}a^2$. That is to say: $\frac{qdA}{dz} + \frac{Adq}{dz} + \frac{Bdp}{dz} + \frac{pdB}{dz} = \frac{qdA'}{dx} + \frac{A'dq}{dx} + \frac{B'dp}{dx} + \frac{pdB'}{dx}$. I say now that this equation will occur if $A' = B$, and $B' = -A - \frac{p}{z}$. Because $Adx + Bdz$ and $A'dx + B'dz$ being exact differentials, it will give $\frac{dA}{dz} = \frac{dB}{dx} = \frac{dA'}{dx}$; and $\frac{dB'}{dx} = \frac{dA'}{dz}$ or $\frac{dB}{dz}$; therefore[7]: $A\frac{dq}{dz} + B\frac{dp}{dz} = A'\frac{dq}{dx} + B'\frac{dp}{dx}$. Finally $\frac{Adq}{dz} + \frac{Bdp}{dz} = AB - BA - \frac{Bp}{z}$ and $\frac{A'dq}{dx} + \frac{B'dp}{dx} = BA - AB - \frac{Bp}{z}$. Therefore the two quantities $\frac{d(qA+Bp)}{dz}$ and $\frac{d(qA'+B'p)}{dx}$ are actually equal. So, etc.

Scholium I

49. It is worth noting that the equation $\frac{d(qA+Bp)}{dz} = \frac{d(qA'+B'p)}{dx}$ would not happen if instead of assuming $A' = B$, it was assumed $A' + B = \lambda$, λ being a constant. Because then $\frac{Adq}{dz} + \frac{Bdp}{dz}$ would be $A(A' + \lambda) + (A' + \lambda)\left(-A - \frac{p}{z}\right) = (A' + \lambda)\frac{p}{z}$; and $\frac{A'dq}{dx} + \frac{B'dp}{dx}$

---

[7] There was a major misprint with the next formula. It was written $\frac{qdA}{dz} + \frac{pdB}{dz} = \frac{qdA'}{dx} + \frac{pdB'}{dx}$. See Mss.50.

would be $= \frac{A'dq}{dx} + \frac{B'dp}{dx}$. Therefore we will not be able to have $\frac{Adq}{dz} + \frac{Bdp}{dz} = \frac{A'dq}{dx} + \frac{B'dp}{dx}$ unless $\lambda = 0$.

Scholium II
50. Before assigning the values of p and q by the conditions that have been found above, it is good to know the values of p and q at the first instant. This research and the comments with which we accompany it will be useful to determine the fluid pressure, and we will show that the values of p and q must have the same conditions in the first instant as in the following ones.

Proposition IX. Problem
51. *Let a body ADCE (Fig. 4.1) be submerged in the midst of a stagnant fluid QGHq and firmly stopped in the middle of this fluid. Then imagine that all parts of the fluid receive from whatever cause any velocity u parallel to the body axis AC. What is the change which the presence of the body must produce in the velocity of the parts of the fluid and in its direction?*

It is clear: 1st, that the particles of the fluid adjacent to the surface *EAD* cannot move parallel to *AC*, they will be forced to change direction and the same will hold for the parts neighboring to those, at least up to a certain distance from body; 2nd, a portion of fluid *FAM* which will be stagnant must necessarily exist at the front of the body (*art. 36*), and therefore that the motion will be suddenly destroyed. Hence it is clear that the fluid particles at the first instant will describe the curves *FaMD*, *OKm*, etc.

Moreover, it will be proved, as in *art. 39*, that the velocity of any point of the fluid depends only on its position; then a velocity parallel to *AC* equal to *Uq* can be assumed in the parts of the fluid, and another perpendicular to *AC* equal to *Up*, *U* being in a given ratio with *u*; so instead of *Uq*, it can write *uq* and *up* instead of *Up*, being *q* and *p* functions of *x* and *z*. It is therefore necessary (*art. 1*) that the parts of the fluid moving by the velocities of tendency *u*, and *−up*, *−uq* are in equilibrium. Now then, the velocity *u* being the same in all them, the parts would be already in equilibrium in virtue of the single velocity of tendency *u*. Therefore they must be in equilibrium in virtue of the singles velocities *−up*, *−uq*. Then if one makes $dq = Adx + Bdy$ and $dp = A'dx + B'dz$, at first $B' = -A - \frac{p}{z}$ will be found, as in the *art. 45*, Now as the channel *NM'm'm* (Fig. 4.5) must be in equilibrium, it is necessary that the pressure of channel *m'm* along *m'm* joined to the pressure of channel *mN* along *mN* to be equal to that of the channel *M'm'* plus that of channel *M'N*. That is to say that the pressure channel *m'm* minus that of channel *M'N* to be equal to the pressure of channel *m'M'* minus that of channel *mN*. Now making $Tt = \beta$ one obtains $NM' = \frac{\beta^2}{qz}$ and $mm' = \frac{\beta^2}{q'z'}$ so that the pressures of the small channels *m'm* and *M'N* are $\frac{\beta^2}{q'z'}p'$ and $\frac{\beta^2}{qz}p$, and their difference will be: $\beta^2 d\left(\frac{p}{qz}\right) = \beta^2 qz\left(\frac{A'dx - B'dz}{q^2z^2} - \frac{pdz}{zq^2z^2}\right) - \beta^2 p\left(\frac{zAdx + zBdz}{q^2z^2} + \frac{dz}{qz^2}\right)$, that is to say (putting for dz its value $\frac{pdx}{q}$ [and $B' = -A - \frac{p}{z}$]) equal to $\beta^2 dx\left(\frac{A'}{qz} - \frac{2pA}{q^2z} - \frac{2p^2}{q^3z^2} - \frac{p^2B}{q^3z}\right)$.

Now, the pressure of $m'M'$ minus that of mN must be equal (*art*. 15) to the pressure of $m'rM'$ minus that of mNR; that is to say to the pressure of rM' minus that of RN, and to the pressure of $m'r$ minus that of mR. Now then the pressure of rM' minus that of RN is: $NR \times \frac{dq}{dz} \times NM' = \frac{\beta^2}{qz} B dx$; and pressure of $m'r$ minus of that mR is: $p'dz' - pdz = \frac{d}{dz}\left(\frac{p^2}{q}\right) \frac{\beta^2 dx}{qz} = -\frac{2p\beta^2 A dx}{q^2 z} - \frac{2p^2 \beta^2 dx}{q^2 z^2} - \frac{p^2 \beta^2 B dx}{q^3 z}$, thus it will give: $dx\left(\frac{A'}{qz} - \frac{2pA}{q^2 z^2} - \frac{2p^2}{q^2 z^2} - \frac{p^2 B}{q^3 z}\right) = \frac{Bdx}{qz} - \frac{2pA dx}{q^2 z} - \frac{p^2 B dx}{q^3 z} - \frac{2p^2 dx}{q^2 z^2}$. From that $A' = B$ results as in *article* 45.

Hereafter it is clear that the quantities p and q are found at the first instant by the same equations as in the following instants. But before determining them, it still rests to us to make some essential remarks.

Remark I

52. The velocity of the particles from F until M in the fluid thread FaM (Fig. 4.3) must be extremely small. Because let V be the velocity of the particle b along bn and let us imagine, as in *art*. 36, the straight channel infinitely small $mqnb$, it is clear that all the particles that compose the channel are assumed to be animated by the velocity u parallel to FA, and the particle nb of the velocity V along nb must be in equilibrium. Now then the particles of the channel are obviously in equilibrium being assumed to be animated by the velocity u which is the same in all them. Therefore the equilibrium is not disturbed by the velocity V along bn, this velocity must be null, or at least so small that it can be considered as null.

That is the rigorous demonstration of this proposition, and we can still be convinced of its truth by the following reflection. In the first instant of the motion, all particles receive a velocity u equal and parallel to AC, and this velocity is suddenly and completely destroyed in the particles that fill the space FAM. Now then it would be shocking that while the particles contained in the FAM space are stopped suddenly, the particles that are on the curve FaM, and which are the limit of this space, had a velocity not infinitely small, since nothing in nature is done by *leaps*, but by insensible degrees; and if the velocity u becomes zero in whatever particle, the velocity of the neighboring particle cannot be more than infinitely small. Hence all convene in ensuring us that the velocity is very small in the curve FaM.

Remark II

53. As curves FaM, MDL (Fig. 4.1) are of different nature, the values of p and q will be different for these two curves, so nevertheless these values are the same at the point M. Moreover, we will not need to know the curve FaM; but it is necessary to observe that the values of p and q are the same for the first instant and for the following ones.

From that and from the *article* 45, it follows that from the first instant of the impulse, the fluid begins to describe the curves $FaMD$, (Fig. 4.1) Km, Sn, etc. and in the following instants it continues to describe them without any change happening either in its direction or its velocity. Therefore not only in the first instant, but in the following ones, the velocity along the curve *FaM* is very small, or must be deemed such. This is what we promised to prove in the *article* 36 *n*° 5.

4.2 On the Fluid Pressure at the First Instant of the Impulse

54. As we shall see below this research is absolutely necessary for determining the quantities p and q.

We have seen that the forces destroyed at the first instant in each particle are u and $-uq$, $-up$; then the pressure resulting from the common velocity u to all particles and parallel to AC will be $\mu\delta u$ (*art.* 23) naming μ the body volume[8] and δ the density of the fluid, and this pressure will be along CA. Now in order to have the pressure that comes from the velocities $-uq$, $-up$, or, which is the same, from the velocity $\sqrt{p^2 + q^2}$ along LDM, let be $PM = A$, $IL = b$. It will be seen (*art.* 28),[9] 1st that this pressure is equal to the integral $u\delta \int 2\pi y dy \int ds \sqrt{p^2 + q^2}$ taken so that it is zero for $y = b$, and that it finishes at the point M or $y = A$. 2nd, the pressure along AC expressed as the amount $A^2 \int ds \sqrt{p^2 + q^2} u\delta$ will need to be subtracted. That said, first it will be remarked that $\int ds \sqrt{p^2 + q^2} = \int pdy + qdx$ because $ds = \frac{pdy+qdx}{\sqrt{pp+qq}}$. Besides $\int 2\pi y dy (\int pdy + qdx) = \Omega$ for $y = b$, where the integral is taken so that it is zero for $y = A$; finally, let us assume gain that $\int pdy + qdx = \Gamma$ for $y = b$ [and zero for $y = A$].[10] Now take the integral $\int pdy + qdx$ so that it is zero for $y = b$, and the integral $\int 2\pi y dy (\int pdy + qdx)$, so that it is also zero for $y = b$. I say that this integral will be $\pi \Gamma A^2 - \pi b^2 \Gamma + \Omega$ for $y = A$. In order to prove it, let us express by $\int pdy' + qdx'$ the integral $pdy + qdx$ taken so that it is zero for $y = b$, and by $\int pdy + qdx$ integral of the $pdy + qdx$ taken so it is zero for $y = A$; it will give $\int pdy' + qdx' = \Gamma - \int pdy + qdx$. Let also express by $\int 2\pi y dy'$ the integral of $2\pi y dy$ taken so that it is zero for $y = b$, and it will give $\int 2\pi y dy' \int pdy' + qdx' = \int 2\pi y dy' (\Gamma - \int pdy + qdx)$.

Now then, first the integral of $2\pi \Gamma y dy'$ when $y = A$, is $\pi \Gamma A^2 - \pi \Gamma b^2$; second, the integral of $2\pi y dy' \int pdy + qdx$, taken so it is zero for $y = b$, it will be $-\Omega$ for $y = A$. Because this integral is evidently equal to $\int 2\pi y dy \int pdy + pdx$, taken negatively. Therefore $-\int 2\pi y dy' \int pdy + qdx = \Omega$. Therefore $\int 2\pi y dy' \int pdy' + qdx' = \pi \Gamma A^2 - \pi \Gamma b^2 + \Omega$. Therefore the value of $u\delta \int 2\pi y dy \int ds \sqrt{p^2 + q^2} = u\delta (\pi \Gamma A^2 - \pi \Gamma b^2 + \Omega)$.

The quantity $\pi A^2 u\delta \int ds \sqrt{pp + qq}$ must be subtracted, that is to say $\pi \Gamma A^2$; finally, $\mu \delta u$ must be added; therefore the pressure at the first instant will $u\delta (\mu + \Omega - \pi \Gamma b^2)$.

---

[8] In the text says mass.
[9] We do not see any sense to call to art. 26, we think that it is a misprint for art. 28.
[10] This is deduced from the context and it is mentioned in the *Mss.* 60.

Corollary I

55. It can be easily proved by the experiment that $\mu + \Omega - \pi \Gamma b^2 = 0$. Because a weight may be found which is capable, by its own mass, of keeping the body ADCE in equilibrium from the first instant of the impulse of the fluid, and of preventing that the body is set in motion by this impulsion. Now then, the action of a weight that is in equilibrium is equivalent to a finite mass animated by an infinitely small velocity. Therefore, the force with which the weight will be in equilibrium will also be infinitely small, thus the quantity $u\delta(\mu + \Omega - \pi \Gamma b^2)$ must be equivalent to a finite mass animated by an infinitely small velocity or an infinitely small mass animated by a finite velocity. So since the velocity u is finite, then it follows that $u\delta(\mu + \Omega - \pi \Gamma b^2)$ should be necessarily infinitely small; that is equal to zero.

Corollary II

56. Let us suppose a body at rest in the middle of a stagnant fluid, and that an equal velocity U parallel to the axis of the body is impressed to all parts of the fluid. We have seen that from the first instant the fluid particles must move along threads which will continue to be the same, while no new force arrives, and that they will always be the same, no matter the impressed velocity U. Let us suppose now that in one of the subsequent instants another velocity U' is impressed to the fluid parts, it is clear that this new velocity will disturb nothing in the threads, because if it had been alone, it would have had to describe them; only the velocity at each point must change in the atio $U + U'$ to U.

This proposition will be very useful to us in the following.

4.3 Method for Determining the Fluid Velocity at Any Point

57. To solve this subject, it is only necessary to determine the quantities p and q by means of the conditions that have been found above (*art.* 45). However, in order to solve this problem more easily, I will begin by solving it with the following hypothesis, which is simpler, that $dq = Mdx + Ndz$, and $dp = Ndx - Mdz$.

Proposition X. Problem

58. *Let* $Mdx + Ndz$ *and* $Ndx - Mdz$ *be exact differentials, we propose to find the quantities* M *and* N.

Since $Mdx + Ndz$ is an exact differential, it follows that $Mdx + N\sqrt{-1}\frac{dz}{\sqrt{-1}}$ will be also an exact differential; in the same way since $Ndx - Mdz$, then $N\sqrt{-1}dx - M\sqrt{-1}dz$, or $N\sqrt{-1}dx + \frac{Mdz}{\sqrt{-1}}$, will also be one. [Hence the sum and the difference of these two quantities will be both exact differentials. Therefore $(M + N\sqrt{-1})\left(dx + \frac{dz}{\sqrt{-1}}\right)$ and $(M - N\sqrt{-1})\left(dx - \frac{dz}{\sqrt{-1}}\right)$ will be exact differentials. So $dx + \frac{dz}{\sqrt{-1}} = du$ or $F + x + \frac{z}{\sqrt{-1}} = u$; $dx - \frac{dz}{\sqrt{-1}} = dt$ or $G + x - \frac{z}{\sqrt{-1}} = t$;

therefore $M + N\sqrt{-1} = \alpha$, and $M - N\sqrt{-1} = \beta$, udu and βdt will be exact differentials. Thus α is a function of u, that is to say, $M + N\sqrt{-1}$ a function of $F + x + \frac{z}{\sqrt{-1}}$, and β is a function of t, that is, $M - N\sqrt{-1}$ a function of $G + x - \frac{z}{\sqrt{-1}}$; from which the value of M and N will be obtained.

Corollary I

59. The functions q and p can be found also by the following method that is a little simpler.[11] Because $\frac{\partial p}{\partial z} = -\frac{\partial q}{\partial x}$ and $\frac{\partial p}{\partial x} = \frac{\partial q}{\partial z}$, then $qdx + pdz$ and $pdx - qdz$ will be exact differentials. Therefore $q + p\sqrt{-1} = \text{fonct.}\left(F + x + \frac{z}{\sqrt{-1}}\right)$ and $q - p\sqrt{-1} = \text{fonct.}\left(G + x - \frac{z}{\sqrt{-1}}\right)$. Therefore $q = \frac{1}{2}\text{fonct.}\left(F + x + \frac{z}{\sqrt{-1}}\right) + \frac{1}{2}\text{fonct.}\left(G + x - \frac{z}{\sqrt{-1}}\right)$ and $p = \frac{\text{fonct.}\left(F+x+\frac{z}{\sqrt{-1}}\right)}{2\sqrt{-1}} - \frac{\text{fonct.}\left(G+x-\frac{z}{\sqrt{-1}}\right)}{2\sqrt{-1}}$.

Hence if one wants p and q to be real quantities, it must be assumed that $G = F$, and we will have $q = \xi\left(x + F + \frac{z}{\sqrt{-1}}\right) + \sqrt{-1}\,\zeta\left(x + F + \frac{z}{\sqrt{-1}}\right) + \xi\left(x + F - \frac{z}{\sqrt{-1}}\right) - \sqrt{-1}\zeta\left(x + F - \frac{z}{\sqrt{-1}}\right)$. With $\xi\left(x + F + \frac{z}{\sqrt{-1}}\right)$ and $\zeta\left(x + F - \frac{z}{\sqrt{-1}}\right)$ designating any functions of $x + F + \frac{z}{\sqrt{-1}}$ [and $x + F - \frac{z}{\sqrt{-1}}$], different from each other if one wishes, but there are not any imaginary constants in them. Similarly we will have

$$p = \frac{\xi\left(x+F+\frac{z}{\sqrt{-1}}\right)}{\sqrt{-1}} + \zeta\left(x + F + \frac{z}{\sqrt{-1}}\right) - \frac{\xi\left(x+F-\frac{z}{\sqrt{-1}}\right)}{\sqrt{-1}} + \zeta\left(x + F - \frac{z}{\sqrt{-1}}\right).$$

It is obvious that in these values of p and q the imaginary quantities destroy themselves.

Corollary II

60. It must be noted that in the previous expressions the letter F is used only to place the origin where one wants in the line AP. Now then as the nature of the problem the origin can be placed where we like, it follows that $F = 0$ can be assumed by placing the origin of x at the convenient point of the line AP,[12] so that the expressions for p and q become simpler.

We will have therefore[13]

$$q = \xi\left(x + \frac{z}{\sqrt{-1}}\right) + \sqrt{-1}\,\zeta\left(x + \frac{z}{\sqrt{-1}}\right) + \xi\left(x - \frac{z}{\sqrt{-1}}\right) - \sqrt{-1}\,\zeta\left(x - \frac{z}{\sqrt{-1}}\right)$$

$$p = \frac{\xi\left(x + \frac{z}{\sqrt{-1}}\right)}{\sqrt{-1}} + \zeta\left(x + \frac{z}{\sqrt{-1}}\right) - \frac{\xi\left(x - \frac{z}{\sqrt{-1}}\right)}{\sqrt{-1}} + \zeta\left(x - \frac{z}{\sqrt{-1}}\right)$$

Thus if it is assumed, for example

---

[11]In the original M and N are written instead of q and p [Mss.55].
[12]Not clear in the text. We follow Mss.56.
[13]In the original the expression for q is missing [Ms.57].

4 On the Pressure That a Fluid Exerts on a Body at Rest and Immersed in It

$$\xi\left(x + \frac{z}{\sqrt{-1}}\right) = a\left(x + \frac{z}{\sqrt{-1}}\right) + b\left(x + \frac{z}{\sqrt{-1}}\right)^2$$

and

$$\zeta\left(x + \frac{z}{\sqrt{-1}}\right) = e\left(x + \frac{z}{\sqrt{-1}}\right) + f\left(x + \frac{z}{\sqrt{-1}}\right)^2 + g\left(x + \frac{z}{\sqrt{-1}}\right)^3$$

It will give

$$p = -2az + 2ex - 4bxz + 2fx^2 - 2fz^2 - 6cx^2z + 2cz^3 + 2gx^3 - 6gxz^2$$

and

$$q = 2ax - 2ez + 2bx^2 - 2bz^2 + 2fz^2$$

Now then from these expressions, I deduce the following method to determine p and q when $dq = A dx + B dz$ and $d(pz) = z B dx - A dz$.

Proposition XI. Problem

61. *To determine p and q for the conditions that $q = A dx + B dz$ and $d(Pz) = zBdx - zAdz$ are both exact differentials.*[14]

Let $p = a'x + b'z + c'x^2 + e'xz + f'z^2 + g'x^3 + h'x^2z + l'xz^2 + m'z^3$ etc. Being a', b', c', etc. undetermined coefficients. Therefore $pz = a'xz + b'z^2 + c'x^2z + e'xz^2 + f'z^3 + g'x^3z + h'x^2z^2 + l'xz^3 + m'z^4$ etc. Therefore $d(pz) = (a'z + 2c'xz + e'z^2 + 3g'x^2z + 2h'xz^2 + l'z^3)dx + (a'x + 2b'z + c'x^2 + 2e'xz + 3f'z^2 + g'x^3 + 2h'x^2z + 3l'xz^2 + 4m'z^3)dz$ etc. So due to $dq = -\frac{dxd(pz)}{zdz} + \frac{dzd(pz)}{zdx}$, we will have $dq = (a' + 2c'x + e'z + 3g'x^2 + 2h'xz + l'z^2)dz + (-\frac{d'x}{z} - 2b' - \frac{c'x^2}{z} - 2e'x - 3f'z - \frac{g'x^3}{z} - 2h'x^2 - 3l'xz - 4m'z^2)dx$. Now then so that this quantity be an exact differential, it is necessary that $2c' + 6l'x + 2h'z = \frac{d'x}{z^2} + \frac{c'x^2}{z^2} - 3f' + \frac{g'x^3}{z^2} - 3l'x - 8m'z$. It gives then $a' = 0$, $g' = 0$, $c' = 0$, $f' = 0$, $l' = 0$, $4m' = -h'$; so $pz = b'z^2 + e'xz^2 + h'x^2z^2 - \frac{h'z^4}{4}$, and $p = b'z + e'xz + h'x^2z - \frac{h'z^3}{4}$. Being b', e', h' undetermined coefficients.

From there the law of the quantity p is rather clear. Because it will give $p = b'z + e'xz + h'x^2z + m'z^3 + n'xz^3 + k'x^3z + r'x^2z^3 + s'x^4z$ etc. An equation in which the values of de m', k', s' can be substituted by their equivalent in h', n', r', etc. respectively,[15] and the unknowns b', e', h', n', r' etc. will remain to be determined by the nature of the curve *AMD*. [Once p is known, the quantity q will be easily determined].[16] QED.

---

[14]In the original A and B are written instead p and q [Ms.58].

[15]The original wording "in which the value of m' in h', of k' in n', of s' in r', etc. can be determined" has been changed, because the sense is to apply the relations between each pair of coefficients.

[16]We think that this sentence adds clarity to the text. From *Mss.* 58: "cognita vero p, determinabimur facillime quantitas q" "But once p is known, we will determine the quantity q very easily".

Corollary I

62. In order to determine now the coefficients b', e', h' etc. it will be noted that introducing y as z in the values of p and q, gives $\frac{dy}{dx} = \frac{p}{q}$. Therefore a certain number of points will be taken on the curve AMD, in which the values of dy/dx, y, and x are known, and then the coefficients b', e', h', etc. will be determined precisely in a similar way as how the integration of a curve is found by approximation, making a line of parabolic type to pass through a number of points of the curve [AMD].

Remark

63. Moreover, after the calculation of these coefficients, one still remains whose absolute value is ignored, and this value will affect the value of all the others. This is obvious, because when the top and bottom of the fraction p/q is multiplied by any quantity m, its value will not change anything. therefore this coefficient must be determined; moreover, the position must be found of the points M, L, (Fig. 4.1) that determine the length of the fluid filet in contact with the curve; or, what is the same thing, the abscissas must be found that have these points. So three new unknowns must be found for the full solution of the problem. For this, it is noted that the velocity at M and L must be very small, or equal to zero (*art.* 52); from where it is deduced $a\sqrt{p^2 + q^2} = 0$ at M and L. Therefore, calling C and D the abscissas of the points M and L, and A, b their ordinates, which are known functions of C and D, we must have: 1st, that has $a\sqrt{p^2 + q^2} = 0$ or $p^2 + q^2 = 0$ putting in p and q, C for x, and A for y; 2nd, that $p^2 + q^2 = 0$, putting in p and q, D for x, and b for y; 3rd, looking at A and B as known as well as C and D, we will have the values for Γ and Ω, assigned above (*art.* 54); now then, these values must be such that $\mu + \Omega - \pi \Gamma b^2 = 0$ (*art.* 55). Therefore this equation, along with the previous two, will serve to determine the three unknowns that remain to us.

Corollary II

64. From the previous it is noticeable, that it is sufficient to know the velocity of the fluid thread which is immediately adjacent to the body surface. Therefore, let the quantities p and q supposedly be found, and put these quantities y in the place of z; moreover let us assume that after this substitution p is divided by q and the quotient is n; it will give $p/q = n$ or $p = qn$ and $pz = qnz$, being always z in the place of y. Therefore making $dn = \lambda dx + \omega dz$, it will give $d(pz) = nzAdx + qz\lambda dx + qndz + nzBdz + zq\omega dz$, now then we have $d(pz) = zBdx - Azdz$; these two values of $d(pz)$ are equal and identical,[17] because the quantities p and qn are equal and identical, so $nzA + qz\lambda = zB$ et $qn + nzB + zq\omega = -zA$. Hence it is deduced: 1st, $A = -\frac{zq\omega + zqn\lambda + qn}{n^2 z + z}$; 2nd, $B = \frac{-nzq\omega - n^2 zq\lambda - qn^2}{n^2 z + z} + q\lambda = \frac{-nzq\omega - qn^2 + q\lambda z}{n^2 z + z}$. Consequently

---

[17] I call identical amounts, those not only equal, but expressed with the same letters: for example $\frac{a^2 - b^2}{a+b} = a - b$ or $(a^2 - b^2) = (a+b)(a-b)$ is an identical equation. But I simply call equal the quantities, which although the same, are expressed by different letters. For example y and $\sqrt{2ax - x^2}$ in the equation $y = \sqrt{2ax - x^2}$. [*Original note*].

4 On the Pressure That a Fluid Exerts on a Body at Rest and Immersed in It

$dq = \frac{-zq\omega dx - zqn\lambda dx - qndx}{n^2z+z} + \frac{-nzq\omega dz - qn^2dz + q\lambda zdz}{n^2z+z}$. Where because of $\lambda dx + \omega dz = dn$ it is $\frac{dq}{q} = -\frac{ndn}{n^2+1} - \frac{ndx+n^2dz}{n^2z+z} + \frac{\lambda dz - \omega dx}{n^2+1}$. Now then we have $\frac{dy}{dx}$ or $\frac{dz}{dx} = n$. Therefore $ndx = dz$, and $\lambda dz = n\lambda dx$; therefore in the filet AMD it will give $\frac{dq}{q} = \frac{-ndn}{n^2+1} - \frac{dz}{z} + \frac{n\lambda d'x - \omega dx}{n^2+1}$. Now then $n\lambda dx - \omega dx = ndn - n\omega dz - \frac{\omega dz}{n} = ndn - \omega dz\left(\frac{n^2+1}{n}\right) = ndn - \omega dx(n^2+1)$. Therefore $\frac{dq}{q} = -\frac{dz}{z} - \omega dx$.

Corollary III

65. It seems at first that nothing is easier than determining q by the equation found in the *previous art.*, since ω is given by n and n is given by the equation of the urve $dy/dx = n$. But with only paying some attention to it, it will be seen that although the equation of the curve or the value of dy/dx is given, n is not given for it. Indeed, n must be equal to p/q; then the ratio dy/dx can be expressed of an infinite number of ways; and among these different expressions, which are not identical although equal, one must be found that is equal to p/q, dq being $Adx + Bdz$ and $d(pz)$ being $zBdx - Azdz$. For example, in the circle it is $\frac{dy}{dx} = \frac{a-x}{y}$ or $\frac{ay-xy}{2ax-x^2}$, or $\frac{a\sqrt{2ax-x^2}-xy}{y^2}$; now then one cannot take as one wishes one of these values for expression of n; it is also required that the equation $p = qn$ be identical.

To see more clearly that n cannot be taken at will; it will be observed that from equation $\frac{dq}{q} = -\frac{dz}{z} - \omega dx$ we obtain $\frac{dq}{q} = -\frac{dz}{z} - \frac{\omega dz}{n}$ and $\frac{dq}{q} = -\frac{dz}{z} - \frac{dn}{n} + \frac{\lambda dx}{n}$; equations from which precisely the same value of q must result. Now if n could be taken at will, let n be taken so that in the first equation n is a function of z only. It will give $-\frac{\omega dz}{n} = -\frac{dn}{n}$ and $\frac{dq}{q} = -\frac{dz}{z} - \frac{dn}{n}$ or $= \frac{\beta}{nz}$, β denoting a constant. Now, let n be equal to a function of x only in the other equation, it will result $\frac{\lambda dx}{n} = \frac{dn}{n}$ and $\frac{dq}{q} = -\frac{dz}{z}$, or $q = \frac{\beta}{n}$, an equation very different from $q = \frac{\beta}{nz}$. Thus etc.

In *art.* 64, we found $nA + q\lambda = B$ and $qn + nzB + zq\omega = -zA$ considering the equation $p = qn$ as identical. Now in general let n' be the value of dy/dx, so that the equation $p = qn'$ is not identical; and let be $dn' = \lambda'dx + \omega'dz$, we will have $n'zAdx + q\lambda'zdx + qn'dz + n'zBdz + zq\omega'dz = Bzdx - Azdz$; therefore (because of $dz = n'dx$) we will find $n'zA + q\lambda'z + qn'^2 + n'^2zB + zq\omega'n' = Bz - An'z$. Therefore by this equation we will have the value of A in B, where $z = y$. But as the unknown B remains to be determined, this method is perhaps not very useful.

4.4 On the Pressure of the Fluid at Each Moment

66. Let us suppose that from the condition equations $Adx + Bdz = dq$ and $d(pz) = zBdx - zAdz$ the functions p and q have been found, as we have taught. Next, when y is placed instead of z in these functions, the velocity at N (Fig. 4.1) will be equal to $a\sqrt{p^2 + q^2}$; from which the pressure at N equal to $\frac{a^2}{2}[1 - (p^2 + q^2)]$ (*art.* 27). As the quantities p and q depend only on the body shape, it is obvious that the pressure at

the point N is proportional to the square of the velocity, and in consequence the pressure upon the entire surface is proportional to the same square.

Moreover, in order for this expression to be exact, it must be assumed that $p^2 + q^2$ is for all smaller than one, that is to say that the velocity along the filet MDL is smaller than a in all the points, or at least it is not greater. However, if after having determined p and q by calculation it would be found that $p^2 + q^2$ was > 1 in some points, then at first one should seek the point where the value of $\sqrt{p^2 + q^2}$ is *maximum*, which will be done assuming $pdp + qdq = 0$. Then naming K the value of $\sqrt{p^2 + q^2}$ at this point, we will have $\frac{a^2}{2}(K^2 - p^2 - q^2)$ (*art. 27*) for the pressure at N.

Remark

67. Some readers perhaps may imagine that the velocity along the thread MDL must be greater than a; they can truly be based on daily experience, by which it seems verified that the fluid accelerates when turning around the body. However if $\sqrt{p^2 + q^2} < 1$ is found by calculation one should not rush to conclude that our theory was contrary to experience. Because in this theory only the thread that touches immediately the body surface is taken; this thread escapes observation, and it may be that threads which are at very little distant from it have much more velocity than it.

Corollary I

68. Let be $K^2 > 1$. In order to have the total pressure we must first integrate $2\pi y dy$ $(K^2 - p^2 - q^2)\frac{a^2}{2}$ (*art. 26*). Furthermore, the pressure at M is $\frac{a^2}{2}K^2$, as well as in L, since the velocity at M and L is zero or it is taken as such. It follows that the part AM will be pressed (*art. 24*) along AC with a force $\frac{a^2}{2}K^2\pi A^2$, and the part LC in the opposite direction with a force $\frac{a^2}{2}K^2\pi b^2$. Therefore it will add the quantity $\frac{a^2}{2}K^2(\pi A^2 - \pi b^2)$ to the pressure.

But if the velocity along the curve AMD was found to be smaller than a, that is to say, if $\sqrt{p^2 + q^2}$ was in all points smaller than one, then instead of $K^2 - p^2 - q^2$, the $1 - p^2 - q^2$ is taken and instead of $a^2K^2(\pi A^2 - \pi b^2)$, the $a^2(\pi A^2 - \pi b^2)$. Because the pressure in F would always be $\frac{a^2}{2}$ and this pressure would act on the arc AM by the channel $TFAM$, and the pressure at L would also be $\frac{a^2}{2}$ because the velocity at L is zero, so that this pressure would act on the arc LC. Thus etc.

In general in all these cases, I called $a^2 \varphi$ the quantity found by the calculation for the total pressure, which as one sees is proportional to a^2, because φ will always be the same regardless of a, since the position of the points L, M and the values of p and q do not depend on a.

Corollary II

69. Let be $KD = \gamma$ the greater ordinate, we will have $\int 2\pi y dy \frac{K^2 a^2}{2} = \frac{K^2 a^2}{2}(\gamma^2 - A^2) - \frac{K^2 a^2}{2}(\gamma^2 - b^2)$. Therefore the total pressure will be reduced in the first case to $-\int 2\pi y dy(p^2 + q^2)\frac{a^2}{2}$, since $\int K^2 2\pi y dy + K^2 A^2 - K^2 b^2 = 0$. For the second case it will also be found that the total pressure will be reduced to $\int 2\pi y dy(p^2 + q^2)\frac{a^2}{2}$.

4 On the Pressure That a Fluid Exerts on a Body at Rest and Immersed in It

Therefore in general, if after having determined p and q by the method of *art*. 62, the integral $-2\pi y dy(p^2+q^2)$ is taken so that it is cero at M and that the value of this integral at L is named φ, it will give $a^2\varphi$ for the total pressure.

Corollary III

70. Let us suppose (Fig. 4.8) that the parts AD, DC are equal to each other, and let the points V be taken, u equally distant from point D. It is clear that the value of n, that is dy/dx, is the same in these points, but of different sign. Therefore if it is assumed that the point K is the midpoint of AC and that the distance from this point K to the origin of the x is h, the value of n must be a function of y and x, such that making $h-x=u$, and taking u as successively negative and positive, n is the same but of different signs. Therefore this function must be such as there is not any term which does not contain an odd power of u or $h-x$. Therefore in the differential $\lambda dx + \omega dz$, ω will be negative [*art.* 64].[18] When $h-x$ will be negative, but will always keep the same value. Therefore in the points V, u the value of ω is the same, but with different signs; from which it can be easily proved that the value of $\int \omega dx$ will be the same in these points and of the same sign. So because of the equation $\frac{dq}{q}=-\frac{dy}{y}+\omega dx$, it follows that the value of q is the same at V and u, etc., thus the value of $p=qn$ is also the same there, but with different signs. Therefore the value of p^2+q^2 will be the same at V and u.

Therefore in general let V and u be any two points equally distant from the point D, so that the velocity at V and u is the same, the quantities $-\int 2\pi y dy(p^2+q^2)$ will be of different signs at V and u, but of equal value.

From above it follows that the arcs LD, DM (Fig. 4.1) cannot be equal; because if they were, then the quantity $-\int 2\pi y dy(p^2+q^2)$ would be equal to zero, so that the body would not undergo any pressure from the fluid, which is against experience. There is nobody that at first glance had not judged that the arcs DL, MD are always equal when the body is composed of four similar and equal parts. Even more, if we stand by the theory alone, it seems to me we would be moved to think, that these arcs must be equal in effect. Hence it is clear how experiments are needed in the present question.

In addition, it is clear that in order that the pressure be directed along AC, as experience teaches, LD must be greater than DM, otherwise $-\int 2\pi y dy(p^2+q^2)$ would be negative.

Remark I

71. Let us imagine (Fig. 4.8) the straight line RT[19] that separates the parts of the fluid where the velocity and direction are not changed from those where in which the velocity and direction are changed; it seems to me that it can be proved by a

---

[18] Reference to *art.* 64 added for clarity.

[19] It is obvious that RT should be a straight line; because the parties that are at the right of this line must have (*hyp.*) a rectilinear motion. [*Original note*].

Fig. 4.8

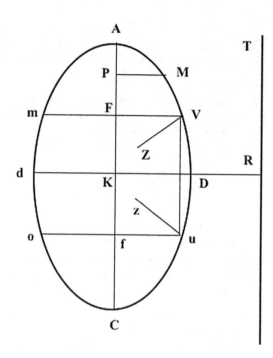

common and simple experiment that the line *RT* is quite close to the body. Let a pendulum be exposed in a fluid current; at first such that it is equally away from the channel walls where the fluid flows, the action of the fluid will move the pendulum apart from the vertical position and the pendulum will rise in a vertical plane along the direction of the fluid current. Then let the same pendulum be exposed again in the same current, but so that it is so much closer to a wall than the other, it will seem to rise to the same level as in the first case and in a vertical plane also along the direction of the flow. Therefore, either the body is placed in the middle of the channel, or much closer to a wall on the other, the pressure on the parties *ADC*, *Adc* (Fig. 4.8)[20] will be equal in both cases, and consequently the velocity in the parts *ADC*, *Adc* as well. Whereby it results that the parts of fluid too near to the body are the only ones where the motion is significantly changed by the effect of the body.

Remark II

72. We can prove the same proposition by means of a body that rises in a vessel full of water; because the body always rises vertically at the same velocity from any part of the vessel where it has been placed and as close to the walls that it is. Whence it follows that it only communicates motion to the parties of the fluid that are quite close to it. Finally, we can still observe that whatever the fluid velocity is, the *RT* line (Fig. 4.8) must always (*art*. 39) be at the same distance from the body. Now then experiment shows that when the velocity is very small the motion and direction

---

[20] In the original says *AMD* y *Amd*, which corresponds with the quote to Fig. 23 in *Mss*. 66.

of the parts of the fluid are not altered until a relatively small distance from the body. So in general, it is quite close to the body, regardless of the velocity a.

4.5 The Resistance of a Plane Figure

73. So far we have considered solids of revolution exposed to the action or to the resistance of a fluid. Let us now imagine a plane figure, or rather to avoid any difficulty, let us imagine a cylindrical body whose cross section perpendicular to its axis is the *EADC* curve (Fig. 4.1) and that it fills exactly the entire width of the channel, which I assume to be a parallelepiped rectangle filled with water, and whose height perpendicular to *qGHQ* is equal to that of the cylinder. It is clear that can be content to consider what happens to one of the layers perpendicular to the axis. Now, keeping the names of the *art*. 45, it will be easily found that $B' = -A$ and $A' = B$; since we will only need to put α/q in the places of *NM* and α^2/qz in the prove of the *art*. 45, and to put zero instead of p/z in that of *art*. 48.

Then in this case we will have $dq = Adx + Bdz$ and $dp = Bdx - Adz$, and the general formula for the value of p and q will be easily found (*art*. 58). But it will not be very easy to apply this formula to the different figures proposed; at least I have not found a method, other than *art*. 61, to choose in this general equation, which is the one that may agree with the equation of the given figure.

4.6 Notes on Our Solution to the Problem of Fluid Pressure

74. The solution that we give here for the problem of fluid pressure etc is based it seems to me, on principles less vague and less arbitrary than all the ones that have been given so far. Everything is rigorously proven, and this is perhaps why it is so difficult to apply the calculation to it and to compare it with the experiment. Because 1st, we only determine by approximation the values of the suitable p and q in each case. 2nd, the analysis by which we propose to find them is so long that it can discourage the most intrepid calculator. However, I do not think a more direct and simple method can be found for determining the resistance and fluids pressure, and I dare even assert that if this method does not agree with what will be found by experiment, we should almost despair of finding the resistance of fluids by the theory and *by the analytical calculation*; since all the physical principles on which our analysis is supported have been demonstrated in rigor. There is only an analytical hypothesis that could be absolutely questioned by us; it is that by which we have assumed that p and q are functions of x and z so that *TFMD*, *OKm*, etc (Fig. 4.1) are curves of the same nature and enclosed in the same general equation. Strictly speaking, that assumption can be disputed to us, but in this case we must give up all hope to determine the fluid pressure by calculation, and consequently by the theory, because since we have proved that the values of the

quantities p and q depend only on the position of the point at which they are, a more general hypothesis could not be made for the calculation than to assume these quantities are functions of x and z.

4.7 Reflections on the Experiments That Have Been Made or That Can Be Made to Determine the Pressure of the Fluids

75. The pressure of a fluid that strikes a body at rest can be determined by experiments in two ways.

The first is to place the body in a fluid current and to find out through experiments the action of the stream on the body; it seems to me that *M. Mariotte* has made it by the simplest method. It consists in placing at first a horizontal axis in a plane perpendicular to the current; after that, in a plane perpendicular to this axis, two rods are attached, which form between them a right angle. At the end of one of these rods, the body whose pressure is to be found is fixed, and the quantity of this pressure will be known by the weight it is necessary to be put at the end of the other rod, so that both are in equilibrium.

Mariotte found by this method that the fluid pressure against a flat surface perpendicular to the stream is equal to the weight of a cylinder of fluid that had this surface as base and whose height was that *due to the velocity* of the fluid. By this method the pressure against a flat surface could be also determined easily that would be exposed obliquely to the fluid stream.

To make this experiment easier and to make the calculations simpler, it is convenient that the two rods be arranged in such a way that, as in the equilibrium one is vertical and the other horizontal. To do that, the axis of the body, whose pressure we wish to determine, must be perpendicular to the rod to which the body it is attached.

In the case where we want to test the pressure of a flat surface situated obliquely in respect to the current, this flat surface can be placed inclined relative to the rod, and maintaining the two rods in their horizontal and vertical situation; or the flat surface will be left in the same plane as the rod and then the two rods will be forced to be inclined.

We can also determine the pressure by means of a pendulum that will be exposed to the fluid stream; since this pendulum moves out of the vertical position and having measured the angle at which it deviates, this proportion will be made*: as the total sine is to the tangent of that angle, so is the weight of the pendulum to the pressure sought*; an analogy so easy to prove that I do not think I need to dwell on it.

This latter method can be hardly employed conveniently except for determining the pressure upon spherical bodies. To this regard, when it is employed to determine the pressure of a rectangular plane or circular or oval, or made as a triangle, or in general as any polygon, it should be noted that the center of pressure or the fluid is or must be supposed to be the center of gravity of the figure. The knowledge of this center is needed to determine the lever arm on which the impulsive force of the fluid acts.

76. The second method to find the fluid pressure by experiment consists in finding out the resistance. We are going to talk about it the following *articles*.

Remark I

77. According to experiments that have been made so far by various authors, we have $\pi b p \delta / 2$ for the pressure on the globe, b expressing the height due to the fluid velocity, δ its density, p the natural gravity, 2π the ratio of the circumference of a circle to its radius, and 1 and the radius of the globe. Which I prove in this manner.

The resistance that a fluid exerts against a body moving therein is equal, as we will prove later, to the pressure that the same fluid, moved with a velocity equal to that of the body, would exercise against the same body at rest. Moreover, following the Proposition 39, Book 2 of Principia Mathematica of *M. Newton*, the resistance of a fluid to a spherical body is to the force with which the complete motion of the body may be destroyed or generated while it traverses the 8/3 of its diameter, as the density of the fluid to that of the body. Now then, let θ be the time during which the globe describe uniformly the 8/3 of its diameter with the velocity $\sqrt{2pb}$, let be 1 the globe the density, and consequently its mass $4\pi/3$; the time θ will be $16/3\sqrt{2pb}$, and the resistance according to *M. Newton* will be $\frac{4\pi}{3} \frac{\sqrt{2pb}}{\theta} \frac{\delta}{1} = \frac{\pi b \delta p}{2}$. Such as this is the formula of the resistance according to *M. Newton*, a formula that he says to have confirmed by a large number of experiments.

In the Memoirs of the Academy of Petersburg Tom. 2*M. Daniel Bernoulli* gave another formula for the resistance of globes, that he confirmed equally by experiments, and which agrees, as will be seen, with the previous one. Here is the proposition of *M. Bernoulli*. Let s be the space that a heavy body traverses freely when falling in the interval of 1 s, na the space that a body would traverse in the same time with uniform speed $\sqrt{2pb}$, p' the weight of a fluid cylinder whose base is the circumference 2π described of radius 1, and whose height is a. *M. Daniel Bernoulli* found that the pressure on the globe, or the resistance of the globe, is $\frac{n^2 a p'}{8s}$ or $p' = \pi \delta a p$; therefore $\frac{n^2 a p'}{8s} = \frac{\pi \delta p n^2 a^2}{8s}$; but as the spaces na and $2s$ are traversed in the interval of a second (*hyp.*) it gives $\frac{na}{\sqrt{2pb}} = \frac{2s}{\sqrt{2ps}}$, or $a^2 = 4bs$. Therefore $\frac{n^2 a p'}{8s} = \frac{\pi \delta b p}{2}$.

Remark II

78. According to *M. Newton*, the resistance of the globe is equal to that of the cylinder; thus the pressure on the latter would be $\pi \delta b p / 2$, but following *M. Daniel Bernoulli*, the pressure on the cylinder is double the pressure on the globe and therefore it will be $\pi \delta b p$. This latter proposition seems to agree with the experiments of *M. Mariotte*, according to which the fluid pressure against a flat surface is equal to the weight of a cylinder whose base would be that surface, and whose height is equal to the line b. Moreover, *M. Newton* does not seem to have sufficiently demonstrated the pretended equality of the two resistances, as has been shown in the Introduction. With respect to f *M. Daniel Bernoulli*, he shows that the

resistances are in the ratio of 1to 2, by the same method that *M. Newton* used in his Principia Math. Book 2, Prop. 34; that method would not have taken place in the case of a continuous fluid. *M. Daniel Bernoulli* ensures that he has tried several experiments on the resistance cylinders, and that they agree with his theory; disregarding the viscosity of fluids, which often contributes to increase the resistance, especially in the cylindrical bodies. That is why, while waiting for new experiments on this subject, we take $\pi\delta bp$ and $\pi\delta bp/2$ for the pressures of the cylinder and the globe.

Considering the fluid particles as small corpuscles of the fluid without elasticity and separated one from another, as I have done elsewhere,[21] it results from the formulas that I have given, that the pressure on the cylinder would be $2\pi\delta pb$; and assuming that the particles of the fluid as small elastic corpuscles the pressure would be $4\pi\delta pb$. *M. Euler*, who, in his treatise titled Scientia navalis, has determined the resistance of fluids by ordinary principles, found the same results, and rightly concluded that the theory on which they are supported should not be very accurate, since it is contradicted by the experiment. Furthermore he observes in the case of elastic corpuscles that the velocity communicated to the fluid particles would be, according to the ordinary laws of motion, greater than the velocity that would remain in the body; and then with this hypothesis a vacuum must be produced in the outer part of the body, between the body and the fluid. From which he rightly concludes that this hypothesis is not consistent with nature; which, together with the reasons given in the Introduction, should determine him to reject it.

Therefore, the wise geometrician we are talking about has then tried to prove by another method that the pressure against a flat surface is *bp*; his reasoning can be summarized here. Let us imagine a vessel full of water up to the height *b* in whose bottom a circular hole is made and that a flat surface is applied to the hole. This flat surface will be pressed by a force *bp*. Now let move this surface away at some distance from the hole, the water will come out with the velocity due to the height *h*, and *it can be assumed that the pressure on the surface will be the same as before*. This last assumption is not true as we will prove in the following. Since the pressure of a fluid stream that comes out from a vessel and that hits a plane is very nearly equal to *2ph*, not to *ph*, as happens when the surface is completely submersed in a fluid. Also the author has not he given any proof of the assumption that we are disputing; and we owe him the justice to say that he seems to have felt, or at least suspected the little accuracy.

Remark III

79. *M. s'Gravesande* in his Elem. of Phys. Math. finds for the resistance of a globe a very different quantity from those we have just given from *M. Newton* and. *Bernoulli*. According to this author, the action of a fluid on a cylinder (excluding the viscosity, the weight, and the friction of parts) is the same that *M. Bernoulli* had found. With respect to the pressure of the globe, it is not as 1 to 2, but as 2 to

---

[21]Treaty of the Equilibrium and Movement of Fluids, Book 3, Ch. 1. (*Original note*).

Fig. 4.9

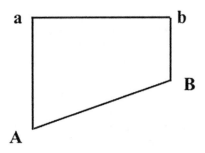

3 (§.1950). *M. s'Gravesande* has confirmed this relation by experiments, and he even undertook to demonstrate it geometrically. The demonstration he gives is the same, as respect to the method, to that by which *M. Newton* and *M. Bernoulli* have found the resistances in a ratio of 1 to 2, but with the difference that the authors quoted have assumed that the action of the fluid perpendicular to each small segment of a curve was in ratio to this small segment, to the square of the velocity and to the square of the sine of incidence; whereas according to the principles of *M. s'Gravesande*, the action was in the compound ratio of the segment of the curve, the square of the velocity and simple sine of the incidence. From this comes the difference of the ratios of 1 to 2 and of 2 to 3.

It is true that the assumption of *M. s'Gravesande* seems contrary to the principle accepted so far by all hydraulic authors, namely, that the action of a fluid which shocks obliquely against a flat surface is as the square of the sine of incidence, being all other things equal. But we must confess that this proposition has so far been poorly proved. Because this demonstration that has been given is based on this single consideration (Fig. 4.9): that the more the surface *AB* is oblique to the flow of the fluid, less are the particles that strike it, since the number of these particles is represented by *ab* perpendicular to the direction of fluid *aA*. Besides this, less is the force with which each of these particles strikes the plate; so that this compound ratio leads to the square of the sine of incidence. Now we would prove by the same reasoning that the oblique pressure that a stagnant fluid would exert against the surface *AB* would be in the ratio of this surface and the square of the sine of incidence; because *aA* is the direction of gravity, it seems that *ab* must represent the number of particles, and the force of gravity acting on *AB* seems to be the sine of the angle *aAB*. However it is known by the hydrodynamic principles that the pressure upon *AB* is for *AB* only, whatever the positioning of this surface is relative to the fluid; because the fluids act equally in any direction, the pressure on the surface *AB* is always the same as if the fluid was perpendicular to that surface.

On the other hand, however, if the fluid pressure on the surface *AB* was assumed to be proportional only to the surface *AB*, so that in this regard a moving fluid is considered as a fluid at rest; it would follow that this pressure would not depend in any way on the position of the surface, which is contrary to experience; because there is no one who has experienced that resistance is even greater when the surface is more directly opposite to the fluid stream. Therefore the sine of incidence must enter into the value of the pressure; but how should it go in? This is what seems to me very difficult to decide. *M. Daniel Bernoulli* in the Tom. 8 of the Mem. Petersburg finds that the pressure of a fluid stream is proportional to the amplitude or width of the flow and the simple sinus of the incident, which, as can easily be seen, means the square of the sine. But he admits that his formula is vague and uncertain, and he promises to make experiments on that. With regard to *s'Gravesande*, that I know, he does not bring any experiment to show that the pressure is simply in the ratio of the sine of incidence, and even if attention is paid to the position of the vanes in windmills, the truth of this proposition can be doubted. For the following experiment, the more advantageous position of the vanes is when they have 54° of inclination, assuming that the pressure is proportional to the square s^2 of sine of the incident s; whereas, if it is assumed proportional to simple sine s, the angle would be found of 45°, which is contrary to the experiment. Indeed, in the case of the square of the sine, the wind effort to rotate the wing is proportional to $s^2\sqrt{1-s^2}$, which has a *maximum* for $1-s^2 = 1/3$; in the hypothesis of simple sine the effort is proportional to $s\sqrt{1-s^2}$ which is a *maximum* for $s = 1/2$.

We have already observed that the quantity of the globe pressure determined by *M. s'Gravesande* appears confirmed by the experiments he reports (§.1495) and that this quantity is very different from that which *M. Newton* and *Bernoulli* have also confirmed by experiments. However *M. Daniel Bernoulli* admits in his Hydrodynamique that experiments give nothing as the theory, the pressure of the globe equal to half that of the cylinder. But what should he respond to the experiment that himself had done? With regard to the cylinder pressure, *M. Mariotte*, *Bernoulli* and s'*Gravesande* find it by the same theory; but besides that it is very difficult to determine this pressure by experiments; *s'Gravesande* admits that those he made on this subject do not agree with his formula. It would therefore be necessary to repeat all these experiments again and to begin by determining the pressure of a fluid that strikes obliquely against a flat surface. This is can be easily performed by means of the method specified in *article* 75.

The experiments of *M. Mariotte*, *Bernoulli*, *Newton* and *s'Gravesande* on the resistance of the globe should be then repeated. But even when in any particular experiment the oblique pressure of a flat surface had been found proportional to the simple sine of the incidence, this would not be a reason to admit the theory of *M. s'Gravesande*. Since this theory has all the defects which we have expounded in the Introduction.

Chapter 5
On the Resistance of Fluids to the Bodies Moving Therein

5.1 General Observations on the Various Classes of Fluids

80. All fluid wherein a body moves is elastic or non-elastic. I called elastic fluid one whose parts can be contracted so they occupy a lesser space than before their compression, and reciprocally they dilate so that they occupy a larger space than before their expansion. I called non-elastic fluid one whose parts can be neither contract nor dilate, but always occupy the same space, whatever the force be that compresses them.

81. If a body moves in a fluid of the latter class and the fluid is either indefinite or is enclosed in a finite vessel and closed on all sides, so that it fills its volume completely; in this case it neither should have and nor can have a vacuum between the parts of the fluid and the surface of the body that moves therein. Because, there could be no empty space, unless the parts of the fluid do not contract themselves, which goes against the hypothesis.

82. It could occur otherwise [the empty space] if the body moves in a non-elastic fluid contained in a vessel which is not closed in all sides. As is for example, stagnant water in a basin and in which a body is immersed that is not far from the upper surface of the water, and that its weight is also equal to the volume of water; I add this condition to be able to exclude the weight of the body and the fluid more easily. When an upward impulse is given to that body towards the upper surface of the standing water, it is clear that by this impulse the fluid is pushed in the front part, which is the part that is between the water surface and the upper surface of the body. Thus, as the parts that are on the surface of the water can freely move upwards, it may occur that the motion impressed to the body indeed compels these parts to move in that way, so that the surface of the water in that place loses its state and its straight and horizontal figure, and it rises above its level. Then nothing prevents that a void could be produced between the rear surface of the body and adjacent parts of the fluid; especially if the motion impressed to the body is large enough so that the

© Springer International Publishing AG 2018
J. Simón Calero (ed.), *Jean Le Rond D'Alembert: A New Theory of the Resistance of Fluids*, Studies in History and Philosophy of Science 47,
https://doi.org/10.1007/978-3-319-68000-2_5

pressure is communicated from the first instant to the surface of the water, and so that the fluid adjacent to the rear part of the body cannot spring forth with enough velocity into the empty space that the body will leave behind.

83. If the fluid is elastic, whether finite or indefinite, it is obvious that the parts of the fluid must necessarily be contracted in the front of the body and dilated at the rear part. In many cases it may even happen that the fluid rushing into the void left behind the body does not completely fill this vacuum, which will happen if the velocity that the fluid must have in virtue of its compression is less than the velocity impressed to the body.

84. We will divide our research on the resistance of the fluids. In the first we will deal with the resistance of non-elastic and indefinite fluids, or, what is equivalent, those contained in an undisturbed vessel and closed on all sides; so that they fill its volume exactly. That is to say (usually speaking) of the fluids resistance in the case where no vacuum between the fluid and the body is produced.

In the second part, we will deal with the resistance of the non-elastic and finite fluids; that is where a vacuum is produced behind the body.

Finally in the third, we will discuss the resistance of the elastic fluids. We destine a separate section[1] to each of these parts, and in these sections we will deduce several important remarks.

5.2 The Resistance of Non-elastic and Indefinite Fluids

85. Before determining this resistance, it is worthwhile to make some necessary observations in order to understand the subsequent calculations.

1st. At first, we will exclude the viscosity and friction of the fluid parts in this research, whose effect we will examine later separately.

2nd. We will also exclude at first the weight, both for the body and the fluid and we will consider later its effect separately.

3rd. If the fluid is compressed by any force other than gravity, we will not pay any attention to this compression; for the reason that however large it may be it could not provide any changes to the resistance in the case where no vacuum is produced behind the body.

Because let a body be at rest in whatever compressed fluid, one half of which is ADC (Fig. 4.8), and let us draw from any point V the line Vu parallel to the axis AC. The fluid (hyp.) being equally compressed on all sides, the points V, u are pressed by equal forces along VZ and uz [perpendicular to the surface ADC][2]; therefore if these forces are changed by others along VF and Vu along uf and uV, it will be easily proved that the forces along Vu and uV are equal between themselves, and that the

---

[1]"Chapter" in the original.
[2]"Ad superficiem ADC perpendiculares", $Mss.$ 77–2°.

forces along *VF* and *uf* are also destroyed by the opposing forces along *FV* and *fu*. Therefore the compression of the fluid cannot produce any motion in the body.

Once that the body is set in motion by any cause, the compression on the points *V* and *u* will always be the same, since (*hyp.*) there is never a vacuum between the fluid and the body, and the force that compresses the fluid always acts equally.

So whatever the fluid compressing force is, it should neither produce any motion in the body, nor any change in its motion. This observation was already made by *M. Newton* and I am surprised that some authors, otherwise very skillful, have thought that the resistance should be zero in an infinitely compressed fluid. Here is their reasoning. If a fluid, they say, is infinitely compressed, the empty space that a body moving therein leaves behind will be filled on the way by fluid particles which rush in with infinite speed. I agree; but I say that for this same reason the resistance in the front part must be much too large, because it is obvious that the compression at the front part is contrary to the body movement; so if the compression at the rear part makes the motion easier in some way, the compression at the front part must retard it in some way; so that the compression at the front part and at the rear always tends to produce equal and directly opposite effects.

Thus the compression of the fluid must be taken for nothing in the case where there is no vacuum is produced between the fluid and the body.

4th. We proved in the *art.* 8[3] that whatever the initial velocity of the moved body is, the fluid particles always describe the same curves. Now it can be proved by a similar reasoning that whatever the initial velocity of the body moved is, the number of parts to which it communicates the motion in the first instant is always the same; and that the parts of the fluid that are moving at the instant when the body is at the end of any space x are always the same, whatever the velocity of this body is at the end of that space. Now then, the experience shows that when a body moves very slowly, only the particles close to the body receive motion, so that the action of a body moving slowly through a fluid extends only until a short distance from itself. Therefore the action of a body moving with any velocity in a fluid must also extend until a short distance from itself, and the same thing results from the *art.* 71 and 72. Moreover, in all the following propositions, as above, we will only need to regard the fluid particles immediately contiguous to the body surface.

Prop. XII. Problem
86. *Let us determine the velocity that a body of whatever shape, moved with whatever velocity, communicates to the parties of a fluid without weight and of any density, when it is moving in such fluid,*

As in *art.* 48, N, B, C, D, (Fig. 4.4) let four fluid particles be arranged so that they form a rectangular parallelogram whose side *NC* is parallel to the trajectory of the body. It is clear that at each moment the velocity of these particles can be regarded as composed of another two; namely, of a velocity equal and parallel to the moving body at that instant, and of another velocity which will be the respective velocity of

---
[3] *Art.* 38 in the original. In *Mss.*77–3° the art. 14 is referred, which corresponds to the art. 8.

these particles with respect to the body. Let u be the rectilinear velocity of the body in any instant and V the relative velocity of the particle N; thus the absolute velocity of this particle will be composed of the velocity u and the velocity V. The first of these velocities u is along CN, parallel and equal to the body velocity, with respect to the second velocity V it can be regarded as composed of the two other velocities, one that I call v, which will be along NC, and the other that I name v', which will be along NB.

Now, when the body is at the end of any space, the absolute velocity of the particle N, must have (*art.* 8)[4] the same relation to the present velocity of the body, whatever it is, and the particle N must have the same position relative to the body and the same direction. Therefore since the absolute velocity of the particle N along NE is $u - v$ and along NB is v', it is clear that the ratio of $u - v$ to u depends on the position of the particle N in relation to the body and space r already traversed by the body. Now as $\frac{u-v}{u} = 1 - \frac{v}{u}$, it follows that the ratio of v to u, and of v' to u depends on the space r traversed by the body and the position of the point N.[5]

In addition, (*art.* 6) $\frac{-du}{u} = \xi dr$, being ξ a function of the space r traversed by the body. So (*art.* 9) r will be a function of u/g. Therefore the ratios of v to u and v' to u depend on u/g and the position of the point N; that is to say on u/g and x and z. Now, in the thread AMD (Fig. 4.5) $\frac{v}{v'} = \frac{dx}{dy}$ is given, that is, equal to a function of x and y. So the expression of the ratio of v/u and v'/u must be such that dividing v by v' and making $z = y$, u/g vanishes and disappears from the ratio. Then let us suppose at first that the quantity u/g is not found in the ratio of v to u and v' to u and let us see what will result from this hypothesis.

Let, as in the *article* 48, be $v = uq$, et $v' = up$, $NB = \alpha$, $NC = \beta$, $NN' = k$ (Fig. 4.6),[6] $dq = Adx + Bdz$, $dp = A'dx + B'dz$, it will be found, as in *art.* 48, $\alpha\beta k = (\alpha - updt + updt + udtB'\alpha)(\beta - uqdt + uqdt + udtA\beta)(k + \frac{kpudt}{z})$. From which $B' = -A - \frac{p}{z}$ will be derived as in the same *article*. Thereupon this equation takes place whether the fluid moves or is at rest. Besides, it will be proved again by the same reasoning as in *art.* 43 (Fig. 4.4) that $uq : uq - qd$ $u - u \times NE \times \frac{dq}{dx} - u \times \frac{NE \cdot p}{q} \times \frac{dq}{dz} :: uqdt : NE$. Therefore $NE = uqdt - qdudt - u^2 qAdt^2 - u^2 pBdt^2$; and it will also be given $FE = updt - pdudt - u^2 pA'dt^2 - u^2 qB'dt^2$. So the particle N, impelled by the forces $\frac{du}{dt} - \frac{qdu}{dt} - u^2 qA - u^2 pB$ along NC and $-\frac{pdu}{dt} - u^2 pA' - u^2 qB'$ along and NB, must be in equilibrium Now then, the force du/dt is the same for all points N, B, C, D, so that the $NBCD$ channel parts impelled by this force are in equilibrium. Therefore the

---

[4] This assertion is in art. 6.

[5] The space traversed in the two previous mentions is written as x, but it is changed to r in the next paragraph and the x passes to be the axis coordinated. We have taken r for the space to avoid ambiguity.

[6] In the original it is Fig. 4.7. Also the values α and β taken here are changed with respect to art. 48, even though this does not affect to the subsequent steps.

5   On the Resistance of Fluids to the Bodies Moving Therein

particle N [*art.* 48] must be in equilibrium driven by single forces $-\frac{qdu}{dt} - u^2qA - u^2pB$ and $-\frac{pdu}{dt} - u^2qA' - u^2pB'$. So substituting B' by its value $-A - p/z$, [and operating as in *art.* 48] it will have $\frac{du}{dt}\frac{dq}{dz} + \frac{d(u^2qA + u^2Bp)}{dz} = \frac{du}{dt}\frac{dp}{dx} + \frac{d\left(u^2qA' - u^2Ap - \frac{u^2p^2}{z}\right)}{dx}$. So since in this equation p and q do not depend on the unknown u, separately we must have $\frac{du}{dt}\frac{dq}{dz} = \frac{du}{dt}\frac{dp}{dx}$, this is $\frac{dq}{dz} = \frac{dp}{dx}$ or $B = A'$, and $\frac{u^2 d(qA+pB)}{dz} = \frac{u^2 d\left(qA' + pA - \frac{p^2}{z}\right)}{dx}$ or $\frac{d(qA+pB)}{dz} = \frac{d\left(qA' + pA - \frac{p^2}{z}\right)}{dx}$. This equation does not differ from that of *art.* 48 $\frac{d(qA+pB)}{dz} = \frac{d(qA'+pB')}{dx}$, assuming $B' = -A - p/z$ and $A = B'$; because we have seen (*art.* 48) that the latter equation is then reduced to zero: thus the conditions already $B = A'$ and $B' = -A - p/z$ satisfy the equation $\frac{d(qA+pB)}{dz} = \frac{d\left(qA' + pA - \frac{p^2}{z}\right)}{dx}$, from which no new condition results. So the equations $B = A'$ et $B' = -A - p/z$ also take place either when the fluid moves, or when the fluid is at rest and the body moving.

Scholium I

87. Had we assumed $\frac{v}{u} = q\varphi\left(\frac{u}{g}\right)$ and $\frac{v'}{u} = p\varphi\left(\frac{u}{g}\right)$, q and p denoting functions of x and z, and $\varphi(u/g)$ any function of u/g, we would arrive at the same equations.

Let us find here the reason why we assumed the function $\varphi\left(\frac{u}{g}\right) = 1$. If we assume $\frac{v}{u} = q\varphi\left(\frac{u}{g}\right)$, we would find, as it is very easy to see, the pressure proportional to $u^2\varphi\left(\frac{u}{g}\right)^2$, which should not be surprising, since in general the resistance R (*article* 9) is proportional to ξu^2, being ξ a function of u/g; therefore the resistance would not be proportional to the single square of the velocity. Now we are going to demonstrate in the next *art.* that effectively it is only proportional to this square.

Scholium II

88. Let be u the variable velocity of the body at every moment, and let us assume that during all the time of the motion of the body the system of fluid and body is carried away in the opposite direction with variable velocity u equal to that[u]. It is obvious that the body will remain at rest, and it will be the fluid which will strike it with a variable velocity u, but by the primitive laws of motion the fluid pressure on the body will not be changed. Now I say that in this case the fluid pressure against the body is proportional to u^2, because if it is assumed that any accelerating or retarding force proportional to kdt acts in every instant upon the parts of the fluid, the threads will not be disturbed (*art.* 56), but the velocity of each particle will be increased or decreased at each instant by a quantity proportional to $kdt\sqrt{p^2 + q^2}$. Therefore, if velocity is u at any instant, the pressure at this instant will be formed: 1st, by a quantity proportional to $\varphi u^2 \delta$ (*art.* 68) that comes from the velocity u; 2nd, by a quantity which comes from velocity of tendency kdt that is $\delta kdt(\mu + \Omega - \pi \Gamma b^2)$ (*articles* 54 and 56). Now then (*art.* 55) this quantity $\mu + \Omega - \pi \Gamma b^2 = 0$, thus the fluid pressure is simply proportional to u^2.

So far all the authors of hydraulics have taken as a principle that the resistance of a moving body in a fluid is equal to the pressure that this fluid, moving with the same velocity, would exert against the body assumed at rest. However, 1st, they have not paid attention to this velocity being variable, the resulting pressure could contain the element du, and in consequence would not be proportional to u^2; 2nd, even considering, as they would have done, this variable velocity to an uniform velocity, they have only proved in a very vague way that the pressure was like u^2; see above *art.* 10. It seems to me that we have fully satisfied all these difficulties, proving that the coefficient of du/dt is zero, and that the coefficient φ of u^2 is always the same, whatever u is.

It is necessary only to remark that in the first instant the pressure is not proportional to u^2, but it is equal (art. 54) to $\delta u(\mu + \Omega - \pi \Gamma b^2)$, that is to say zero.

Then let a body be pushed into a stagnant fluid with an initial velocity U, and this body after a time t has a velocity u. It is clear that it can be considered as driven along CA (Fig. 4.1) at every moment by the force of $+\frac{du}{dt}$. Let us apply to the system of fluid and body the initial velocity $-U$, and at each successive instant dt the velocity $+\frac{du}{dt}$ along AC; it is clear that the body will be at rest, and however it will continually pushed along AC by a force $+\frac{du}{dt}$, which will be balanced by the pressure of the fluid.

Propos. XIII. Problem

89. *The same things being supposed ast in the previous article, let us determine the resistance of the fluid.*

The force which tends to move the body in the instant dt is $+du/dt$. Let μ be the volume of the body and Δ its density; therefore $\mu\Delta$ will be its mass and $\mu\Delta du/dt$ will be the force along CA. This force must be balanced (*art.* 1) by the pressure of the fluid, that is to $\delta\varphi u^2 - \frac{\delta du}{dt}\left(\mu + \Omega - \pi\Gamma b^2\right)$. So since $\mu + \Omega - \pi\Gamma b^2 = 0$, it will give $\mu\Delta\frac{du}{dt} + \delta\varphi u^2 = 0$.

Corollary

90. So $-\frac{du}{u^2}\mu\Delta = \varphi\delta dt$ is the formula to find the velocity of a body which moves in a fluid. Hence it is obvious that the resistance of the fluid, all other things being equal, is proportional to $u^2\varphi\delta$, that is to say it is equal to the pressure that this fluid would exert upon the body assumed at rest, if this fluid came to strike it with velocity u. As we have said this proposition so far was recognized to be true, but it does not mean that there was no less need to prove it. Because the pressure of a fluid uniformly moved with the velocity a upon a body at rest is $a^2\varphi\delta$, instead of the pressure of a fluid at rest upon a body that moves therein with a variable speed u is $u^2\varphi\delta - \frac{du}{dt}\delta(\pi\Gamma b^2 - \Omega - \mu)$, however this quantity cannot be reduced to $u^2\varphi\delta$, unless $\pi\Gamma b^2 - \Omega - \mu = 0$. This did not seem easy to prove, because of the difficulty of expressing the quantities Γ and Ω analytically; but fortunately we have reached

the end for the consideration of the primitive velocity of the body, without needing to know these quantities.

Propos. XIV. Problem

91. *With everything the same, let us find the resistance of a body moved in a fluid, having regard to the gravity of the fluid and the body, and assuming that the body rises in the fluid.*

The weight is not something other than a force which acts equally on all the particles of the fluid along parallel lines; it is easy to prove by the same reasoning as the *art.* 48 and 49 that we will have $A' = B$ and $B' = -A - p/z$. Besides this, the fluid pressure which comes from its relative velocity and from the force $-du/dt$ must be increased in the pressure $g\mu\delta$ which comes from the gravity of the body, and it must be reduced by the quantity $gM\Delta$ which comes from the effort of the fluid under its own weight, and which acts from below upwards. So the pressure $\varphi u^2 \delta$, found in the *art.* 89, must be increased by the quantity $g\mu\delta - g\mu\Delta$. So it will give $-du = \frac{\varphi u^2 \delta dt}{\Delta \mu} + \left(\frac{g\delta}{\Delta} - g\right) dt$.

Propos. XIV-B. Problem[7]

92. *With everything the same, let us find the resistance of the fluid having regard to the viscosity and the friction of the parts.*

1st. The friction of the fluid over the body may only come from the relative velocity of the fluid respect to the body.

2nd. The experiments made by the famous *Musschenbroek* teach us that the friction is proportional to the velocity, hence it follows that if the relative velocity at any point is called U, the friction is proportional to nU, n denoting a coefficient to be determined by experiment, and which is the resistance coming from the friction when the velocity U is equal to 1.

3rd. The equations $B' = -A - \frac{p}{z}$ and $A' = B$ still take place here. Since the forces lost by each fluid particle, defined in the *art.* 48, must be decreased by the forces unp and unq; because the friction that decreases the velocity can be represented by a force which would act in the opposite sense to the direction of the velocity. Therefore the force $-unp$ must be subtracted from the force lost along NB (Fig. 4.4) and the force $-unq$ from the lost force along NC. Then, it will give (*art.* 86)[8] $-nu\frac{dq}{dz} + \frac{dudq}{dtdz} + u^2 \frac{d(qA+Bp)}{dz} = -nu\frac{dp}{dx} + \frac{dudp}{dtdx} + u^2 \frac{d(qA'+B'p)}{dx}$, and as neither p nor q depend on u, we will have $\frac{dq}{dz} = \frac{dp}{dx}$ separately and $\frac{d(qA+Bp)}{dz} = \frac{d(qA'+B'p)}{dx}$; from which $A' = B$ and $B' = -A - \frac{p}{z}$. is obtained as in the *article* 48. Therefore the pressure from A to C found in the previous calculations, must be decreased in (*article* 54) $un\delta \int 2\pi y dy' \int p dy' + q dx' - un\delta\pi\Gamma A^2$; that is, we need to add (*article* 54) $un\delta(\Omega - b^2\Gamma\pi)$. Consequently we will have for the general equation[9]

---

[7]In the original this propositions repeats the same number as the previous one, therefore I have called it XIV-B.

[8]Said 86 in the original.

[9]This equation and the successive ones derived from it have a misprint in the place of *dt*.

$-\mu\Delta\frac{du}{dt} = \varphi u^2 \delta - g\delta\mu + g\Delta\mu - un\delta\Omega + un\delta\Gamma\pi b^2$. And as $-\Omega + \Gamma\pi b^2 = \mu$ (*article 55*), it will give $-\mu\Delta\frac{du}{dt} = \varphi u^2 \delta dt + g\delta\mu - g\Delta\mu + un\delta\mu$.

Remark

93. The resistance proportional to velocity, which we talk about here, is due to friction[10] of the parts of the fluid and body. This friction comes from the unevenness of the body surface. But there is also another resistance that comes from the viscosity[11] of the parts of the fluid; and, as far as we can conjecture from all experience, it can be regarded as a constant force. Because, 1st, there are many bodies that, although having a greater specific gravity than water, do not descend into the water. Now then, as this descent is only prevented by the viscosity of the water parts, it follows that the viscosity is necessarily in a finite ratio with the gravity. Indeed, any body being just slightly heavier than water would always descend if the viscosity was proportional to some power of velocity; because making the velocity equal zero, the viscosity would be zero, and so it would not oppose the first descent of the body. 2nd, anybody has observed this viscosity in the drops of water, as it often prevents the drops from falling when they are adhered to the lower surface of any body. Therefore viscosity, whether it comes from a compressive force or from the attraction of parts, is a constant force such as gravity, though very small compared to it.

The only objection that can be made against this reasoning is that any pendulum, provided it is a little heavier than water, always adopts the vertical position in the water, and it returns to it when it is moved out slightly; this would not happen if the viscosity was an opposing force. Because let be g gravity and α the viscosity force, it is clear that the pendulum should stop in any case where it will make an angle or smaller or equal to α/g with the vertical; which nobody until now has noticed, as I know. But as the angle α/g is very small, and consequently not easy to observe, and also any extraneous motion coming from the air or from the surrounding bodies can disturb this experience. I do not consider that the objection in question is large enough to make me reject a truth that seems consistent with reason, and which is supported by infinity of experiments. In the work that we have already mentioned *M. s'Gravesande* finds (§. 1911) that the pressure of a fluid in motion against a body at rest is proportional in part to the single velocity, due to the fluid viscosity, and in part to the square of the velocity due to the force of inertia. The intensity of these two pressures against a cylindrical body appears to be in a ratio of 20 to 39 following the experiments he made, §.1930 and 1945. The first, which comes from the viscosity, is independent, as *M. s'Gravesande* says, of the shape of the body (§.1916); but it is not the same with the second, because in a globe it is the 2/3 of that of the cylinder, and in a right-angled cone it is as the half diameter of the base is to the side, §.1917, 1918 etc.

---

[10] In the original "frottement".
[11] In the original "ténacité".

5 On the Resistance of Fluids to the Bodies Moving Therein

When the fluid is at rest and the body in motion, then according to *M. s'Gravesande* the total resistance is again composed of another two: one part is constant, the other is due to the square of the velocity (§.1975). He proves that a part of the resistance is constant because a body moved in a fluid finally stops; which does not happen if the resistance simply dependent on the velocity (§. 1963). This test reinforces what we have already given about the same proposition at the beginning of this article. In effect, when the resistance is assumed to be as u^2, or as u or as $u^2 + u$ we will always find that t, that is the time of the body motion, is equal to infinity; although t is finite when a constant term enters in the expression of the resistance.

But how is the force that comes from the cohesion of the parts proportional to the velocity in the case of the moving fluid, and constant in the case of the fluid at rest? This is what *M. s'Gravesande* tries to explain, §.2065 et *seq.,* but using an argument that seems to me very obscure. Besides, I do not see that he has made any experiment to verify the difference; or rather, in the same experiments that he made on the pressure of a fluid in motion, he remarks (§.1914) that the theory does not agree with experiment when the velocity is very small, but that the pressure given by the experiment (§.1911) is greater than what is found by the theory. This proves, it seems to me, that the fluid pressure that comes from the viscosity is not strictly proportional to the velocity.

However, I must admit that considering the opinion itself of *s'Gravesande*, independently of the obscure and insufficient evidences in which he has sought to support himself. This opinion may seem founded to some extent, at least at first glance. That is to say, it can be thought that the pressure of a moving fluid against a body at rest, and the resistance that a moving body experiences in a fluid at rest are not the same in the case where the viscosity of the parts of the fluid is considered; if the viscosity is understood as the difficulty of separating the fluid particles. Indeed, when a body moves, it is clear that the difficulty in separating the particles is an obstacle for it, which must necessarily make it lose velocity. But when the body is at rest, and it is the fluid which strikes, one cannot see distinctly how the viscosity of the parties increases the pressure. Because this viscosity seems to be a simply passive force capable of resistance rather than action.

However, considering this issue more carefully, it is soon perceived that the viscosity is a force by which the particles of the fluid resist their division; so that if the fluid particles have precisely only a justly low velocity so they could not be detached from each other, these parts would move in virtue of that velocity as it an absolutely solid body would; and the fluid would move together with the body, so that the fluid particles would not have any relative velocity in relation to the body. To clarify our thoughts better, in a fluid let us suppose a body at rest slightly heavier than the fluid, but that remains suspended therein due to the adherence of parts of the fluid; the whole system will therefore remain at rest. Let us now give a velocity equal and contrary to that with which the body tends to descend; it is obvious that the body and the fluid will be transported with this velocity in the same manner as if they formed a solid body; and they are transported in the opposite direction to which the body tends to move along. Thus it is seen how the viscosity of the particles may be reduced to the action of a force tending to move the body in an opposite direction to which it moves along. The viscosity can be further reduced to an active force, considering that when a body, that is slightly heavier than the same volume of fluid, remains suspended due to the viscosity of the fluid; it is in the same case, if the viscosity is excluded, as if the weight of fluid was increased in such a quantity so the fluid and the body were in equilibrium. From that it also follows that the viscosity can be supposed equal to a constant force, since the effect of the viscosity is equivalent to that which would result from an increase of the weight in the fluid.

It seems to me that we have distinguished very well founded three classes of resistance: the first is constant, coming from the viscosity of the fluid particles, that is to say, the resistance that the particles oppose to being divided; the second proportional to the velocity, and coming from the friction that the fluid particles undergo when they slide on the body surface by virtue of their relative velocity; the third proportional to the square of the velocity, coming from the force of inertia. The constant resistance does not depend on the body figure or velocity, or even on its width. Because this resistance comes mainly from parts of fluid that are found in the axis AC extended to A (Fig. 4.1),[12] and that the body is forced to separate in order to move itself, now then, the number of particles to separate is as the distance traversed; therefore the *live force* lost is proportional to this space. Indeed, this resistance can be compared with sufficient likelihood to the effect of the gravity or to the force of a spring wire, which will be always the same.

Corollary

94. So if it is assumed that g' is the part of the resistance that must be constant and one considers the weight of the fluid and the body as much as the viscosity and the friction of the parts, it will give $-\mu\Delta\frac{du}{dt} = \varphi u^2 \delta + g\delta\mu - g\Delta\mu + un\delta\mu + g'\delta$. This is the general equation of motion of a body in a fluid. It is reduced to $\frac{-du}{\alpha u^2 + \beta u + \gamma} = dt$, being α, β and γ constants. Now, the integration of this equation has no difficulty. Let $u+k$ et $u+k'$ be the two roots of the factor of the quantity $u^2 + \frac{\beta u}{\alpha} + \frac{\gamma}{\alpha}$, we will have $\alpha t = -\frac{1}{k-k'}\log\frac{(u+k)(g+k')}{(u+k')(g+k)}$, indicating g the initial velocity of the body.

Therefore $\frac{(u+k)(g+k')}{(u+k')(g+k)} = c^{-\alpha t(k-k')}$; which gives the value of u in t.[13] Now if k is imaginary as well as k', these expressions can easily be reduced to real quantities by the method I explained in the Memories of Academy of Sciences of Prussia 1746. Thus the solution of the problem is now reduced to a pure difficulty of analysis. That is why I go on to other researches, contenting myself with making the observation that I arrived at this formula by an entirely new method.

---

[12] It seems that it refers to Fig. 4.1.

[13] We notice that the letter c corresponds with number e.

5.3 On the Use of Pendulum Experiments to Determine the Resistance of Fluids Whose Velocity Is Very Small

95. At first one can easily ensure by the experiments of the pendulum if the resistance is roughly as the square of the velocity when the velocity is very small. To find it, let us imagine a pendulum that describes very small arcs; let be p the natural gravity, A the point where it is assumed that the body starts to move at any instant (Fig. 5.1),[14] $AM = y$, $DM = s$, $QP = x$, $AD = B$, $CD = a$, u the velocity at M, m the mass of the pendulum, f the resistance that the fluid would produce upon the mass m moved with a velocity $\sqrt{2ph}$. It will give $pdx - \frac{fu^2 dy}{2phm} = udu$ in the hypothesis of the resistance as the square of the velocity; from which it follows $u^2 = 2px - \int \frac{fu^2 dy}{phm}$ 0; now then $x = \frac{B^2 - s^2}{2a}$ and $s = B - y$; from which it is clear that u^2 is very nearly equal to $\frac{p(2By - y^2)}{a} - \int \frac{fdy}{phm}\left(\frac{2pBy - py^2}{a}\right)$. Therefore, $u^2 = \frac{p(2By - y^2)}{a} - \frac{f}{Rhm}\left(By^2 - \frac{y^3}{3}\right)$. So to find the point A'' up to where the body rises, $u = 0$ must be taken, which will give $ADA' = 2B - \frac{2fB^2}{3phm}$. Therefore $AA' = \frac{2fB^2}{3phm}$. Now assuming the resistance as the square of the velocity, $\frac{phm}{f}$ must be equal to a constant n. So $' = \frac{2B^2}{3n}$.

96. Now it is easy to see that from whichever point A the body starts to move, the very short time that it uses to make a vibration, that I call τ, will always be about the same. So looking $A'A''$ as an infinitely small quantity and equal to $-dB$, and also assuming $\tau = dt$, we will have $-dB$ proportional to $\frac{2B^2 dt}{3n}$. Therefore if it is assumed that the first arc traversed by the body is equal to B', we will find that $\frac{3n}{2B} - \frac{3n}{2B'}$ is proportional to t. In order for the resistance to be as the square of the velocity, $\frac{B' - B}{B'B}$ must be a constant quantity, t expressing the time that the pendulum is moving. Let

Fig. 5.1

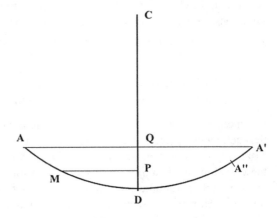

---

[14]The symbol a is used with two meanings, and in the next article α is also changed. To avoid any mistake we have introduced A', A'' in the places of the points a and α in this article and τ for α in the next maintaining them until §.98 inclusive.

Fig. 5.2

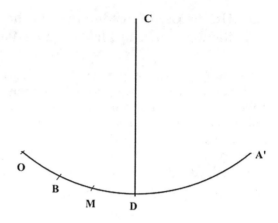

us see now how it will be ensured by the experiment, if the resistance is as the square of the velocity.

97. A pendulum of arbitrary length *CD* (Fig. 5.2) will be made to move in the fluid whose resistance we want to determine. Care should be taken that the suspension in *C* is such that the friction can be counted for nothing; for that we can use a rope cord attached between two plates of very polished copper or steel from which the pendulum will be suspended. Next we will mark the point *O* from which we let the pendulum fall and we will be note the points *B*, *M* to which it ascends after two times t, t'. I say that the resistance will be as the square of the velocity if $\frac{OB}{DB} : \frac{OM}{DM} :: t : t'$.

Assuming that this proportion is indeed found to be true, or very nearly true, it will be easy to know the constant n. For this, let be τ the time of one vibration of the pendulum, which can be determined with great accuracy by counting the number of vibrations it makes in a given time; it is clear that it can be assumed $\frac{f}{pm} = \frac{n}{T}$, where T denoted a time constant, but unknown. Therefore we will have $-dB = \frac{2dtB^2}{3Th}$ and $\frac{3h}{2B} - \frac{3h}{2B'} = \frac{t}{T} = \frac{tf}{pm\tau}$; so $\frac{1}{B} - \frac{1}{B'} = \frac{2t}{3\tau n}$; so $n = \frac{3tB'B}{2\tau(B'-B)} = \frac{3t}{2\tau} \frac{BD \times DO}{OB}$.

98. If it is found that the resistance is not proportional to the square of velocity, in this case the resistance will be assumed equal to $+\frac{fu^2}{2ph} + \frac{ku}{\sqrt{2ph}}$, as we did in *art.* 94; but in this case it will be very difficult to determine by experiments with pendulum the coefficients *f, g, k*, because the formula that would give the value of *B* at *t* will be extremely complicated. So I think in this case the analysis in order to determine these coefficients is almost an insuperable difficulty because the almost impossibility of finding a simple and convenient equation between the times and the spaces traversed. However, it seems to me that from all assumptions that can be made on the fluid resistance, the more truthfully and less subject to dispute is the one given here. It would be wished that a way to easily compare it with experiments could be found.

Moreover, let us see here in the most convenient way that is possible, the method I have devised to reduce the experiments to the calculus.

5 On the Resistance of Fluids to the Bodies Moving Therein

We will have $u^2 = \frac{p}{a}(2By - y^2) - \frac{f}{ahm}\left(By^2 - \frac{y^3}{3}\right) - \frac{2gy}{m} - \frac{2k}{m\sqrt{2ah}}\int dy\sqrt{2By - y^2}$.

Now, if π is called the ratio of the circumference of a circle to its radius, wet will find $'A'' = \frac{2fB^2}{3phm} + \frac{2ga}{pm} + \frac{2k a \pi B}{2pm\sqrt{2ha}}$. Making $\frac{f}{pm} = \frac{\tau}{T} = \frac{dt}{T}$ we will have $dB = \frac{2B^2 dt}{3Th} + \frac{2agdt}{fT} + \frac{ak\pi B dt}{fT\sqrt{2ah}}$ or $\frac{dt}{T} = -\frac{\frac{3hdB}{2}}{B^2 + \frac{k\pi 3\sqrt{2ha}}{f}B + \frac{3hga}{f}}$.

It can be therefore assumed $\frac{dt}{T} = \frac{-3hdB}{2(B+G)(B+A)}$, G and A being imaginary or real; from where $\frac{1}{T} = \int \frac{3h}{2(A-G)}\left(\frac{dB}{B+A} - \frac{dB}{B+G}\right)$ is drawn and $\frac{2(A-G)t}{3hT} = \log\frac{B+A}{B+G}\frac{B'+A}{B'+G}$. So t is proportional to $\log\frac{B+A}{B+G} - \log\frac{B'+A}{B'+G}$. From this it follows that by three observations A and G are known, and therefore $\frac{h}{f}$ and $\frac{g}{f}$. First of all let us observe the arc B' in the first descent, and after let us observe the three arcs β, β', β'' in the times t, t', t'' which are in arithmetic progression, it will give this continuous geometric progression $:-: 1 : \frac{\beta+A}{B'+A}\frac{B'+G}{\beta+G} \cdot \frac{(\beta'+A)(B'+G)}{(B'+A)(\beta'+G)} : \frac{(\beta''+A)(B'+G)}{(B'+A)(\beta''+G)}$. Hence the values of A and G will be obtained, by a very long indeed calculation; further, instead of T putting its value $\frac{pm\tau}{f}$, we will have $\frac{2(A-G)tf}{3hpmn} = \log\frac{B+A}{B'+A}\frac{B'+G}{B+G}$. Therefore $\frac{pm}{f}$ or the relation of f to pm is also known.

Remark

99. *M. Daniel Bernoulli*, in tome 3 of the Memories of Petersburg, proposed himself to determine by theory the movement of a heavy body in a medium whose resistance is in part constant by the viscosity, and in part proportional to the square of the velocity. Having applied the calculation to the experiments, he found that the viscosity with which water resists a globe whose weight in water would be a grain[15] would be equivalent to about1/4 the weight of the globe; a result that seemed suspect to *M. Bernoulli* as giving a too big value for the resistance that comes from viscosity. He observes moreover, that according to the experiments of *M. Newton*, Book 2, Prop. XL, Sch., this resistance occurs only in very slow movements and that in the others the resistance is more like the square of the velocity.

In tome 5 of the same Memories, this great geometrician continues treating the same subject. He first applies the calculation to the experiments of pendulums made by *M. Newton*, Book 2, Prop. 31, Sch. and he find, after having weighed and discussed all the circumstances, 1st that in pendulums whose motion is not too slow the resistance is almost like square of the velocity, 2nd that in slower motions a constant force joins that resistance. 3rd finally, in extremely slow motions, it seems very difficult to determine accurately enough the law that follows the total resistance, because the experiments do not agree then with the theory. However, *M. Bernoulli* believes that even in this case the theory should not be completely rejected, because the experiments are so delicate that it seems difficult to conclude anything certain and positive from them. Perhaps, moreover, they would better agree with the theory if in the case when the motions are very slow the resistance is

---

[15] A grain is equivalent to1/72 livres (0.0531 g).

imagined proportional to $fu^2 + ku + g$, as we have done. Moreover, it seems to me that the formula that has been given, p. 135, Tome 5 of the Mém. of Acad. of Petersburg, does not accurately represent the difference in the arcs traversed in the hypothesis of resistance proportional to $fu^2 + g$; this is what can be made sure by comparing our two methods.

The author finds that by calling t the arc of the first descent, the ascended arc will be $t - \frac{4}{3}nt^2$, the resistance being as the square of the velocity, which is consistent with what we have found. From that he concludes that after a number of oscillations equal to l, the ascended arc will be approximately represented by the geometric progression $-\left(\frac{4}{3}nl\right)t^2 + \left(\frac{4}{3}nl\right)^2 t^3 - \left(\frac{4}{3}nl\right)^3 t^4$, etc., whose sum is $\frac{t}{1+\frac{4}{3}nlt}$; which is still consistent with our calculation, as we can easily see even although we have used a different method and denominations.

But it is not the same in the case of resistance as the square of the velocity plus a constant; because as from the excess of the descended arc over the next ascended arc, that is $\frac{4ma}{g} + \frac{4}{3}nt^2$ as the author has found and we do as well, I do not see how the author concludes that $\left(t - \frac{4mla}{g}\right) : \left(1 + \frac{4}{3}nlt\right)$ will be the arc ascended after a number of oscillations equal to l. On the contrary it seems to me that following the very short and very simple method that we have used, it will give (keeping the names given by the author) $= \int \frac{-dt}{\frac{4ma}{g}+\frac{4}{3}nt^2}$, which is very different from the value of l which would be derived from the equation $\left(t - \frac{4mla}{g}\right) : \left(1 + \frac{4}{3}nlt\right) = r$, r expressing the ascended arc.

5.4 Examination of an Hypothesis Which Would Lead to Strange Paradoxes on the Resistance of Fluids

100. All the authors who have so far treated the motion of fluids enclosed in vases have taken as hypothesis that all parts of the fluid placed in the same horizontal line have the same vertical velocity. This hypothesis, that is, or at least it appears to be, confirmed by experiment, had so captivated me that I had thought at first to derive from it the theory of fluids resistance. But having paid attention to the resulting calculations, I noticed that there were many cases in which the fluid resistance will be null according to this theory, and that it lead to a lot of other consequences very contrary to experience. Maybe it will not be useless to explain this more at length.

101. Let a body be *ANB* (Fig. 5.3), which I will take, for convenience, as a flat surface and that here I will consider only one half because the other half is assumed similar and equal to this one. Let this body be immersed in a fluid, which is enclosed in a cylindrical vessel,[16] *QV* being one of the walls. The body moves from *B* to *A* and the line *PNV* is perpendicular to *AP*. It is obvious that the point *N* will come at

---

[16] It should be understood as prismatic instead of cylindrical.

5 On the Resistance of Fluids to the Bodies Moving Therein

Fig. 5.3

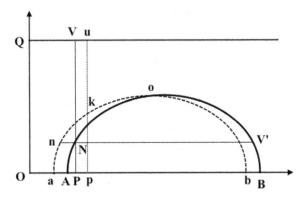

n, so that the space $OANVQ$ decreases by an amount equal to $AaNn = aA \times PN$. So the parts of the fluid contained in NV must necessarily move towards ku; so if the velocity of all these parts parallel to Vu is assumed the same, all parts NV will keep themselves in the situation ku parallel to NV and it will give $NVuk = ANKa$. Therefore $Vu = \frac{PN \cdot Aa}{NV}$. Let u be the velocity of the body along Aa, $PN = y$, $PV = a$ and v the velocity of the particles NV, it will give $v = \frac{uy}{a-y}$. Having said that, I look for the fluid resistance in the following way.

102. Let be M the mass of the body, V the velocity imprinted at the first instant, V' the actual velocity that it must have because of the resistance of the fluid. In the first instant the particles placed in any line NV will move parallel to Vu with the velocity $\frac{V'y}{a-y}$. So (*art.* 1) the pressure of these particles is the same as if they tended to move with the same velocity parallel with uV, now then in this case the pressure would be (*art.* 23) $\delta \int y dx \frac{V'y}{a-y}$; therefore $M\Delta(V - V') = \delta V' \int \frac{ydx \, y}{a-y}$; so $V' = \frac{M\Delta V}{M\Delta + \delta \int \frac{y^2 dx}{a-y}}$. So since $M = \int y dx$, it will give $V' = \frac{MV}{\int \frac{ydx}{a-y}}$ assuming $\Delta = \delta$. But here is the drawback according to this formula: in the first instant of the motion following the experiment $V = V'$ is; but following the formula $V = V'$ this happens only in the case which a is infinitely large. Because then $\int \frac{ydx}{a-y} = \int y dx = M$; in other cases, it is $\int \frac{ydx}{a-y} > M$, and therefore $V > V'$, and the smaller a is, so V' will be smaller in respect to V. But I do not know any experiment that proves that the velocity lost at the first instant is greater when the vessel is narrower. It even appears that the figure of vase contributes nothing, or almost nothing, to the resistance, because, as it has been proved above, the motion that the body communicates to the fluid particles extends up to a very short distance around it (*art.* 71 and 72).

103. I will prove now that in the next instants, the fluid resistance would be absolutely null if the parts contained in the line NV all had the same velocity parallel to AP. In effect being $\frac{uy}{a-y}$ the velocity of the particles NV in any instant dt, these particles, when they reach in the next the instant the situation ku, will have the velocity $\frac{u'y'}{a-y'}$, y' denoting the line pk and u' the velocity of the body in this second

instant. So the pressure will be the same, as if all parts of the line NV were impelled parallel to uV by a force equal to $\frac{1}{dt}\left(\frac{u'y'}{a-y'} - \frac{uy}{a-y}\right)$. Now $u' = u + du$, $y - y' = pk - PN = (Nn + Vu)\frac{dy}{dx} = \left(udt + \frac{uydt}{a-y}\right)\frac{dy}{dx}$. So the pressure along AB will be $= \delta \int ydx \frac{ydu}{(a-y)dt} + \delta \int \frac{au}{(a-y)^2 dt}\left(udt + \frac{uydt}{a-y}\right)\frac{dy}{dx}ydx = \frac{\delta du}{dt}\int \frac{y^2 dx}{a-y} + \delta \int \frac{au^2 ydy}{(a-y)^2} + \delta \int \frac{au^2 y^2 dy}{(a-y)^3} = \frac{\delta du}{dt}\int \frac{y^2 dx}{a-y} + \delta au^2 \int \frac{ady}{(a-y)^3} = \frac{\delta du}{dt}\int \frac{y^2 dx}{a-y} + \frac{a^2 \delta u^2}{2}\left(\frac{1}{(a-y)^2} - \frac{1}{a^2}\right) = \frac{\delta du}{dt}\int \frac{y^2 dx}{a-y}$, because the second part is equal to zero when $x = AB$.[17] So $\frac{\delta du}{dt}\int \frac{y^2 dx}{a-y}$ is the fluid pressure; therefore if the vessel is very wide, this pressure will be null or may be regarded as null; and if the vessel has not a large extension it will give $\frac{M\Delta du}{dt} = \frac{\delta du}{dt}\int \frac{y^2 dx}{a-y}$, which is absurd. 1st, because in an infinity of cases $M\Delta$ will not be $= \delta \int \frac{y^2 dx}{a-y}$, as $\int \frac{y^2 dx}{a-y}$ depends on the figure of the vessel; now then, the shape of the vessel has not any influence on the amount $\frac{M\Delta}{\delta}$, which depends only on the shape of the body, its density and that of the fluid. 2nd, when even it would occur in some cases by chance that $M\Delta$ was equal $\delta \int \frac{y^2 dx}{a-y}$, then du may be taken as one would like and the problem will be undetermined, which is still absurd; since the fluid and the body are given, the quantity of resistance and velocity at every instant is necessarily determined and not arbitrary.

Scholium I

104. In the *previous art.* we have only considered the velocity of the fluid particles parallel to AP. But we would find also the resistance null if we had considered the actual velocity of the fluid particle that are immediately contiguous to the body surface. Let us suppose, for convenience, that the curve ANB is composed of two equal and similar parts AO, OB, and let be $AN = s$, and $\frac{ds}{dx} = Y$, this quantity will be the same at the corresponding points of the arcs AO, OB.

Now, since the absolute velocity of the particles of fluid parallel to AP is $\frac{uy}{a-y}$, and the velocity of the body ANB in opposite direction is u; it follows that the velocity relative to the body will be $\frac{uy}{a-y} + u = \frac{au}{a-y}$, as this velocity is parallel to AP. Therefore the velocity of these particles along the surface of the body will be $\frac{uaY}{a-y}$. So 1st, the body is pressed by the fluid with a force equal to $-\frac{\delta du}{dt}\mu$, which acts along AB. 2nd. The velocity lost at the point N is $\frac{u'aY}{a-y} - \frac{uaY'}{a-y'} = \frac{aYdu}{a-y} + \left(\frac{audY}{(a-y)dy} + \frac{auY}{(a-y)^2}\right)\left(udt + \frac{uydt}{a-y}\right)\frac{dy}{dx}$, and for the point V directly opposed to N, it will be $-\frac{aYdu}{a-y} - \left(\frac{audY}{(a-y)dy} + \frac{auY}{(a-y)^2}\right)\left(udt + \frac{uydt}{a-y}\right)\frac{-dy}{dx}$. Hence it is easy to see that the pressure is the same as if the points N and V were impelled

---

[17] This formula has an error in the third term of the third equality, which must be $\int \frac{ady}{(a-y)^3}$, whose integral between 0 and y is $\int \frac{ydy}{(a-y)^3} = -\left(\frac{1}{a-y} - \frac{1}{a}\right) - \frac{a}{2}\left(\frac{1}{(a-y)^2} - \frac{1}{a^2}\right)$. However, the final result does not change because its value between 0 and AB is also zero. The error comes from the *Mss*.90.

by the only force $\frac{adu}{dt}\frac{Y}{a-y}$. So the resulting pressure will $\frac{adu\delta}{dt}\int\frac{dy}{ds}\int\frac{Yds}{a-y}$. Therefore it will give $-\mu\Delta du = -\mu\delta du + a\delta du \int\frac{du}{ds}\int\frac{Yds}{a-y}$, from where $du = 0$ is deduced, and consequently the velocity u is constant, and the resistance null. Which is absurd. Thus etc.

Scholium II
105. It should not be concluded from the latter that the hypothesis, which has just been rejected for finding the resistance of fluids, has also to be rejected when it comes to determining the motion of a fluid in a vessel. The experiment, that must be our guide here, proves that in the latter case the assumption in question gives an analytical result fairly consistent with the observation, whereas in the first case the result of the calculation gives null resistance, which is absolutely contrary to experience. Moreover, we shall have occasion later to examine by our principles the laws of motion of a fluid in a vase.

5.5 About the Resistance of Non-elastic and Finite Fluids

106. We have seen (*art.* 82) that this resistance occurs when a body, submerged in a fluid close to the upper surface of this fluid, rises up and tends to move towards this surface. Because if the body was deeply immersed in the fluid, and the fluid was not elastic, then there would be no vacuum behind the body, however great its velocity was. Indeed, no matter what the velocity is, the communication of motion is always made (*art.* 85 *no.* 4) to the same number of parts; now when the velocity is very small, the experiment proves that the parts to which the body communicates motion are few, and they are very close to the body. Therefore the motion communicated to the fluid in the case in question would not be extended to the upper surface of the fluid; then in order to produce a vacuum behind the body, when the fluid is not elastic, it is necessary that the motion is extended up to the surface and that the part of this surface, which is perpendicularly above the body, rises somewhat above the level.

From this it follows that even in cases when there is a vacuum behind the body, the parts to which the motion is communicated are not far distant from the body. Thus for this reason only it could be enough just to regard the contiguous parts to the body surface, if this did not follow otherwise from *art.* 1, 21 and 22.

107. Now then, before determining the resistance in this case, we notice that the number of particles to which the body communicates motion here is not always the same, as in the case when no vacuum is produced behind the body. Because the proposition from which we have deduced this theorem in *art.* 8 only takes place when the particles of fluid are not impelled by any accelerating force, so that its effect is null; which happens when there is no vacuum behind the body, because then (*art.* 85 n°. 3) the effect of compression is null. It is not the same when a vacuum is produced, because then the velocity with which the parts of the fluid tend

Fig. 5.4

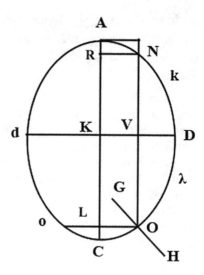

to move due to their compression must be combined with that which the body communicates to them, either directly or by means of other parts.

108. Now, in order to determine the resistance it must be observed: 1st, that the vacuum, that will be behind the body, will be much greater as the body velocity will be greater in relation with that which the particles of the fluid would erupt in an empty space due to their compression; 2nd, that these particles, being compressed along perpendicular lines to the body surface, should jump in the same direction; so that if the body moves, for example, along *CA* (Fig. 5.4), the particle placed in *O* behind the body will be moved along *OG* perpendicular to the body surface. Thus the velocity along *OG*, when the particle does not leave the body, is to the body velocity along *CA* as dy is to ds. Then, if *V* is the velocity that the compression of particles of the fluid must give to them, there will be *V* to *u* as dy to ds, and the vacuum will begin to be produced behind the body in the point where $\frac{V}{u}$ will be equal to $\frac{dy}{ds}$. From this it follows that there will not be any vacuum if $V = u$ or if $V > u$; and that it will be only when $V < u$; and that this vacuum will be much lower, all other things being equal, as *u* will be smaller. Because the smaller *u* will be, the more the ratio $\frac{V}{u}$ will approach to unity; therefore also the point where $\frac{dy}{ds} = \frac{V}{u}$ will be near the point *C*, since $\frac{dy}{ds} = 1$ at *C* and equal to zero at *D*.

Propos. XV. Problem

109. *Find the resistance of a heavy fluid, compressed, of finite extension and non-elastic.*

Let *O* be the point where $\frac{dy}{ds} = \frac{V}{u}$; $DK = b$; $KV = Z$, the parts *ADC* and *Adc* are assumed similar and equal, finally, let ψ be the force that compresses [all][18] the fluid particles. The compression upon the part *odADO* along *AC* will be $\psi \delta \pi Z^2$; now

---

[18]*Mss*.104.

it can be assumed $\psi = p\xi$, that is to say, equal to the pressure of a fluid whose gravity would be p and the height ξ. In this case, by the laws of hydrodynamics, we will have $V = \sqrt{2p\xi}$; because a fluid compressed by a force $\psi\delta = p\delta\xi$ escapes with a velocity of $\sqrt{2p\xi}$, that is to say with the velocity that a body of weight p will acquire falling from a height ξ. Besides that, the pressure along CA coming from the weight of the fluid will be $\delta g(ArN + NVODN)$ that must be subtracted from the weight $g\mu\Delta$ of the body.[19]

With respect to the part of the resistance of the fluid that comes from the inertia, we will prove, as was done in the *article* 55, that the part $-\frac{du}{dt}$ will be multiplied by a coefficient equal to zero, and that only the part $\varphi u^2 \delta$ will remain. Therefore neglecting the friction and viscosity of the parts, we will have $-\mu\Delta\frac{du}{dt} = \varphi u^2 \delta - \delta g(ArN + NVODN) + \mu\Delta g + \pi p\xi\delta Z^2$, an equation[20] from which the value of u cannot be deduced as function of t, because the quantities ArN, $NVODN$ and Z are variables and they depend on the position of the point O, which itself depends on u. But at least it can always have the value of t as function of u, which comes to be about the same.

Scholium I

110. If pressure ψ only comes from the weight of the parts of the fluid, then the term $\pi p\xi Z^2 \delta dt$ should be deleted in the above equation.

Scholium II

111. See here how the quantity φ will be determined. It is obvious that it is equal to the value of $\int 2\pi y dy (1 - p^2 - q^2)$ when $y = OL$, this value being equal to zero at any point of Fig. 6.1. Now to find this point let A and B be the abscise and the ordinate that respond to it; we will seek the quantity which must multiply $-\frac{du}{dt}$ in the formula of the fluid pressure, and as this quantity must be zero, then we will have the unknown A and B; and consequently the value of φ.

5.6 On the Resistance of Elastic Fluids

112. The fluids we have treated so far have been supposed to be unable to occupy a larger or smaller space by the action of any force whatever this is. So that if a body moves in such a fluid of an indefinite extension, it will never produce vacuum behind the body and the fluid will always maintain the same density behind the body.

113. It is not the same when the fluid is elastic. Because when a body moves in such a fluid, the fluid is expanded at the rear part of the body and condensed at the front part; and the compression of a part and the expansion of the other are greater in the measure that the body moves with higher velocity. Furthermore, neither

---

[19] Let note that by *Arn* and *NVODN*, we understand here the solid generated by the revolution of these figures around the *AC* axis. (*Original Note*).

[20] In the original there are several dt misprinted and in §.92, 94 as well.

compression nor expansion are the same in all points of the surface, be in the front or in the back. For example, a globe which moves in an elastic fluid, and let us imagine a line parallel to the direction of the globe passing through the center of the globe; it is obvious that the compression of the fluid is greatest at the front end of this diameter and that the expansion is the greatest at the rear end. From this it follows that the compression and dilatation will be smaller in any point of the surface, in the measure this point will be further away from the ends of the assumed diameter; so that in the line which is in the middle between the front and back surface, that is to say, in the great circle which is separated 90° from these ends there are neither any compression nor any dilatation; which we may also prove as follows. Since the compression of the fluid is produced at the front of the body and the dilatation takes place at the rear part, and that nothing happens in nature other than by insensible degrees, the fluid contiguous to the body must pass from the state of compression to the dilatation through insensible degrees. Therefore, after the point where the compression is the greatest, it must decrease to a point where compression is changed to dilatation, and consequently the fluid must not be dilated or compressed in this latter point.

114. Given this, the fluid dilates in the rear part because the motion of the body leaves an empty space behind the body into which the fluid rushes with more velocity in the measure its compression is greater.

So, as *M. Robins* has noted the first, if the velocity of the body is greater than that with which the fluid can jump into an empty space, then an empty space behind the body will necessarily remain. From this consideration *M. Robins* rightly concludes that the laws of the resistance for elastic fluids must be very different from those of the resistance for non-elastic fluids, especially in the cases when a vacuum is produced behind the body. Because, as *M. Robins* noted very well, if the resistance of a non-continuous fluid, composed of parts separated from each other, is greater than that of a continuous fluid, this is because in a continuous fluid a reflux of particles is produced behind the body by a kind of circular motion, which helps to reduce the resistance of the fluid upon the front surface. From this he concludes that in an elastic fluid, when a vacuum is produced behind the body, the resistance is much greater than in a continuous fluid, because the motion and the circular reflux of the parts cannot take place then. Now, according to *M. Newton*, the resistance that a non-continuous fluid causes to a cylinder moving therein is four times the resistance that a continuous fluid causes to the same cylinder; moreover, the resistance that a continuous fluid causes to a globe is equal to that of the same fluid to the cylinder. Finally the resistance that a non-continuous fluid causes to a globe is a half that the same fluid causes to a cylinder.

Therefore, *M. Robins* concludes, the intensity of the air resistance to a cannon ball when a vacuum is formed behind the ball, is twice as large (excluding the velocity) than in those cases in which no vacuum is produced.[21] Besides, as in the

---

[21]The original is confused and unclear. In Mss.115 "in iis casibus in quibus nullum fit vacuum": "In those cases in which no vacuum is produced".

case when there is a vacuum, the fluid is condensed much at the front part, *M. Robins* judges that the resistance increases also by this circumstance; so that, according to him, the intensity of the air resistance upon a cannon ball moving very rapidly is three times what it would be if the ball was moving with a medium velocity, so that no vacuum was left behind it. This proposition, or, if it is preferred this conjecture, seems to be confirmed by the experiments that *Robins* has made. But as he has not given another theory on this subject, I thought that it would not be useless to show some views here on this matter. Since the air is the only elastic fluid we know, I will deal here only with the air resistance.

Observations

115. [1st]. The air in its natural state is compressed by a force equal to that of a column of water about 32 pieds.[22] Now, the air is about 800 times lighter than the water. So the air in its natural state is compressed by a force corresponding to an air column of approximately 32×800 pieds. So if the air compressed by this force rushes into an empty space, its velocity would be that which a heavy body would acquire in falling from a height of 32×800 pieds. Now a heavy body traverses 15 pieds per second, therefore in a second it would traverse 30 pieds with a uniform motion. So the air moving with the aforementioned velocity would traverse in one second a space of $30 \times \frac{\sqrt{32 \times 800}}{\sqrt{15}} = 2\sqrt{15} \times \sqrt{8 \cdot 8 \cdot 4 \cdot 100} = 2\sqrt{15} \cdot 8 \cdot 2 \cdot 10 = 328\sqrt{15} = 320\sqrt{16-1} = 320\left(4 - \frac{1}{8}\right) = 1280 - 40 = 1240$ pieds. Therefore in order to produce a vacuum behind the body, its velocity must be greater than 1240 pieds per second.

2nd. Let δ' be the density of the air in its natural state, δ its density in another state, experience makes us see that the compression of air in direct ration to its density can be assumed fairly accurately; therefore the compression will be $32 \times 800 \cdot \delta/\delta'$. But the velocity with which the air would erupt in a void space will always be equal, regardless of the density δ, to which a heavy body would acquire falling from the height of 32×800 pieds, which will always be 1240 pieds per second. Because when the air has a density δ, its compression is equal to the weight of a column of 32×800 pieds and of the density δ.

3rd. Let there be a body *DAdC* (Fig. 5.5) which moves in an elastic fluid, so that it goes from *DAdC* to *D'a'd'c*. It is obvious from what has been said in *art*. 113, that the largest compression of the fluid will be at *A*, the largest dilatation in *C*, and at *D* there will be neither compression nor dilatation. Therefore if *Nn* is drawn parallel and equal to *Aa*, and *NV* perpendicular to *AD*, it is obvious that the compression will be smaller in the measure that the line *NV* will be smaller. Besides this, the compression at *A* is greater in the measure that the velocity is greater. Therefore, naming as above δ' the density of the air in its natural state, u the velocity of the body, that is to say of point *A*; it seems to me that it will not deviate far from the truth, assuming that the density at *A* is $\delta'\left(1 + \frac{nu}{G}\right)$, designating *G* some known

---

[22] A pieds of Paris equals to a 0.325 m. The air density at sea level and at 15° is 1.219 kg/m³, therefore 1/820 of the water. The air pressure in the same conditions is 101.3 kPa.

Fig. 5.5

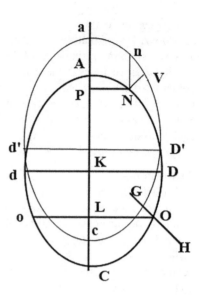

velocity that makes the density $\delta = \delta' + n\delta'$. For the same reason the density at N is will be $\delta'\left(1 + \frac{nu}{G}\frac{NV}{Aa}\right) = \delta'\left(1 + \frac{nu}{G}\frac{dy}{ds}\right)$; and this density, which is greater than the natural density in the part DAd, will become lower in the part DCd, where $\frac{dy}{ds}$ is negative.

4th. The air particle which is at O tends to move, whatever its density δ is, with a velocity of 1240 pieds per second. So to find the part OCo that the fluid does not touch, we have only to search for the point O where $-\frac{udy}{ds} = 1240$ pieds; a problem easy to solve, especially when the figure is a globe.

This all well understood, here is how we seek the resistance of elastic fluids in the case when no vacuum is made behind the body; and in those when a vacuum is produced.

5.7 Principles Necessary for Determining the Pressure of an Elastic Fluid[23]

116. The parts of the fluid that move in the front surface DAd can be, as in the case of the non-elastic fluids, regarded as having simultaneously two velocities. One of which, that I call u, is equal and parallel the velocity of the body and the other is composed of two respective velocities uq, up, one of which is parallel to AC, and the other to Dd; being p and q functions of x and z. Besides that, the density δ is still a function of x and z. So considering here, as in Fig. 4.6, the points N', N, B', B, C', C,

---

[23] The content of the next articles is related to the previous one. However, in the Table of Content it is listed as a separate entry, for this reason we have named it as an additional section.

D', D which form an infinitely small rectangular parallelepiped and which are near the surface of the body, the density after the point N reached at n will become $\delta + \frac{d\delta}{dx} uqdt + \frac{d\delta}{dz} updt$ and the rectangular parallelepiped $NN'B'BDD'C'CN$, whose mass is $\alpha\beta k\delta$, will turned into another whose mass will be $\left(\alpha + \frac{adp}{dt}udt\right)\left(\beta + \frac{\beta dq}{dx}udt\right)\left(k + \frac{kpdt}{dz}\right)\left(\delta + \frac{d\delta}{dx}uqdt + \frac{d\delta}{dz}updt\right)$; now then this second parallelepiped must be equal in mass to the first. So it will give $\delta\left(\frac{dp}{dz} + \frac{dq}{dx} + \frac{p}{z}\right) + \frac{qd\delta}{dx} + \frac{pd\delta}{dz} = 0$, that is to say $\frac{d(\delta p)}{dz} + \frac{d(\delta q)}{dx} + \frac{\delta p}{z} = 0$.

Besides this, it will be found (*art.* 86) that the force along NC that must be destroyed is $\frac{du}{dt} - \frac{qdu}{dt} - u^2 Aq - u^2 pB$, and the force that must be destroyed along NB is $-\frac{pdu}{dt} - u^2 pA' - u^2 qB'$. So (*art.* 19 and 20) we will have: $\frac{du}{dt}\frac{d(\delta - \delta q)}{dz} - \frac{u^2 d(\delta qA + \delta pB)}{dz} + \frac{du}{dt}\frac{d(\delta p)}{dx} + \frac{u^2 d(\delta qA' + \delta pB')}{dx} = 0$.

117. In order to use this equation, we have only to assume $\delta = \delta(u, x)$,[24] that is to say equal to a function of x and u, so that $\frac{d\delta}{dz} = 0$, and the remaining equations will be exactly similar to those of the *art.* 86; so that we will need no more than to determine δq and δp by the same method used in *art.* 61 to determine p and q. Ones these quantities are found, it will be noted that $\delta = \delta'\left(1 + \frac{nudy}{ds}\right)$ and putting for $\frac{dy}{ds}$ its value at x that I suppose ξ, which is given by the equation of the curve, we will have $\delta = \delta'(1 + nu\xi)$; therefore $\delta(u,x) = \delta'(1 + nu\xi)$, so knowing δ, and having found δp and δq, it will give p and q. Using these quantities and the method of *art.* 66 the fluid pressure will be determined at every instant and therefore its resistance.

In the case when a vacuum must be produced behind the body a method analogously to the Section 5th will be employed.

Moreover, other hypotheses more realistic about the value of δ would make the calculation even more complicated; and all this is no more than a cursory attempt.

---

[24] In the original, $\delta = u, X$. We understand it as $\delta = \delta(u,x)$.

Chapter 6
Oscillations of a Body Floating in a Fluid

6.1 Rectilinear Oscillations

118. Let there be a body *DAd* (Fig. 6.1) consisting of two equal and similar parts placed on either side of the axis *AC*, and which we consider for more convenience as a plane figure. Let us imagine that this body is placed on the surface of a fluid at rest, so that the axis AC is vertical and the immersed part *KAN* in the fluid is a little less weighty than an equal volume of fluid. We ask, What is the law of oscillation of bodies?

1st. We will find by the method of *art*. 86 that the parts of the fluid, besides the velocity which is common with the body, will have a respective velocity composed of two partial velocities and *uq* and *up*. 2nd. The coefficient of *du/dt* in the formula of the fluid pressure upon the body will be null, for the same reasons that have been expounded. Then there will only remain in the formula of the pressure the term that comes from the weight of the body and the one that will contain the square u^2. Now as the velocity is very small here, because the oscillations are very small, it is permissible to ignore this term. So that if the fluid density is called δ, $\Delta\mu$ the body mass, *P* the part which is submersed in the instant *dt* and *p* the gravity; we will be have $\frac{du}{dt} = \frac{\mu\Delta p - P\delta p}{\mu\Delta}$. Using this formula it is possible to solve the problem easily. In Vol. 4 of his works *Bernoulli*[1] has given a solution which can be consulted; moreover it will also be found in the next article. But it was necessary for the correctness of this solution to prove that the fluid pressure in this case comes only from the gravity and that the inertia [of the parts of the fluid] must be counted for nothing[2]; what no one had proven yet.

---

[1]Johann Bernoulli, father of Daniel and his *Opera Omnia*.

[2]In the *Mss*.107, "and ignoring the resistance that comes from the inertia of the fluid parts": And I ignore the resistance derived from the inertia of the parts of the fluid."

Fig. 6.1

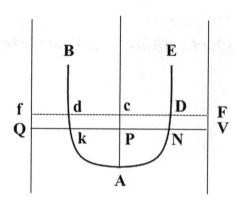

6.2 Curvilinear Oscillations

119. The oscillations of a floating body in a fluid are not rectilinear except in one case, namely where the center of gravity of the total mass and the one of the submerged part are in the same vertical straight line. If they are not in this straight line then the action of the fluid in order to lift the body, which acts along a line that passes through the center of gravity of the submerged part, no longer passes through the center of gravity of the body. Thus, according to the principles of dynamics, the center of gravity must rise up in a vertical line, while the body rotates about the same center. To make this more clear, let there be a power acting along the line *gf* (Fig. 6.2); I say that the center of gravity of the body will move along a line parallel to *gf* with the same velocity that it would have if the direction *gf* of the power passed through the center of gravity, and the body will turn at the same time around its center of gravity with the same velocity that it would have if the center was fixed and the power had the direction *gf*.

120. Then let C be the center of gravity of the body, BOD the submerged parts, $BA = b$, $AD = a$, E the midpoint of BD, G the center of gravity of the part BOD, $CF = \beta$, α the quantity of space that the center C traverses vertically. We will find that $AE = b - \frac{a+b}{2}$, and $EI = b - \beta - \frac{a+b}{2}$. Let N also be the weight[3] of the submerged part BOD; this weight diminishes the amount $\alpha(a+b)$ when the center C travels upwardly the space α, so that the center of gravity G passes to another point g, so that $EG : Gg :: N : \alpha(a+b)$; therefore Ii or $Ff \times N = EI \times \alpha(a+b)$; therefore $Ff = \frac{\alpha(a+b)}{N} \times \left(\frac{b}{2} - \frac{a}{2} - \beta\right)$.

121.[4] Now, let suppose that the body rotates around the center C from D to Q, so that the angle described by the unitary radius in the time t is equal to ε. Then Cg being the almost vertical line, it is clear that in this rotation the center g will advance horizontally a quantity equal to $\varepsilon(CA + GI) = \varepsilon(e+f)$, naming $CA = e$ and $GI = f$.

---

[3]This is not weight, but a volume or a two-dimensional surface. The same is applied to the next time the term weight appears.

[4]There are some misprints corrected. Besides, the symbols b, N and Q are used with two meanings.

6 Oscillations of a Body Floating in a Fluid

Fig. 6.2

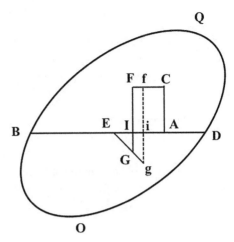

Fig. 6.3

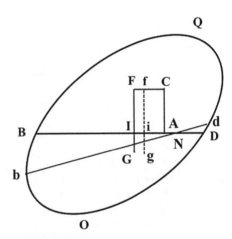

In addition (Fig. 6.3), let the angle $ACa = \varepsilon$, $Ca = CA - \alpha$ or CA, what comes to the same thing here, and bad perpendicular to Ca; BDd will become the submerged part. Then let be $BQ = 1/3\, b$, so that Q is the center of gravity of sector BNb, and if one makes $Ai = \beta'$, the distance from the center of gravity of the part bND to the line CA will be found to be very close to $\beta' - \frac{2}{3}b\frac{eb^2}{2N}$, because $\frac{eb^2}{2}$ represents the sector BNb. Also due to the dND area, the distance from the center of gravity of the BDd part to the AC line will be $\beta' - \frac{2}{3}b\frac{eb^2}{2N} - \frac{2}{3}b\frac{ea^2}{2N}$: So since (Fig. 6.2) $\beta' = Ai = CF - fF = \beta - \frac{a(a+b)\left(\frac{b}{2}-\frac{a}{2}\right)}{N}$, it follows that when bd (Fig. 6.3) is in the horizontal position, the distance from the center of gravity of the part bOd to the line CA is $\varepsilon(e + f) + \beta - \frac{a(a+b)\left(\frac{b}{2}-\frac{a}{2}\right)}{N} - \frac{eb^3}{3N} - \frac{ea^3}{3N}$, and as in this expression α and ε are variables, we can be put y instead of α and x in the place of ε. Now the force which raises the body up is $p\delta\left[N - y(a+b) - \frac{b^2x}{2} + \frac{a^2x}{2}\right]$; because when the center C has traversed the space y vertically and the body has turned the quantity x, the submerged part is $p\delta\left[N - y(a+b) - \frac{b^2x}{2} + \frac{a^2x}{2}\right]$; now the body weight $p\Delta M$ must be subtracted from this force Therefore it will give the first equation $\frac{M\Delta d^2y}{dt^2} = p\delta\left[N - y(a+b) - \frac{b^2x}{2} + \frac{a^2x}{2}\right] - p\Delta M$.

122. Moreover let ΔG be the sum of the products of each particle of the body by the square of their distance to the center of gravity C, it will give $\frac{\Delta G d^2x}{d t^2}$ equal to the product of the force which tends to raise the body and the distance from CA to the center of gravity of the submerged part; because the direction of this force passes through this center of gravity. We will therefore have $\frac{\Delta G d^2x}{d t^2} = p\delta N\left[\beta - \frac{y(a+b)\frac{b-a}{2} - \frac{b^3x}{3} - \frac{a^3x}{3}}{N} + x(e+f)\right]$; second equation, which, with the previous one, will serve to determine the quantities x and y, as we will see in a moment.

123. Let α be the space that a heavy body traverses in the time θ, and $N\delta - \Delta M = k\Delta M$; it will give $d^2y = \frac{2\alpha dt^2}{\theta^2}\left(k - \frac{\delta y(a+b)}{\Delta M} - \frac{\delta b^2 x}{2\Delta M} + \frac{\delta a^2 x}{2\Delta M}\right)$, and $d^2x = \frac{2\alpha dt^2}{\theta^2}\frac{\delta}{\Delta G}\left[N\beta - y\frac{b^2-a^2}{2} - \frac{b^3x}{3} - \frac{a^3x}{3} + Nx(e+f)\right]$.

Before coming to the integration of these equations, we notice that in the solution he gave to this problem the celebrated *Johann Bernoulli*, has only attended to the case when the center C is motionless, or $b = a$, and wherein the submerged part always has the same volume. Whence $y = 0$, $k = 0$, $d^2x = \frac{2\alpha dt^2}{\theta^2}\frac{\delta}{\Delta G}\left[N\beta - \frac{b^3x}{3} - \frac{a^3x}{3} + Nx(e+f)\right]$. Now then, a pendulum of length l would have a motion determined by the equation $d^2x = \frac{2\alpha dt^2}{\theta^2}(\varphi - x)$, being φ the initial angle of the pendulum with the vertical; and a pendulum which had a length l, and is isochronous with the oscillating body, would have as equation of its movement $l d^2x = \frac{2\alpha dt^2}{\theta^2}(\pi - x)$ or $d^2x = \frac{2\alpha dt^2}{\theta^2}\left(\frac{\pi}{l} - \frac{x}{l}\right)$, from which it is derived $\frac{1}{l} = \frac{\delta}{\Delta G}\left[\frac{b^3+a^3}{3} - N(e+f)\right]$; which agrees, as can easily be seen, with the formula of *Bernoulli*, in which $e+f$ is negative, the quantity that he calls $\int \delta r^2 p$ is here ΔG, and that he calls gV is $gN\delta$.[5] But it is clear that our formulas are much more extensive and they can generally be used to determine the very small oscillations of a floating body. I say *very small*, because the oscillations can be quite large although the initial distance CF (Fig. 6.2) is very small; for example it is what would happen if the body QDO was an ellipse whose major axis was almost vertical to the surface of the fluid.

124. Now for integrating the two equations that give the oscillatory motion, it is proposed to integrate these two generic ones $d^2x + Axdt^2 + Bydt^2 + Mdt^2 = 0$, and $d^2y + Cydt^2 + Dxdt^2 + Pdt^2 = 0$, which are more general and where M and P are constants, or functions of t, and A, B, C, D any constant coefficients. The second of

---

[5]The g now represents the gravity.

these equations will be multiplied by an undetermined coefficient ν, then it will be added to the first, then $x+\nu y$ will be assumed proportional to $Ax+D\nu x+By+C\nu y$, that is to say $A+D\nu=\frac{B+C\nu}{\nu}$. From here an equation will be obtained that will provide two values for ν, which I call ν' and ν''. Now let $x+\nu' y=u$ and $x+\nu'' y=z$, whereby the equations will change to these two $d^2 u+(A+D\nu')u dt^2+\Gamma dt^2=0$ and $d^2 z+(A+D\nu'')z dt^2+\rho dt^2=0$; being Γ and ρ functions of t or constants.

Now it is easy to integrate each of these equations by known methods. See the Memoirs of the Paris Academy of 1745 and those of the Prussia of 1748. That is why I do not delay more on this subject, contenting myself to have reduced the problem to the calculus.

Scholium I

125. If b and a were approximately equal, then the equations will become much simpler, because we will have $d^2 y=\frac{2adt^2}{\theta^2}\left(k-\frac{\delta y(a+b)}{\Delta M}\right)$ and $d^2 x=\frac{2adt^2}{\theta^2}\frac{\delta}{\Delta G}\left[N\beta-\frac{b^3 x}{3}-\frac{a^3 x}{3}+Nx(e+f)\right]$, equations that will integrated separately. If in the second of these equations the coefficient of x is positive, that is to say, if $N(e+f)>\frac{b^3+a^3}{3}$, that is to say $>\frac{2a^3}{3}$, the value of x will no longer contain arcs of a circle and the oscillations will not be infinitesimal.

For that reason, an ellipse whose major axis is nearly vertical to the surface of the fluid will not make small oscillations. Let suppose at first that this ellipse is a circle and $b=a$, it will give $N(e+f)=\frac{2a^3}{3}$, as it is easy to prove by the static principles. If the figure is an ellipse whose minor axis is to the large one as ρ to 1, and the major axis is very close to the vertical, then we will have $N(e+f)=\frac{2a^3\rho}{3\rho^3}$, and consequently $N(e+f)>\frac{2a^3}{3}$. On the contrary, if it is the minor axis that is almost vertical, then it gives $(e+f)=\frac{2a^3\rho^3}{3\rho}$ and $N(e+f)<\frac{2a^3}{3}$. So then the oscillations are small and the solution is only good for this case.

Scholium II

126. I have assumed so far that the fluid was indefinite, so that its surface does not rise with that of the body, but it still remained at the same level. But if the fluid was contained in a finite vessel, here is how it would necessary then to solve the problem.

It was found that if the fluid surface was still always at the same level, the submerged part of the solid at the end of time t would be $-y(a+b)-\frac{b^2 x-a^2 x}{2}$. So if the width of the fluid at the surface is called k', the fluid must be lowered at the end of time t by an amount equal to $\frac{y(a+b)+\frac{b^2 x-a^2 x}{2}}{k'-a-b}$, therefore the sunken part will become $N-y(a+b)-\frac{b^2 x-a^2 x}{2}-=\frac{y(a+b)^2-\frac{x(a+b)^2(b-a)}{2}}{k'-a-b}=N-\frac{k' y(a+b)}{k'-a-b}-\frac{k' x(b2-a^2)}{k'-a-b}$; and the distance from the center of gravity to the line CF will be reduced by the amount $\frac{y\left(a+b^2\frac{b-a}{2}\right)}{(k'-a-b)N}+\frac{\frac{x}{2}(a+b)^2\frac{(b-a)^2}{2}}{(k'-a-b)N}$; whence it follows that this distance will become $x(e+f)+\beta-\frac{y(b^2-a^2)k'}{2N(k'-a-b)}-\frac{x}{4}\frac{(b^2-a^2)^2}{N(k'-a-b)}-\frac{b^3 x+a^3 x}{3N}$. Therefore it

will give $d^2x = \frac{2adt^2}{\theta^2} \frac{\delta}{\Delta G} \left[N\beta - \frac{y(b^2-a^2)k'}{2N(k'-a-b)} - \frac{b^3x}{3} - \frac{a^3x}{3} - \frac{x}{4} \frac{(b^2-a^2)^2}{N(k'-a-b)} + N(e+f) \right]$ and
$d^2y = \left[k - \frac{\delta y(a+b)k'}{\Delta M(k'-a-b)} + \frac{\delta k'x(b^2-a^2)}{2\Delta M(k'-a-b)} \right] \frac{2adt^2}{\theta^2}$.

If the body must make only rectilinear oscillations, in this case $x = 0$, and $d^2x = \frac{2adt^2}{\theta^2} \left[k - \frac{\delta y(a+b)k'}{\Delta M(k'-a-b)} \right]$.

Scholium III

127. So far we have considered only plane figures. Now let us see what the oscillations of a solid should be, and at first let us take the solids of revolution. In the first place it is easy to see that the center of gravity of the solid and that of the sunken part will always be in the same plane passing through the axis of the body and perpendicular to the fluid surface. Once that is put, the problem will not have more difficulty than the one we have solved for the plane figures. Here is only what must be observed.

Let *QBOD* (Fig. 6.3) be the cut of the solid by a plane perpendicular to the fluid surface and wherein the oscillation must be made. Retaining the names given above, we will simply introduce: 1st, in the place of $a + b$, the entire surface that is the common section of the solid and fluid surface. 2nd, in the place of *N*, the sunken; solid part and in the place of *M*, the entire solid. 3rd, in the place of *G*, the sum of the products of the particles by the square of their distances to a horizontal axis perpendicular to the plane *QBOD*. 4th, in the place of $\frac{b}{2}$ and $\frac{a}{2}$, the distance from the line *CA* to the center of gravity of two parts of the horizontal surface that have *AD* and *AB* for abscissa, whose distances, in the case in question, are equal or deemed such, because the body is a solid of revolution. 5th, in the place of $\frac{\varepsilon b^2}{2}$ and $\frac{\varepsilon a^2}{2}$, we can set εqb^3 and εpa^3, assuming that $4Dqb^3$ and $4Dpa^3$ are the solids that these portions of surface would form rotating about their ordinates, *D* being taken to designate the right angle $\left[\frac{\pi}{2}\right]$. 6th, finally in the place of $\frac{2}{3}b$ and $\frac{2}{3}a$, we will put the distance from the line *CA* to the centers of gravity of these solids, which is roughly the same for both and that will be called *r* and *s*, so that $r - s$ will be an infinitely small quantity, or supposed as such.

Let *A* and *B* be the two parts of the surface that have *AD* and *AB* for abscissas, *h* and *l* the distances of their centers of gravity to the line *CA*, by the principle of P. Guldin, known by the geometers it will give $\varepsilon qb^3 = Ah\varepsilon$, $\varepsilon qa^3 = Bl\varepsilon$; therefore it will give $d^2y = \frac{2adt^2}{\theta^2} \left[k - \frac{\delta y(A+B)}{\Delta M} \right]$ and $d^2x = \frac{2adt^2}{\theta^2} \frac{\delta}{\Delta G} [N\beta - 2rxhA\varepsilon + Nx(e+f)]$, for the case where the surface of the fluid does not elevate with the body; and in case it elevates, the fluid surface will be called K', and it will give $d^2x = \frac{2adt^2}{\theta^2} \frac{\delta}{\Delta G} [N\beta - 2rxhA + Nx(e+f)]$ and therefore $d^2y = \frac{2adt^2}{\theta^2} \left[k - \frac{\delta y(A+B)K'}{K'-A-B} \right]$. We assume here for convenience $B = A$, and $h = l$.

6.3 Oscillations of a Body of Irregular Shape

128. The problem becomes much more difficult when the body is of irregular shape. To solve it, at first I imagine the two vertical lines *CA* and *GI* (Fig. 6.4) through which, in the first instant, pass the centers of gravity of the body and of the submerged part. I make a plane to pass through these lines which forms with the body the vertical section *QBOD* perpendicular to the fluid surface. Next the horizontal line *Cp* is drawn through the center *C*; let us imagine that this line *Cp* turns around the fixed point *C* in any plane inclined as desired with the surface of the fluid, but in such way that when changing situation the plan *QBOD* remains perpendicular to the surface of the fluid. The movement of the line *Cp* may be regarded as composed of two movements, one in a plane perpendicular to the fluid surface, the other in a plane parallel to this same surface and which will be around a vertical passing through C.[6]

129. As the motion of the axis *Cp* is very small and the line *AI* is also very small, it is clear that the latter of these two motions will only produce an infinitesimal rotation of second order in the center of gravity *G* that can be ignored, and besides this same motion will not make any part of the body either to emerge or to sink.

But it is not the same with the motion of the line *Cp* perpendicular to the surface of the fluid. For let *BZDY* (Fig. 6.5) be the common section of the body and the fluid surface, *ZY* perpendicular to *BD*, and γ, γ' the center of gravity of the wedges formed by the solid parts *ZDY*, *ZBY* turning around the ordinate *ZY*. Naming the angle of this rotation η, it is easy to see that the sunken part will become $N - \eta q b^3 + \eta p a^3$, and the center of this part will be: 1st it will go backward horizontally and parallel to *DB* in a vertical plane in the quantity $\eta (e + f)$; 2nd, it will advance in the

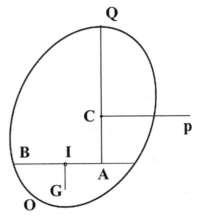

Fig. 6.4

---

[6]For coherence with the former we assume that the following can be classified as a new Article, with the same reasoning as in Chap. 5, Sect. 5.7.

Fig. 6.5

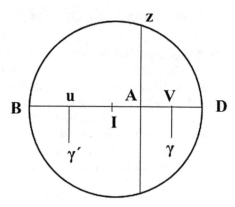

Fig. 6.6

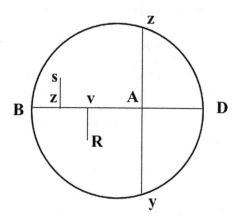

same plane parallel to *DB* by the quantity $uA \times \frac{nqb^3}{N} + VA \times \frac{nqa^3}{N}$; 3rd, it will advance horizontally and parallel to $V\gamma$ by the quantity $-u\gamma' \times \frac{nqb^3}{N} + V\gamma \times \frac{nqa^3}{N}$; 4th finally, it will be also rise vertically by some quantity not useful for our solution.

130. Now, in order to have the total body motion, it will be necessary, as I have done elsewhere,[7] to imagine a section perpendicular to *QBOD* (Fig. 6.4) passing through *QA*, which rotates about the axis *Cp* with an angular motion *P*.[8] Due to this motion: 1st, the center of gravity will be moved horizontally in an opposite direction to the angle *dP* at a velocity equal to $P(e+f)$. 2nd, if *R* and *S* (Fig. 6.6) are taken as the centers of gravity of the wedges formed by the parts *DYB*, *DZB* rotating about *DB*, and these wedges are called $q' \times AY^3 \times P$ and $p' \times AZ^3 \times P$, we will see that the center of gravity will advance in the direction of the angle *dP* by a

---

[7]See *Recherches sur la précession des Equinoxes*, article 26 et seq. (*Original note*).

[8]According with the context of the entire article, it seems to refer to an angle *P* rotating with a velocity dP/dt.

6 Oscillations of a Body Floating in a Fluid

Fig. 6.7

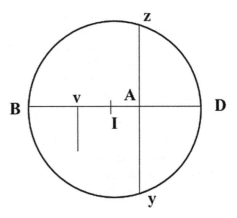

Fig. 6.8

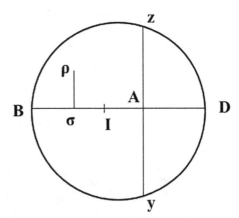

quantity equal to $\frac{P \cdot q' AY^3}{N} \times yR + \frac{P \cdot p' AZ^3}{N} \times SZ'$ and it will advance parallel to AD by a quantity equal to $\frac{q' AY^3}{N} \times P \times yA - \frac{p' AZ^3}{N} \times P \times Z'A$.

131. In addition, let a (Fig. 6.7) be the center of gravity of the area $BZDY$; it will be found that while the body rises perpendicular by the quantity α, the center of gravity of the part submerged advances in an opposite sense to AY by the quantity $\frac{\alpha \times BZDY}{N} \times ab$, and it advances in the sense of AD by the quantity $\frac{\alpha \times BZDY}{N} \times bA$.

132. So by joining the different quantities that we have calculated, in the vertical line $\sigma\rho$ (Fig. 6.8) it will give the point ρ where the center of gravity of the submerged part after the time t is found, so that $A\sigma = AI + \eta(e+f) - \frac{uA \cdot \eta q b^3}{N} - \frac{\gamma A \cdot \eta p a^3}{N} - \frac{q' AY^3 \cdot P \cdot yA}{N} + \frac{p' AZ^3 \cdot P \cdot Z'A}{N} - \frac{\alpha \cdot BZDY}{N} \times bA$, that I call $\beta - \omega$; and $\rho\sigma = u\gamma' \times \frac{\eta q b^3}{N} - V\gamma \times \frac{\eta p a^3}{N} - \frac{P \cdot q' AY^3}{N} \times yR - \frac{P \cdot p' AZ^3}{N} \times SZ' + \frac{\alpha \cdot BZDY}{N} \times ab$, that I call z.

133. Finally it is confirmable that after time t the sunken parts will be $N - \alpha \cdot BZDY + \eta p a^3 - \eta p b^3 - AY^3 \times Pq' + AZ^3 \times Pp'$, that I call $N-k$ for abbreviate.

134. Let us imagine now, following the method taught in my *Recherches sur la précession des Equinoxes*, that while the body has its various motions, there are: 1st, let G be the forces that have to be destroyed parallel to the plane QBD (Fig. 6.4) and to the fluid surface, and that their distance to this plane is χ and its distance to the surface of fluid ξ; 2nd, let F be the forces that must be destroyed at every instant parallel to the surface and perpendicular to the plane QBD, that their distance to the vertical plane passing through QA and perpendicular to plane QBD is θ and the distance to the surface of the fluid ζ: finally, 3rd, let the vertical forces to be destroyed be π', and their distance to the vertical plane AY (Fig. 6.6) is ν' and their distance to the plane QBD is μ'. These forces must be balanced with the vertical forces of the body, namely: 1st, with the forces $g\delta M$ and $-g\delta(N-k)$,[9] one of which is applied, or considered as applied, to A, and the other at a distance of $YA = \beta - \omega$, and at a distance of $BD = z$; 2° with the vertical force $+\frac{M\Delta d^2x}{dt^2}$ applied at A. The sum of these three forces is $g(N-k) - g\delta M - \frac{M\Delta d^2x}{dt^2}$, and it can therefore be reduced to a single one that I call π'', which joined with the force π' will be $\pi'' + \pi'$, I which I call π, and let whose distance to the line AY be named ν and let the distance to the line AD be named μ. So now we have three powers G, F, π whose positions are given and they must be in equilibrium. This condition gives

$$F\zeta - \pi\mu = 0.$$
$$G\xi - \pi\nu = 0$$
$$F\theta - G\chi = 0$$

And for the principles of the statics it will give, $\pi\mu = \pi'\mu' + g\Delta Nz$; et $\pi\nu = \pi'\nu' + g\Delta\nu(\beta - \omega)$.

Now, let be $K/2$ half the sum of the products of each particle by the square of its distance to the axis Cp (Fig. 6.4), and J[10] to the sum of the products of each one by the square of its distance to a plane vertical passing through GA, ϵ the angle described during the time t by the projection of the axis Cp on a horizontal plane, y the cosine of the angle that Cp makes with the horizon. It gives,[11] 1st. $G\xi - \pi'\nu' = \frac{K}{2}\left[-yd\left(\frac{ydy}{\sqrt{1-y^2}}\right) - 2yd\epsilon dP - ydP^2\sqrt{1-y^2} - yd\epsilon^2\sqrt{1-y^2}\right] + J(-d^2y\sqrt{1-y^2} + yd\epsilon^2\sqrt{1-y^2}) - \frac{K}{2}(d^2y\sqrt{1-y^2} - y\sqrt{1-y^2}dP^2) - Jyd\left(\frac{ydy}{\sqrt{1-y^2}}\right)$.

---

[9] In the original the symbol Δ is used both for the body and fluid density. We have maintained Δ for the fluid and introduced δ for the body.

[10] In the original, the letter M is used both the body volume and the moment of inertia. We have maintained M for the volume and introduced J for the second. Also, the moment of inertia should be respect an axis perpendicular to the plane mentioned.

[11] These equations are derived from formulas found in my *Recherches sur la précession des Equinoxes*, for determining the rotation of a body animated. (*Original note*).

6 Oscillations of a Body Floating in a Fluid

2nd.

$$F\theta - G\chi = J(2ydyd\varepsilon + y^2d^2\varepsilon) + \tfrac{K}{2}\left(-2ydyd\varepsilon + (1-y^2)d^2\varepsilon + d^2P\sqrt{1-y^2}\right) + \tfrac{K}{2}d^2\varepsilon - \tfrac{K}{2}\tfrac{2ydydP}{\sqrt{1-y^2}} + \tfrac{K}{2}d^2P\sqrt{1-y^2}.$$

Finaly 3rd.

$$F\zeta - \pi'\mu' = \tfrac{K}{2}\left(\tfrac{2y^2dyd\varepsilon}{\sqrt{1-y^2}} - yd^2\varepsilon\sqrt{1-y^2} - yd^2P\right) + J(2ydyd\varepsilon\sqrt{1-y^2} + yd^2\varepsilon\sqrt{1-y^2}) - \tfrac{K}{2}(2dydP + yd^2P).$$

Substituting these values in the above equations, it will give three new equations, of which comparing the second with the third,[12] $Kd^2P = -d\left(d\varepsilon\sqrt{1-y^2}\right) + g\delta Nzdt^2$ will be found; and as y is very little different from 1 and ε very small, it will give simply $d^2P = g\Delta\delta Nzdt^2$. Also if the first equation is examined, it will be seen that, neglecting all other terms that are infinitely small, it can be reduced to $g\delta N(\beta - \omega) = (K + J)\left(-yd\left(\tfrac{ydy}{\sqrt{1-y^2}}\right) - d^2y\sqrt{1-y^2}\right) = \left(\tfrac{K}{2} + J\right)d^2\eta$, assuming $y = \cos\eta$. With respect to the second equation, it will be reduced to $(M - \tfrac{K}{2})d(y^2d\varepsilon) + \tfrac{K}{2}d\left(dP\sqrt{1-y^2}\right) = 0$, or as y is almost equal to 1, $\left(J - \tfrac{K}{2}\right)d\varepsilon = 0$.

135. Then it will give

1st $\tfrac{M\Delta d^2x}{dt^2} = g\delta(N - k) - g\delta M$
2nd $g\delta N(\beta - \omega)dt^2 = \left(\tfrac{K}{2} + J\right)d^2\eta$
3rd $Kd^2P = g\Delta\delta Nzdt^2$

So introducing in the place of β-ω and z their values in P and x, and giving analytical values to the constants AY (Figs. 6.5, 6.6 et 6.7), AZ, yA, $Z'A$, uA, VA, $BZDY$, bA, $u\gamma'$, $V\gamma$, yR, SZ', ab, we will arrive at three equations of this form,

$$d^2x = (Hx + LP + K\eta + \Omega)dt^2$$
$$d^2\eta = (H'x + L'P + K'\eta + \Omega')dt^2$$
$$d^2P = (H''x + L''P + K''\eta + \Omega'')dt^2$$

Whose integration can be easily completed by the method which I have already mentioned above, and have explained at length in the 4th volume of the Memoirs of the Royal Academy of Sciences of Prussia, in the year 1748.

136. It is seen in this solution that 1st, that since $d\varepsilon = 0$, the line Cp, that we have taken as the body axis, has only an almost imperceptible motion parallel to the surface of the fluid, and the body has only proper rotation in two planes

---

[12] In the next and subsequent formulas, the moment of inertia K is missing but we have included it.

perpendicular to each other and both vertical. This is not surprising if it is considered that the forces acting here on the body are merely vertical. 2°. It follows from the first equation (*art.* 134) that force π'' is equal to $-g\delta M + g\delta(N-k) - \frac{M\Delta d^2 x}{dt^2} = 0$. In addition, we will easily find that the forces G, F, π' are also each one equal to zero, for this we only require to look for the expression of each of these forces that are in my *Recherches sur la précession des Equinoxes*, and to remember that for the property of the center of gravity $\int \mu \sin X = 0$, $\int \mu \cos X = 0$, $\int \mu(a-b) = 0$. Because it is a known law and a proven static, that whatever plan is made to pass through the center of gravity of a body, the sum of the products of each particle by its distance from the plane is zero. 3rd, it is easy to know by completing the calculations we just indicated here, that in any case the solid will only make infinitely small oscillations, that is to say, it will tend to restore its equilibrium state.

Chapter 7
On the Action of a Fluid Stream That Exits from a Vessel and Strikes a Plane

137. About this question, that has some relation to the theory of fluid resistance, I thought it would be good to deal with it here, not only because its solution is easily deduced from my principles, but also because it will give me the opportunity to make some new observations about this matter, and consistent with the experiment.

I will remark at first with *M. Daniel Bernoulli*, that whenever a fluid stream (water for example) strikes perpendicularly on a plane, all water particles leave the plane along lines parallel to the direction of the plane. This is better understood (I use here the terms of *Bernoulli*) in Fig. 7.1, wherein *AB* marks the axis of the stream flow that strikes the *EF* plane. It is clear that the threads that make up the stream bend at a small distance from the plane; so that at *E* and *F* where they leave the plane, their direction become parallel to the plane or perpendicular to the axis *AB*.

138. Let us suppose that *AB* (Fig. 7.2) is the opening of a vessel from which the waters flow with a uniform velocity to strike the plane *CD*. We only consider here a half of the plane *CD* and the orifice *AB*, because the other side will be exactly the same. 1st. It is clear by all that has been said, that the velocity of the particles at *D*, in the way that it is parallel to *AC*, will be zero. So if the velocity parallel to *AC* is expressed by a function q of $CP(x)$ and $PN(z)$; q must be a function such that it becomes zero when x is zero, that is to say that all it terms are multiplied by x. 2nd by the same reasoning as in *art.* 36 it can be proved that the velocity is constant in the extreme and external curve *BMD*.

Because *Mm*, *mm'* being two small segments of the curve described by the particle *M* in equal and consecutive instants, and let *mn* be equal in straight line with *Mm*. Moreover let the velocity *mn* be regarded as composed by the actual velocity *mm'* and a velocity *m'n* that has to be destroyed. It is clear by the hydrostatic principles that *m'n* must be perpendicular to the curve *BMD*; therefore $mn = mm'$. So the velocity in the curve *BMD* is constant.

139. Moreover, as the hole *AB* is supposed quite small and all the parts of the slice *AB* have the same vertical velocity, the supposition that all the parts of any

Fig. 7.1

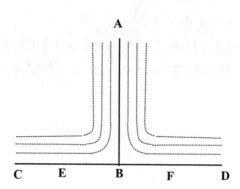

Fig. 7.2

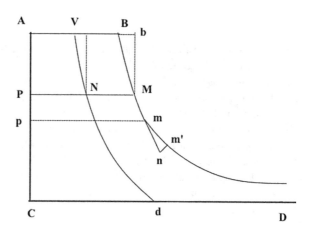

slice *PM*, parallel to *AB*, also have the same vertical velocity, which will not be far from the truth. So that if *Pp* is the distance traversed by the particle *P* in the instant *dt*, *PMmp* is constant, that is to say proportional to instant *dt*. Now then, we have just proved that taking the instant *dt* constant, *Mm* must be constant, because the velocity in the curve *BMD* is constant. Therefore *Mm* must be proportional to *PMmp*, from where the curve equation is obtained as follows.

140. Let $AP = x$, $PM = y$, $GS = s$, $AB = a$, we will have $ydx = ads$, because when $x = 0$, it is $dx = ds$ and $y = a$; therefore $y^2 dx^2 = a^2 dx^2 + a^2 dy^2$. So $dx = \frac{ady}{\sqrt{y^2 - a^2}}$.

Now Let *v* be the particle velocity at *A*, the velocity at *PM* will be $\frac{va}{y}$, and if the lines *AB*, *PM* would be of equal width, the pressure at any point of *PM* would be $\frac{y}{dt} \int \frac{vady}{y^2} dx = y \int \frac{vady}{y^2} dx \frac{va}{ydx}$, since $dt = \frac{ydx}{va}$; therefore the pressure would be $y \int \frac{v^2 a^2 dy}{y^3} = v^2 a^2 y \left(\frac{1}{2a^2} - \frac{1}{2y^2} \right)$ if *AB* and *PM* were equal. But *AB* not being equal to *PM*, the pressure that would come from the part *BbM* must be subtracted from the previous quantity. Now, the vertical pressure at any point of the curve *BM*, would be $v^2 a^2 \left(\frac{1}{2a^2} - \frac{1}{2y^2} \right)$ which multiplied by dy and integrated next will give the quantity that is necessary to subtract of the previous one. It gives consequently $\int v^2 a^2 dy \left(\frac{1}{2a^2} - \frac{1}{2y^2}\right) = \frac{v^2}{2}(y-a) + \frac{v^2 a^2}{2y} - \frac{v^2 a^2}{2a}$. So let $v^2 = 2ph$, and the pressure at PM will be $phy - \frac{pha^2}{y} - phy + pha - \frac{pha^2}{y} + pha = 2pha - \frac{2pha^2}{y}$.

141. Some readers may imagine that the pressure on PM should be the same as if AB was equal to PM, because following the hydrostatic principles, if a fluid is given whose all parts are impelled by any force parallel to AC, which is the same in all points of the same slice PM; this fluid exerts the same pressure as if all ordinates PM were equal. But it should be noted that in addition to the force $-\frac{dv}{dt}$ there are other forces here. Because 1st, in the curve BMD the destroyed force is perpendicular to the curve. Therefore this force is composed of two others, one parallel to AC equal to $-\frac{dv}{dt}$ and the other perpendicular to AC. The same must be said for the other points with the ordinate PM, which describing the curve lose not only a force $-dv/dt$ parallel to AC, but another force perpendicular to AC. Hence it follows that the pressure at M, for example, which would equal to $-\int \frac{dv}{dt} dx$ if there was only the force $-\frac{dv}{dt}$ which will be null. Because this pressure would be the same as the pressure of the channel BM which must be null (*art*. 27) because the equality of velocities at B and M; and the pressure at another point, for example N, is equal to those of the channel BMN. So as the pressure of the channel BM is zero, this pressure is the same as if it came from the only part MN; instead of if there was no more than the force $-\frac{dv}{dt}$ and the fluid was enclosed in a vessel $ABDC$, the pressure at N would be the same as if it came from the only column βN. Because in that case the weight of BM would not be zero, but the weight of βN. Thus it is not without reason that we have determined the pressure as we have done in *art*. 140, without regarding PM and AB as equals.

142. Let see here the conclusions that can be drawn from the above theory. It is clear by the equation $dx = \frac{ady}{\sqrt{y^2 - a^2}}$ that when $x = AC$, dx/dy is never zero, unless y is assumed infinity. Now then, this assumption cannot be made physically, because from this it is follows that the direction of the fluid, when it reached the CD plan is not exactly parallel to that plane, but it makes an angle as acute to the plane CD, as this plane CD is longer in relation to the orifice AB. However, as this length is quite large let b be the length of the plane, and $2pha - \frac{2pha^2}{b}$ will almost be given for the fluid pressure. Therefore b being much greater than a, it is clear that the pressure will be a little lower than $2pha$, which fits perfectly with the experiments made by M. Krafft, and reported in Volume 8 of the Memoires of Petersburg; because, according to these experiments, the action of a fluid stream that strikes upon a plane is a little less than the weight of a fluid cylinder whose base is a and the height $2h$. This is to say, whose base is the opening of the hole and the height is equal to twice the *height due to the velocity* of fluid.

Scholium I
143. In previous articles, for ease of calculation we have assume that the vessel was a parallelogram rectangle; but if it was considered as a cylinder, then it would be $y^2 dt = a^2 ds$ and $dx = \frac{a^2 dy}{\sqrt{y^4 - a^4}}$, equation that cannot be reduced to logarithms, as the

equation $dx = \frac{ady}{\sqrt{y^2-a^2}}$, but can be reduced the rectification of conic sections. See Memories of the Academy of Science Prussia year 1746, Vol. 2. With regard to the pressure, taking 2π for the ratio of the circumference to the radius, it will be found in the following manner.

The velocity at *PM* is $\frac{va^2}{y^2}$; therefore if *PM* was equal to *AB*, the pressure would be $\pi y^2(2v^2a^4)\left(\frac{1}{4a^4} - \frac{1}{4y^4}\right) = \pi y^2 4pha^4\left(\frac{1}{4a^4} - \frac{1}{4y^4}\right)$; now, if from this quantity $\int ph\left(1 - \frac{a^4}{y^4}\right)2\pi y dy = \pi phy^2 - \pi pha^2 + \frac{ph\pi a^4}{y^2} - ph\pi a^2$ is subtracted, the pressure at *PM* will be $2\pi pha^2 - \frac{2\pi pha^4}{y^2}$; expression that agrees again with the experiments of *Krafft*.

Scholium II
144. If the weight of parts of the fluid is taken into account, then the vertical velocity can be assumed the same in all parts of one slice *PM*, but the velocity at the curve *BMD* is not constant. Now then, in this case the lost velocity $m'm$ must be combined in such a way with the weight, which acts vertically, so that a single force perpendicular to the surface of the curved results. Therefore taking *PMpm* constant and *p* for gravity, it is necessary that *Mm* grows by the quantity $pdt^2 \frac{dx}{ds}$; that is to say that due to $dt = \frac{ydx}{va}$ it will give $d^2s = \frac{py^2dx^2}{v^2a^2}\frac{dx}{ds}$; thus $\frac{ds^2}{2} = \frac{py^2dx^2}{v^2a^2}x + \frac{py^2dx^2}{v^2a^2}h$.

Therefore $dx^2 + dy^2 = \frac{y^2dx^2}{2pha^2}(2px + 2ph)$, that is to say (making $h + x = n$) $dn^2 + dy^2 = \frac{y^2ndn^2}{ha^2}$; equation difficult to integrate, but the integral can be found at least by approximation as follows.

Obviously we would have $dx = \frac{ady}{\sqrt{y^2-a^2}}$, if *p* was zero; so that we would have $x = a\log\frac{y+\sqrt{y^2-a^2}}{a}$. Therefore introducing this value for *x* in the second member of the equation $dx^2 + dy^2 = \frac{y^2dx^2}{2pha^2}(2px + 2ph)$, we would have $dx = \frac{dy}{\sqrt{\frac{y^2}{a^2}-1+\frac{y^4}{ah}\log\left(y+\frac{\sqrt{y^2-a^2}}{a}\right)}}$;

equation that represents almost exactly the curve *BMD*, above all in the points that are not too close to *D*.

Now, to determine the pressure in *PM*, it must be noted that the force destroyed in every particle *PM* in the instant dt is $d\left(\frac{va}{y}\right) + pdt$: from where we easily conclude that the pressure determined in the *art. prev.* must be increased by a quantity equal to the weight of the entire vein *ABDC*. Now the weight of the fluid *ABDC* is about the same as if $dx = \frac{ady}{\sqrt{y^2-a^2}}$; therefore $\int pydx = \int \frac{apydy}{\sqrt{y^2-a^2}} = pa\sqrt{y^2 - a^2} = pa\sqrt{b^2 - a^2}$, so the total pressure will be $2pha - \frac{2pha^2}{b} + pa\sqrt{b^2 - a^2}$. This expression does not seem to agree with the experiments of *Krafft*; at least in the cases when $\sqrt{b^2 - a^2} > \frac{2ha^2}{b}$, which can happen often. But it must be noted that in the experiments of *Krafft*, the water came out of the vessel through a vertical hole following a horizontal direction, whence it follows that its weight had no share in the effect of pressure. It might be necessary to make new experiments on fluid pressure coming out of a vase in a vertical direction. But these experiments are difficult to perform; whatever they may be it is sufficient for us that all those that

7 On the Action of a Fluid Stream That Exits from a Vessel and Strikes a Plane 107

have been made to date on this matter, and whose accuracy can be counted, are consistent with our theory.

Scholium III

145. The expression that we have just given for the pressure of a fluid flowing from a vessel is a little different from that given by the famous *M. Daniel Bernoulli* in the Vol. 8 of the Memoirs of the Academy of Petersburg. According to him, the pressure of a fluid that comes out from a vessel is equal only to the quantity $2pha$. Now then, we find it to be a little smaller.

In order to know where the difference comes from, we make some observations on the method of *Bernoulli*.

He assumes at first that the curves described by each thread of fluid may be considered as channels in which a body moves. Therefore let *BMD* (Fig. 7.3)[1] a channel in which a small body, that I call m, moves and naming v the height *due to the velocity* at M. Let us seek with *Bernoulli* the sum of all the momentary powers parallel to the axis.[2]

Let the tangential power at $M = p$[3] and variable according to any law one wishes; the resulting force parallel to AC will be $\frac{pdxdt}{ds}$. Besides this, the centrifugal force at $M = \frac{2mvdt}{R}$, being R the radius of curvature at M, and the resulting for parallel to AC is $\frac{2vm}{R}dt\frac{dy}{ds}$; and as $dt = \frac{ds}{\sqrt{2v}}$ this latter force will be $\frac{mdy\sqrt{2v}}{R}$; therefore the sum of the two pressures $\frac{mdy\sqrt{2v}}{R} + \frac{pdxdt}{ds}$; now then making ds constant, we have $R = -\frac{dyds}{d^2x}$, and more $fdt = -\frac{mdv}{\sqrt{2v}}$. Therefore the pressure will be $-\frac{d^2x\sqrt{2v}}{ds} - \frac{mdvdx}{ds\sqrt{2v}}$, whose integral will be $-\frac{mdx\sqrt{2v}}{ds} + m\sqrt{2k}$, meaning for $\sqrt{2k}$ the initial velocity at B.

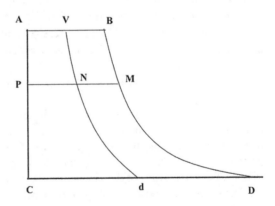

Fig. 7.3

---

[1]This description does not correspond with the former Fig. 7.2, but with the Fig. 29 of the *Manuscript*, that we include as Fig. 7.3.

[2]We notice that now v is the height due to the velocity and in this definition the gravity is missing. The same happens in the *Mss*.98. Likely he is following the Bernoulli's text, which is written in that way.

[3]The letter p is now used as the tangential force at M.

Therefore if $\frac{dx}{ds} = 0$, as occurs at the point where the fluid reaches the plane, the pressure at these points is $m\sqrt{2k}$; and if the velocity of fluid is assumed constant, then the pressure will be $m\sqrt{2k}(1 - \frac{dx}{ds}) = m\sqrt{2k}$ when $dx/ds = 0$. So in all cases, the sum of the momentary pressures from B until D is $m\sqrt{2k}$.

Now let $\sqrt{2A}$ be the uniform velocity of the exiting water, let whatever time t be taken at will, and let us assume that during this time a quantity of water m comes out; let p[4] be the power that supports the plane. Therefore it will give $t = m\sqrt{2A}$; or $p = \frac{m\sqrt{2A}}{t}$. Now then (*hyp.*) the mass m comes out uniformly during the time t with the velocity $\sqrt{2A}$ through the hole 1: so $1 \times \sqrt{2A} \times t = m$; therefore $t = \frac{m}{\sqrt{2A}}$; and $p = 2A$. According to *Bernoulli* that is the pressure of the water; where he concludes that it is equal to the weight of a water cylinder whose base would be the hole, equal to 1, and height $2A$.

It seems to me that this theory is reduced to the following propositions.

That in the first instant $d\theta$ of a time whatever t, through the orifice AB n particles flow and consequently the number of them is $\int n$. It is obvious that if we assume the hole AB divided in very small portions $d\alpha$, we will have $n = d\theta d\alpha \sqrt{2A}$; because at the same time $d\theta$ the particle n that comes out will be greater, as the velocity $\sqrt{2A}$ will be greater. So for the same reason in whatever time t, or $\int d\theta$, the number of particles that come out will be $\int t d\alpha \sqrt{2A}$.

Now, after they have come out from the orifice AB, each of the n particles reaches the plan CD describing any curve with any velocity, with the only circumstance that its motion becomes parallel to the plane CD when it arrives at this plane. Then the sum of the pressures of any particle $d\alpha$ since it goes through the hole AB until it reaches CD will be $d\alpha\sqrt{2A}$, and the sum of the pressures of all the parts $d\alpha$ that come out simultaneously from the orifice AB will be $\sqrt{2A} \int d\alpha$; therefore in time t the pressure will be $\sqrt{2A} \int t d\alpha \sqrt{2A}$. So that the pressure *Bernoulli* considers as instantaneous will be $2A \int d\alpha$, that is to say equal to the product of the width of the orifice by $2A$.

It seems pretty clear from these propositions that the pressure determined by *Bernoulli* is the sum of the pressures that the particles coming out of the vase at the same time, exert upon the plane from the instant they come out of the hole until the time they reach the plane CD. But it seems to me that the sum of these pressures does not represent the true pressure in question here. Because the sum of these pressures acts only in a finite time, that is to say in the time that the particles employ to reach from the orifice of the pipe to the plane CD. Now then, what is asked here is the instantaneous pressure that all fluid particles filling the space ABCD in that instant, exert in the same instant upon the plane CD. This pressure, if I am not mistaken, is different from that of *M. Bernoulli*. Since we consider the particles that describe the curve BMD, as fully covering this curve in any instant, and as we seek the pressure that they exert in this instant upon the plane we will find, using the same method of *M. Bernoulli*, 1st, that the pressure coming from the centrifugal

---

[4] Now p has the meaning of a force.

force is $\frac{2v}{R}ds\frac{dy}{ds} = \frac{-2vd^2x}{ds}$; 2nd, that the pressure coming from the tangential force is $-\frac{dv}{\sqrt{2vds}}ds\frac{dx}{ds} = -\frac{dvdx}{ds}$; therefore the pressure of any particle is $-\frac{2vd^2x}{ds} - \frac{dvdx}{ds}$, whose integral is $-2v\frac{dx}{ds} + 2k + \int \frac{dvdx}{ds}$. Now then, when dv is negative, that is to say when the velocity decreases from A to B, this amount is less than $-2v\frac{dx}{ds} + 2k$. Thus the pressure in each curve is $2k - P$, P being a positive quantity $-\int \frac{dvdx}{ds}$. Therefore, since the number of the curves is equal to the number of points of the orifice AB, it follows that if we call this orifice 1, the pressure will be equal to $1 \times (2h - P)$, that is to say less than that of *M. Daniel Bernoulli*; and more consistent with the one we have given.

We must admit, however, 1st, that when the velocity increases from A to B, the latter formula would provide a pressure greater than $2k$, which is contrary to experience; 2nd, that this same formula agrees with that of *Bernoulli* in the case $dv = 0$, that is to say when the speed is supposed constant in all the curves. But this latter hypothesis, as well as the method itself, seems subjected to some difficulties.

Because, let be ANC and BMD (Fig. 7.4) two curved or channels infinitely close one to other, and let be formed the channel $ANMB$. It is clear that due to be constant the velocity (*hyp.*), the pressure in the parts AN and BM is null (*art.* 27) as well as in the part AB. But in the channel MN there is some pressure that comes from the centrifugal force of the parts, therefore the channel $ANMB$ could not be in equilibrium; this is contrary to *art.* 18; from where it follows that the velocities at N and M cannot be equal. So since the velocity in channel BMD is necessarily constant, it follows that it cannot be in the inner channel ANC. Besides, if the fluid velocity in each curve BD and Vd (Fig. 7.2) was constant, then by the general formula for the pressure, found in *art.* 27, it follows that the pressure at D and d would be null, and thus the plane would not support any effort, which is contrary to experience. With regard to the method which we have used to determine the fluid pressure as equal to $2k - P$, it is faulty that the centrifugal force ought not to multiply the two forces by dy/ds and dx/ds. This is a result of everything that has been said in this book on the laws of fluid pressure.

It seems to me that we approach much closer to the truth with the hypothesis we have made that all the parts of the fluid in the same slice PM have the same velocity parallel to AC. We must admit however, that this assumption cannot be perhaps rigorously true, as it can be concluded from what has been said in *art.* 100.

Before finishing this research, I must warn that following *M. Daniel Bernoulli*, the experiments he has made agree perfectly with his theory. I preferred, however, the experiment of K*rafft* that, it seems to me, are more numerous, and they all agree with the pressure a little lower than $2a$. Perhaps to draw any true conclusions on this subject, it would not be useless to recommence the ones and the others.

Scholium IV

146. Moreover, the method that I explained in this book could be applied to the research of the pressure of a fluid stream. But the calculation would be difficult. Indeed, let QAq (Fig. 7.5) be the circular plane exposed to the fluid thread and FA the central thread; it will be proved, as it was done in *art.* 36, no. 3 and 5, that the

Fig. 7.4

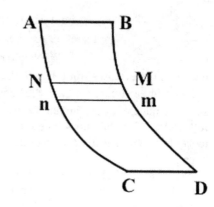

Fig. 7.5

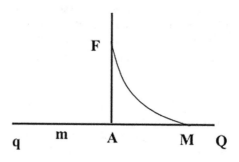

fluid is stagnating in a space *FAM* and that the velocity along *FM* is very small. Hence it follows: 1st, if we call *a* the fluid velocity and we make $Am = AM$, the pressure on *Mam* will be equal to $a^2/2$ multiplied by the circle whose diameter is *Mam*. 2nd, the values of *p* and *q* must be determined by methods similar to those of equations *art*. 45 and 48, and if *AQ* is called *y*, the pressure at *Q* will be $\int 2\pi y dy$ $\frac{a^2}{2}(1 - p^2 - q^2)$; this integral being taken so that it is zero where $AQ = AM$. 3rd, that the value of *q* is expressed as a function of *x* and *z*, and taking the origin of *x* in *A*, this function must be null making $x = 0$, because the velocity *q*, perpendicular to the *FA*, is zero along the plane *AM*. Therefore, the quantity *q* must contain *x* in all its terms. It is easy to find an infinite number of values of *q* which satisfy these conditions, especially if the plane is considered supposed as a single line; because then $dq = Adx + Bdz$ and $dp = Bdx - Adz$. But the problem remains undetermined anyway. That is what I obliged me to look for another route perhaps less rigorous and less direct in order to find the pressure of a fluid stream against a plane.

Scholium V
147. What we have said here about the action of a fluid stream against a flat surface, can also be applied to the action that a current exerts against a plan submerged in it. The values of *p* and *q* seem to me indeterminate in these cases, or rather indeterminable; in such a way that it is as impossible to compare the theory with

7 On the Action of a Fluid Stream That Exits from a Vessel and Strikes a Plane

the experiment, even in this case that seems the simplest of all. Besides, it seems to me that, all things being equal, the pressure of a fluid stream coming out of a vase and that acts against a plane must be greater than that of a fluid in which the plane is fully immersed. Because in the first case it is only the anterior surface of the plane, which is exposed to the action of the fluid; whereas in the second case the fluid acts on the posterior surface of the plane and partially counterbalances, by the pressure it exerts, that which supports the anterior surface. All this is according to experience, whereby, in effect, the pressure in the first case is greater than the pressure in the second (*art.* 75 y 142).

Chapter 8
Application of the Principles Outlined in This Essay in the Research of the Motion of a Fluid in a Vessel

148. As the principles I have given in this book to determine the laws of fluid resistance seem to me to have a lot of scope, I thought it would not be useless to show in some way how to apply them to the research of the movement of fluids in any vessels or channels. But as these researches are not directly related here to my subject, I content myself by only stating the principles.

At first let us imagine a vessel of any figure and with an indefinite length *HGLI* (Fig. 8.1), in which a quantity of fluid *ABFE* is enclosed, which is standing in the vessel supported by the *FE* bottom, and that suddenly the bottom *FE* is taken away. We ask what the fluid motion should be.

To make the calculus easier, at first we consider the vessel as a plane figure, and we will take the origin of x at C, the vertical being x and y or z the horizontal. If the vessel was cylindrical, it is obvious that the fluid would fall like the ordinary heavy bodies, so that naming g the natural gravity, t the time since the beginning of the fall, and u the velocity at the end of time t, we will have $u = gt$. But the curvilinear shape of the vessel must completely change the value of u, so that after a time t the horizontal and vertical velocities must be a function of t, x, z. Now then in the first place, these velocities must be between them such as making $z = PM = y$, the relationship they have with each other is equal to the function of x and y, which represents the ratio of dx to $-dy$ at the point M; and this condition has to take place whatever the time t is. Therefore if the vertical velocity is called Q, and horizontal P it must be $\frac{Q}{P} = \frac{dx}{-dy}$, placing in Q and P y instead of z. So it is necessary that t vanishes entirely in the division of Q by P; which cannot happen unless we assume $Q = \theta q$, $P = \theta p$, being θ a function only of t, and q, p functions of x and z.

149. That said, let $d(\theta q) = qTdt + \theta Adx + \theta Bdz$, and $d(\theta p) = pTdt + \theta A'dx + \theta B'dz$, we will easily find a similar method to the *art.* 48: 1st, that: $\theta B' = -\theta A$; 2nd, that the horizontal accelerating force that must be destroyed is $-\theta B'p - \theta A'q - pT$, and the vertical force $g - B\theta p - A\theta q - qT$, hence it follows that it will give

Fig. 8.1

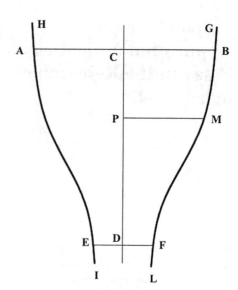

$\frac{\partial}{\partial z}(g - B\theta p - A\theta q - qT) = \frac{\partial}{\partial x}(-\theta B'p - \theta A'q - pT)$; equation which will be satisfied assuming $A' = B$; as in the *art.* 48. Therefore we will have:

$$dq = Adx + Bdz$$
$$dp = Bdx - Adz$$

From these equations the general form of the quantities p and q will be determined.

150. Now, in the beginning of the motion, when the time is $t = 0$, the surfaces AB, EF are horizontal, the force lost must be perpendicular to these surfaces; from which it follows that p must be zero when $x = 0$, and when $x = CD$, whatever value that z has. Moreover, if the walls of the vessel are not perpendicular to the lines AB, EF at A, B, F, E, it must be that $q = 0$ when $x = 0$ and $z = CB$, and when $x = CD$ and $z = DF$. Because the motion of the particles A, B, F, E cannot be take place other than following the walls of the vessel, $p = 0$ can only take place in those points that $q = 0$ as well.

It should also occur that in the beginning of the motion, when $t = 0$, the pressure in the channel CD is null, which gives $\int \left(g - \frac{d(\theta q)}{dt}\right)dx = 0$,[1] the integral is taken so that it is null when $x = 0$, and when $x = CD$. From this we can conclude that the θ function should be such that making $t = dt$, we have $\frac{d\theta}{dt} = 1$. So $\theta = t$.[2]

---

[1] The differential symbol is missing in $d(\theta q)$, and also in the next formula.
[2] The equation should be $d\theta/dt = g$, that is $\theta = gt$. This affects the next formulas.

8 Application of the Principles Outlined in This Essay in the Research...

Fig. 8.2

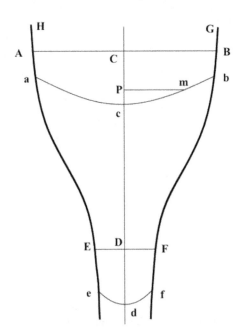

151. Through all the above conditions and the curvature of the walls *BM* that is known, and in which $-dx/dy$ must be equal to q/p, we will have the value of the coefficients to enter in q and p.

152. Let us suppose that after any time t, the two surfaces of the fluid are *acb* and *edf* (Fig. 8.2), and let us call a to Cc and b to Cd. I say at first, that assuming that a and b are known, it will be possible to find the curves *acb* and *edf*, because in each of these curves the force lost must be perpendicular to the curve from which it follows that if we call s to Pm, it must be $\frac{-ds}{dx} = \frac{g-q-Aqt-Bpt}{-p+Bqt-Apt}$, putting s for z in p and q, what will give the two curves. These two curves are known, and closing a and b in their equation, we see here how we determine the quantities a, b. 1° It should be noted that the mass of fluid is given *acbfdea*; first equation. 2nd, the pressure in the CD channel must be zero, hence $\int (g-q-tAq-tBp)dx = 0$ is obtained when $x = a$, and when $x = b$. From which we will have the value of a and b in t, and the problem will be completely solved.

Remark I
153. If the vessel was considered not as a plane figure, but as a solid body, then $B' = A$ should not be assumed; but $B' = -A - \frac{pt}{z}$ as in *art.* 48 Otherwise the calculation will be the same as in the previous article.

Remark II
154. If the fluid, instead of flowing always inside the vessel, escapes from it, then the calculation will again be the same as in the previous *art.* with the difference, that whereas in the former case (*art.* 152) the mass of the fluid *acbfec* was constant, it

Fig. 8.3

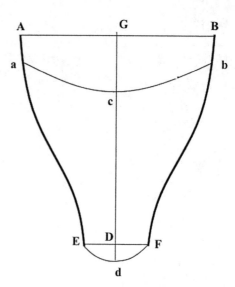

will no longer be here; this condition has to be changed to another, as that in the curve *edf* let $s = DF$ when $x = CD$ (Fig. 8.3). This is the only change we require to make to the calculation.

155. This method of determining the motion of a fluid is much more rigorous than the one I used in my *Traité de l'Equilibre et du mouvement des Fluides*, but the calculation is so difficult that we must almost give it up. Moreover, the experiment seems to agree fairly well with the theory that we have established in the work cited.

Chapter 9
Application of the Same Principles to Some Research on Streams in Rivers

156. Let be Bm (Fig. 9.1) the bottom of the river, CM its upper surface, and let it be assumed that the river flows from C to M, and that its riverbed is of equal width everywhere. Drawing at will the horizontal Qo and the vertical AQ, it is true that we can express the horizontal and vertical velocities of any point o by p and q, that is to say by the functions of QA, as x, and Qo, as z. In addition, it will be found by the methods already explained, that if $dq = Adx + Bdz$, we will have $dp = Bdx - Adz$. Therefore, 1st, [It is possible to] know the quantities q and p with a coefficient, as in art. 62.

2nd. The horizontal and vertical forces to be lost at the surface CM must produce a single one which, combined with the weight, is perpendicular to the surface; from this condition we will obtain the differential equation of the CM surface. Assuming that we know the depth of the river at two points C and M, the quantities of lines AC, AP, PM known, together with the equation of the curve CM, they all will be used to find the unknown coefficient of the quantities p and q.

Remark I
157. The problem is even easier to solve when it is not assumed that the bottom Bm is given, but that it is assumed as a figure at will. Then let be $TV = u$, we have only to take $\frac{dx}{du} = \frac{q}{p} = \frac{\sqrt{-1}\left[\Delta\left(x + \frac{u}{\sqrt{-1}}\right) + \Delta\left(x - \frac{u}{\sqrt{-1}}\right)\right]}{\Delta\left(x + \frac{u}{\sqrt{-1}}\right) - \Delta\left(x - \frac{u}{\sqrt{-1}}\right)}$

Remark II
158. Behold the problem generally solved, but here a comment to be made is presented. As it is assumed that the fluid reaches a steady state, it is clear that the CM surface always remains the same, and therefore all points of this surface move along the same surface, so that $\frac{q}{p} = \frac{dx}{dy}$ is at the surface CM; now then $\frac{dx}{dy} = \frac{g - Aq - Bp}{Bq - Ap}$ is as well, therefore $gp - Bp^2 = Bq^2$ and $qdy = pdx$ must be at the same time the equation of the curve AM.

Fig. 9.1

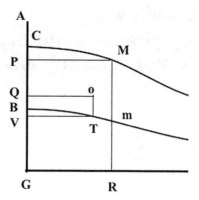

Fig. 9.2

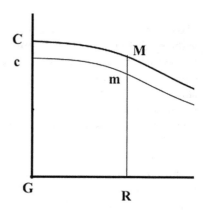

Let us see here, I think, how these two conditions can be met. Let us suppose the points A and C are given, and the curve B*m* as well. Two very general functions of *x* and *z*, with a large number of undetermined coefficients, will be taken for *q* and *p*. The curve CM will be assumed as represented by the equation $gp - Bp^2 = Bq^2$, the coefficients always remaining undetermined. Let us imagine next that the curve CM is drawn, and having taken in both curves CM and B*m* jointly as many points as coefficients to be determined, these coefficients will be found by the equations $\frac{q}{p} = \frac{dx}{dy}, \frac{q}{p} = \frac{dx}{du}$ et $gp - Bp^2 = Bq^2$.

Remark III
159. Instead of taking the origin of coordinates in A, if deemed more convenient we may take the origin at G; it will only be necessary to put *y* instead of *x*, $-dy$ instead of *dx*, etc. Everything else remaining as before.

Remark IV
160. If the surface CM is not required as permanent, but that it changes at every moment; in this case $\frac{q}{p} = \frac{dx}{dy}$ should not be assumed, but instead of this equation we would have another; in effect, the equation of the surface is in general $\frac{g-Aq-Bp}{Bq-Ap} = \frac{dx}{dy}$.

Now then, when the surface CM changes to cm (Fig. 9.2), q becomes $q + Aqdt + Bpdt$; p to $p - Apdt + Bqdt$; A to $A + \frac{dA}{dx}qdt + \frac{dA}{dz}pdt$; B does to $B + \frac{dB}{dx}qdt + \frac{dB}{dz}pdt$; finally dx becomes $dx + dt(Adx + Bdy)$ and dy does to $dy + dt(Bdx - Ady)$. From it $2A\frac{g-Aq-Bp}{Bq-Ap}dt + Bdt - B\frac{(g-Aq-Bp)^2}{(Bq-Ap)^2}dt = d\left(\frac{g-Aq-Bp}{Bq-Ap}\right)$ is obtained, placing for dA, dq, dB, dp, its values $q\frac{dA}{dx}dt + p\frac{dA}{dz}dt$, $Aqdt + Bpdt$, $q\frac{dB}{dx}dt + p\frac{dB}{dz}dt$, $-Apdt + Bqdt$. There will be an equation which, assuming q and p are taken at will with undetermined coefficients, will state only finite quantities because the term dt, found in all terms, can be eliminated. This equation will be used to determine the coefficients q and p by mean of the two curves CM and Bm. I have only indicated the method here, because the details would lead me too far.

However, I must note here that some time ago a handwritten theory on river currents fell into my hands. The method that the author employs, though less simple and less accurate than mine, nevertheless has something, it seems to me, to do with it, but I am able to prove that I had found the principles on which my method is supported by the end of 1749, that is to say, more than a year before the Memoire in question fell into my hands, and more than 8 months before it could fall into them. It would not be impossible that the method outlined in my book was unknown to the author of the Memoire I am talking about, and that it would have helped him in his research on the flow of rivers.

Chapter 10
Appendix

This Appendix will include some reflections on the laws of the Equilibrium of Fluids that I have not thought necessary to include in the body of the work in order not to interrupt the sequence of matters, but they seem to me worthy of being submitted to the judgment of the wise; and besides they have a fairly immediate relation with the subject of this book.

10.1 Reflections on the Laws of the Equilibrium of Fluid

161. The equation $\frac{d(\delta Q)}{dx} = \frac{d(\delta R)}{dy}$ found in *art*. 19 may be also found by some other method, which I will explain here, because it will give me the occasion to make some quite important observations on the laws of the equilibrium of Fluids.

Let M, N, m, O, (Fig. 10.1) be four points of the fluid and such: 1st, that the forces that impel the points M, m, are directed along the lines MN, mO both perpendicular to Mm; 2nd, that MN is to mO as the force along mO multiplied by the density at M. It is obvious that in the infinitely small straight channel $MNOm$, the small columns MN, mO will be in equilibrium between them. Thus the small channels Mm, NO must be also in equilibrium between them. Now then, as all the points of the channel Mm are impelled (*hyp*.) by perpendicular forces to Mm, the weight of this channel is null. Therefore the weight of the channel must also be null, that is to say, the forces acting on the points N, O must be perpendicular to NO at the points N, O. Now I will prove that for this condition to take place, it is necessary that $\frac{d(\delta Q)}{dx} = \frac{d(\delta R)}{dy}$.

Let be $Mm = ds$, the force along MN will be $\sqrt{R^2 + Q^2}$, and it can be assumed $MN = \frac{d\zeta}{\delta\sqrt{R^2+Q^2}}$, being $d\zeta$ an undetermined[1] quantity but infinitely small. Now let be

---

[1] In the original is "determinée" and in the *Mss*. 21 "indeterminata".

© Springer International Publishing AG 2018
J. Simón Calero (ed.), *Jean Le Rond D'Alembert: A New Theory of the Resistance of Fluids*, Studies in History and Philosophy of Science 47,
https://doi.org/10.1007/978-3-319-68000-2_10

Fig. 10.1

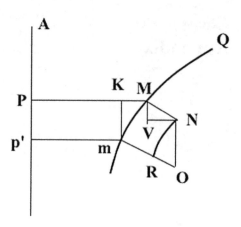

the force at *m* along *mO* be equal to $\sqrt{R'^2+Q'^2}$, it will give (hyp.) $mO \times \delta' \times \sqrt{R'^2+Q'^2} = MN \times \delta \times \sqrt{R^2+Q^2}$. Therefore $mO = \frac{d\zeta}{\delta'\sqrt{R'^2+Q'^2}}$; and carrying *NR* parallel to *Mm*, it will give $RO = \frac{d\zeta}{\delta'\sqrt{R'^2+Q'^2}} - \frac{d\zeta}{\delta\sqrt{R^2+Q^2}}$; now then $R' = R + Pp \times \frac{dR}{dx} - KM \times \frac{dR}{dy}$ (I write -*KM* because *AP(x)* increases, *PM(y)* decreases); even more, Pp or $mK = \frac{Mm \times Q}{\sqrt{R^2+Q^2}}$; because due to the similar triangles *MmK*, *MVN*, it gives $mK : Mm :: VN : MN :: Q : \sqrt{R^2+Q^2}$. Therefore $Pp = \frac{Qds}{\sqrt{R^2+Q^2}}$; it can be found as well $KM = \frac{Rds}{\sqrt{R^2+Q^2}}$; so $R' = R + \frac{dR}{dx}\frac{Qds}{\sqrt{R^2+Q^2}} - \frac{dR}{dy}\frac{Rds}{\sqrt{R^2+Q^2}}$; more $\delta' = \delta + \frac{d\delta}{dx}\frac{Qds}{\sqrt{R^2+Q^2}} - \frac{d\delta}{dy}\frac{Rds}{\sqrt{R^2+Q^2}}$; finally, for the same reason it will give $Q' = Q + \frac{dQ}{dx}\frac{Qds}{\sqrt{R^2+Q^2}} - \frac{dQ}{dy}\frac{Rds}{\sqrt{R^2+Q^2}}$. So

$$\frac{1}{\delta'\sqrt{R'^2+Q'^2}} = \frac{1}{\delta\sqrt{R^2+Q^2}} + \frac{1}{\delta(R^2+Q^2)^{\frac{3}{2}}}\left(-\frac{RdR}{dx}\frac{Qds}{\sqrt{R^2+Q^2}} + \frac{RdR}{dy}\frac{Rds}{\sqrt{R^2+Q^2}} - \frac{QdQ}{dx}\frac{Qds}{\sqrt{R^2+Q^2}} + \frac{QdQ}{dy}\frac{Rds}{\sqrt{R^2+Q^2}} + \frac{QdQ}{dy}\frac{Rds}{\sqrt{R^2+Q^2}}\right) - \frac{1}{\delta^2\sqrt{R^2+Q^2}}\left(\frac{d\delta}{dx}\frac{Qds}{\sqrt{R^2+Q^2}} - \frac{d\delta}{dy}\frac{Rds}{\sqrt{R^2+Q^2}}\right).$$ So as *RO* was found above to be equal to $d\zeta\left(\frac{1}{\delta'\sqrt{R'^2+Q'^2}} - \frac{1}{\delta\sqrt{R^2+Q^2}}\right)$; it will give $\frac{RO}{RN}$, that is to say, the angle $ONR = \frac{RO}{ds} = \frac{d\zeta}{\delta(R^2+Q^2)^{\frac{3}{2}}}\left(-\frac{RQdR}{dx} + \frac{R^2dR}{dy} - \frac{Q^2dQ}{dx} + \frac{QRdR}{dy}\right) - \frac{d\zeta}{\delta^2(R^2+Q^2)}\left(\frac{Qd\delta}{dx} - \frac{Rd\delta}{dy}\right)$.

Now let *Mm*, *Mμ* be two adjacent and equal sides of the curve *QMm*, *Nr* and *Nr* parallel to these sides (Fig. 10.2); and let *MN* be extended to *G*; it is obvious that *MN*, perpendicular (hyp.) to the curve *MN* in *M*, divides into two equals the angle *μMm*. Therefore it will also divide into two equals the angle *RNr*; moreover, let *μo* be to *MN* as the force $\sqrt{R^2+Q^2}$ along *MN* multiplied by the density at *M* is to force along *μo* multiplied by density at *μ*; for *ro* we will have the same value as *RO* but negative. Therefore the angle $RNO = rNo$. But we have proved that the force that the point *N* impels must be perpendicular to the curve *ONo*; so if *Ng* is the direction of this force, we will have the angle $ONg = gNo$; $ONo = RNr$ and $\frac{RNr}{2} = \frac{ONo}{2}$, that is

10 Appendix

Fig. 10.2

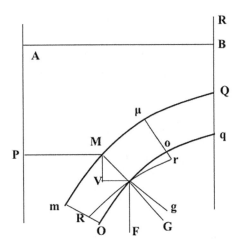

to say $RNG = \frac{ONo}{2} = ONg$. So $RNG = ONg$ and $GNg = ONR$. Now Q/R is the tangent of the angle NMV, and Q'/R' of the angle FNg let $Q' = Q+k$ and $R' = R + m$, the difference of the angles NMV, FNg, that is to say the angle GNg will be $\frac{Rk-Qm}{R^2+Q^2}$, as it is known by the geometricians (because, if y/x is the tangent of an angle, the differential of this is angle will be $\frac{xdy-ydx}{x^2+y^2}$); now we have here $k = \frac{dQ}{dy} \times VN + \frac{dQ}{dx} \times MV = \frac{dQ}{dy} \frac{d\zeta}{\delta\sqrt{R^2+Q^2}} \frac{Q}{\sqrt{R^2+Q^2}} + \frac{dQ}{dx} \frac{d\zeta}{\delta\sqrt{R^2+Q^2}} \frac{R}{\sqrt{R^2+Q^2}}$; and $m = \frac{dR}{dx} \frac{d\zeta}{\delta\sqrt{R^2+Q^2}} \frac{R}{\sqrt{R^2+Q^2}} + \frac{dR}{dy} \frac{d\zeta}{\delta\sqrt{R^2+Q^2}} \frac{Q}{\sqrt{R^2+Q^2}}$. Therefore the angle $GNg = \frac{d\zeta}{\delta(R^2+Q^2)^2} \left(\frac{R^2 dQ}{dx} + \frac{RQdQ}{dy} - \frac{QRdR}{dx} - \frac{Q^2 dR}{dy} \right)$. Therefore since we have just proved that the angle $GNg = ONR$, and we found above the value of the angle ONR, by comparing these two values, we will have the equation $-\frac{R^2+Q^2}{\delta} \left(\frac{Qd\delta}{dx} - \frac{Rd\delta}{dy} \right) + \frac{R^2 dR}{dy} - \frac{Q^2 dQ}{dx} = \frac{R^2 dQ}{dx} - \frac{Q^2 dR}{dy}$. Therefore transposing and multiplying by $\frac{\delta}{\sqrt{R^2+Q^2}}$, it becomes $\frac{\delta dQ}{dx} + \frac{Qd\delta}{dx} = \frac{\delta dR}{dy} + \frac{Rd\delta}{dy}$, that is to say, $\frac{d(\delta Q)}{dx} = \frac{d(\delta R)}{dy}$.

Scholium I

162. In the second method by which we demonstrated (previous *art.*) the equation $\frac{d(\delta Q)}{dx} = \frac{d(\delta R)}{dy}$ a rather important observation is presented. If we assume, with the authors that have so far dealt with this matter, that the density is constant in each layer QMm, ONo in particular, but that it varies at will from one layer to the other, for the law of equilibrium we will find only the equation $\frac{d(Q)}{dx} = \frac{d(R)}{dy}$, that is to say, the same that would be found if the density δ was constant all over. However, if in the same case where the density is not uniform, the equation that results from the equilibrium was sought by the method of *art.* 19, it would be $\frac{d(\delta Q)}{dx} = \frac{d(\delta R)}{dy}$. Then, how can these two equations can take place at the same time in the present case?

I answer that the density being constant in each layer (*hyp.*), it will give $dx\frac{d\delta}{dx} + dy\frac{d\delta}{dy} = 0$; therefore since $dy = -\frac{Rdx}{Q}$, we will be find $\frac{Qd\delta}{dx} - \frac{Rd\delta}{dy} = 0$. Hence the equation $\frac{d(\delta Q)}{dx} = \frac{d(\delta R)}{dy}$ is reduced to $\frac{\delta dQ}{dx} = \frac{\delta dR}{dy}$, that is to say, $\frac{dQ}{dx} = \frac{dR}{dy}$. But it should be noted that the equation $\frac{dQ}{dx} = \frac{dR}{dy}$ takes place in this case only for layers *QMm, ONo* in which the direction of gravity is perpendicular, while the equation $\frac{d(\delta Q)}{dx} = \frac{d(\delta R)}{dy}$ takes place on whatever layer one wishes, whether perpendicular or not to the direction of gravity. From this I conclude that the method of *art.* 19 is the only really general one to determine the laws of the equilibrium of fluids. Whereupon see the *Théorie de la Terre,* by M. Clairaut.

Scholium II

163. Moreover, in *art.* 161 I assumed the density variable, even in every curve especially in *QMm, Ono,* and I do not believe that this assumption has anything absurd. As long as the equation $\frac{d(\delta Q)}{dx} = \frac{d(\delta R)}{dy}$ takes place, and that the force of gravity is perpendicular to the first layer *QMm,* the mass of the fluid will be always in equilibrium whatever the law of the density is in each layer in particular. So I think I can advance in general, that any heterogeneous mass of fluid will always be in equilibrium, providing that the previous equation is observed. It is true that experience seems to contradict this assertion, because it shows us that fluids of different density cannot mix together. But the reason which prevents this mixing is that, gravity being *the same* for all these fluids, the $\frac{d(\delta Q)}{dx} = \frac{d(\delta R)}{dy}$ cannot take place until they are mixed.

Scholium III

164. It should also be noted that equation $\frac{d(\delta Q)}{dx} = \frac{d(\delta R)}{dy}$ only takes place assuming δ, R and Q as variable functions of x and y. But I see no reason to be limited to this assumption. Effectively, in order to facilitate the calculation, let us assume that the density δ is constant everywhere; why would not we assume R and Q functions not only of x and y but also of a third variable z,[2] represented for example by any line RQ, Rq (Fig. 10.2) which would be variable for the different layers *QMm, Ono,* and constant for the same layer? That said, the same value as above would be found for the angle *ONR* (*art.* 161); but in the expression of the angle *GNg* k must be increased by the quantity $d\zeta\frac{dQ}{dz}$, and m by the quantity $\frac{dR}{dz}d\zeta$. Therefore the value found above for the angle *GNg* would be increased by $\frac{d\zeta}{R^2+Q^2}\left(\frac{RdQ}{dz} - \frac{QdR}{dz}\right)$; so comparing the two values of the angles *GNg, ONR,* we will have $\frac{dQ}{dx} - \frac{dR}{dy} + \frac{RdQ}{dz} - \frac{QdR}{dz} = 0$.

Now, let R and Q be functions of x and z only, so that $\frac{dR}{dy} = 0$; it will give $\frac{dQ}{dx} + \frac{RdQ}{dz} - \frac{QdR}{dz} = 0$. Thus taking for Q any function of x and of z, R will be found

---

[2] In the original the symbol for this variable is ζ, which has been used already with another meaning and also in this article. To avoid such confusion we have introduce z in its place, which affects the entire article.

easily. Because we will find $\frac{dQdz}{Q^2dx} = \left(\frac{QdR}{dz} - \frac{RdQ}{dz}\right)\frac{dz}{Q^2}$. So treating x as constant, it will give $\int \frac{dQdz}{Q^2dx} = \frac{R}{Q} + \xi$, being ξ any function of x; and $\int \frac{dQdz}{Q^2dx}$ the integral of $\frac{dQdz}{Q^2dx}$ treating z as variable.

As the force R is assumed perpendicular to the curve at Q, q, etc. we must be careful to take the function Q, so that in all points Q, q, etc. of the line RQ this function is equal to zero. Therefore, taking AB perpendicular to AP and RQ, and making $RB = a$, Q must be a function of x, a, z, so making $x + a = z$, this function becomes zero. This can easily found by infinite ways.

It is easy to see that the latter formula would be $\frac{dQ}{dx} - \frac{dR}{dy} + \frac{RdQ}{dz} - \frac{QdR}{dz} = 0$, in which the density δ is assumed constant, this is more general than the formula $\frac{dR}{dy} = \frac{dQ}{dx}$, found in *art.* 162 for the same case. However, the equations $\frac{dR}{dy} = \frac{dQ}{dx}$ and $\frac{d(\delta R)}{dy} = \frac{d(\delta Q)}{dx}$ are the only ones that we have used in this work, because they are the only ones to which the calculation seems to be able to be applied in the research for the resistance of fluids.

Scholium IV
165. It is clear by the previous scholium that all parts of fluid contained in whatever layer Ono are equally pressed by the fluid that is above, since the weight of the columns MN, mO, μo is the same. As soon as there is equilibrium in the fluid, each inner layer Ono, to which the direction of gravity is perpendicular, is pressed equally in all its points. Can we not conclude from this with likelihood, that the pressure must be equal at all points of the first layer or outer surface $mM\mu$? In this case the forces along μo, MN, mO should be equal among them; but in the lower layers it would not be necessary that the inherent force in each particle was the same, it would be sufficient that each particle was equally pressed by the weight of the above column.

Besides, if we consider the particles of the fluid in the $mM\mu$ layer as small globules that press each other, and ignoring the lower layers, by static principles we can easily find that for these globules to be in equilibrium it is necessary that at any point M the force that acts along MN is inversely proportional to the radius of that evoluted in M. If this proportion takes place, and, besides this, the force must be constant in M according to what we just observed, it follows that in the outer surface $mM\mu$ all osculatory radii should be equal; and thus a fluid may not be in equilibrium, unless its outer surface is flat or spherical.

However it can be shown by the following reasoning that the surface of a fluid in equilibrium is not subject to either of these two figures.

Let us imagine a fluid whose parts are in motion; it is obvious that in countless cases its surface will not be flat or spherical. So let $OPQR$ (Fig. 10.3) be this surface in any instant, P, Q, two points or corpuscles placed upon this surface, whose velocities are u, v, and that these velocities in the next instant are changed to u', v'; finally, let the velocities u, v be considered as composed by u', u'', v', v''. It is obvious (*art.* 1) that if the points P and Q were impelled to move with the singles velocities u'', v'', they would remain in equilibrium. Therefore as the velocities u, u', v, v' are given, or assumed as given, it follows that forces can always be found that

Fig. 10.3

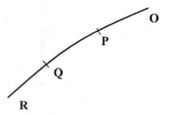

would maintain the points *P, Q*, in equilibrium on the surface of fluid *OPQR*; and it is the same with the other points. So whatever the figure the surface *OPQR* of a fluid has, there is always a possible system of forces which would maintain it in equilibrium.

From the foregoing results at least this first consequence: that the principle of the equality of forces in the outer surface and the [inverse][3] proportionality of these forces with the osculator radius cannot be both true. Moreover, it must be admitted that neither one nor the other is supported well by solid foundations. Because at first, for the principle of [inverse] proportionality of the forces with the osculator radius was true, it should have been proved not only that the particles of fluids are globules, which is very uncertain, but also that the effort of the globules acts only on those contiguous to them in the same layer, and nothing on those that are behind; which is not true. In respect to the principle of the equality of the forces, it is obvious that if it was accepted, all the theories which have been given about the shape of the Earth should be regarded as false, considering it as a fluid, and having regard to the attraction of the parts and to the axis rotation. I do not pretend to decide; I just want to show that if the principle of the equality of pressure on the outer surface is rejected, we must necessarily agree that the equality pressure of the inner layers is only, so to speak, an accidental property and not a fundamental law of fluid equilibrium.

Also *MacLaurin*, the first to speak of these layers *Ono* (Fig. 10.4)[4] to which the weight is perpendicular and that he called *level surfaces*, did not deduce the law of equilibrium from the equality of pressure on these surfaces. But after taking in the interior of the fluid a column *Pp* of any direction whose weight is equal to that of the column *AO*, he was content to derive, by a simple corollary, that the surface *Op* through all points *p* and through the point *O* will be a *level surface*. We have thought that we must follow his method fully in this regard.

Scholium V
166. Moreover, the same method, by which we have proved that the outer surface of a fluid can always be in equilibrium with a suitable system forces, can also be used to prove that an heterogeneous fluid, or several fluids of different densities, can

---

[3]See previous paragraph.

[4]In the original this Fig, is numbered as 49. We have changed to 10.4 in order to maintain the order of appearance in the text. The same is applied to Figs. 47 and 48, changed to 10.5 and 10.6 respectively.

Fig. 10.4

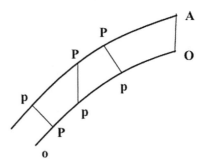

always be in equilibrium whatever way these fluids are mixed and arranged, provided that a suitable system of forces is assumed. We do not need, to be convinced, more than imaging a heterogeneous fluid whose parts are mixed as we like, assuming next that these parts have any motion and to apply here the reasoning of *art*. 165.

So we have reason to suppose that in a fluid in equilibrium, each level surface is not necessarily of a uniform density throughout its extent.

Scholium VI

167. We have just to show that it is not necessary that the level surfaces be of uniform density throughout their extent. We will now see that if a fluid is made up of different layers, each of which is of uniform density, it is not necessary for the equilibrium that these layers be level layers.

To prove this, let *DAEF* (Fig. 10.5) be a fluid mass whose parts are impelled by whatever forces, it is obvious that all the forces acting on each particle P can be reduced to two, one acting along PC, the other along a perpendicular to PC. For simplicity of the calculation let us assume that the second force is very small compared to the first; each layer *EADF* will differ very little from a circle.

That said. Let $CA = r$, the angle $ACP = z$, $CP = r + \alpha\rho Z$, being α a constant very small and that is the same for all layers, ρ a function of $CA(r)$ and Z a function of the angle z, or rather of its sinus, etc., it is obvious that $\frac{P'\pi}{P\pi} = \frac{\alpha\rho dZ}{rdz}$. Even more, the force at P along PC is $\rho' + \alpha\rho''Z'$ (ρ', ρ'' being functions of r, and Z' a function of z) and the force perpendicular to CP is $\alpha\rho'''Z''$; it is clear that the force acting along PC can be divided into two, one perpendicular to the layer *ADP* at P, the other in the same direction of this layer, and this latter will be $\rho' \frac{\alpha\rho dZ}{rdz}$. With respect to the force $\alpha\rho'''Z''$ perpendicular to PC, the resulting force along PP' is also $\alpha\rho'''Z''$, because $P\pi$ and PP' differ only in an infinitely small amount of the third order. So the force along $P'P$ is $\left(\alpha\rho\rho'\frac{dZ}{rdz} - \alpha\rho'''Z''\right)$, and this function multiplied by $P'P$ or rdz and by the density of the layer *APD*, which I call δ, it will give the expression $\alpha dz\left(\delta\alpha\rho\rho'\frac{dZ}{rdz} - \delta\alpha\rho'''Z''\right)$ for the force of the small particle $P'P$. Now, the force along CP is $\rho' + \alpha\rho''Z'$, and it is $Pp = dr + \alpha Zd\rho$; as the density is δ, it will give the force on pP along pP equal to $\delta(\rho' + \alpha\rho''Z')(dr + \alpha Zd\rho)$. Therefore the force of $p'P'$ minus that of Pp will be $\alpha\delta dr \frac{\rho''dZ'}{dz} + \alpha\delta\rho' d\rho\frac{dZ}{dz}$. Now then (*art*. 17) it is required that

Fig. 10.5

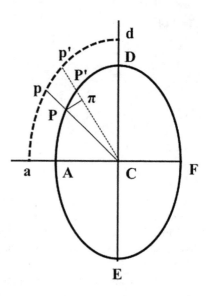

the channel $pp'P'P$ is in equilibrium, that is to say, that the force of $p'p$ minus of that of $P'P$ is equal to the force of $p'P'$ minus of the pP; So $\frac{dZ}{dz}d(\delta\rho\rho') - Z''d(\delta r\rho''') = \delta\rho''dr\frac{dZ'}{dz} + \delta\rho'd\rho\frac{dZ}{dz}$ will be the general equation of equilibrium.

For the layers to be of level, it would be required that the force along $P'P$ was equal to zero, that is to say that $\rho\rho'\frac{dZ}{rdz} - \rho'''Z'' = 0$; hence it follows that making $\frac{dZ}{dz} = \pm Z''$, as is necessary, would give $d(\delta\rho\rho') \mp d(\delta r\rho''') = 0$. However this equation is much less general than the previous one, which is easy to see, but to prove this let us suppose at first $\frac{dZ}{dz} = \pm Z'' = \pm\frac{dZ'}{dz}$, the previous equation will give $d(\delta\rho\rho' \mp \delta r\rho''') = \pm\delta\rho''dr \pm \delta\rho'd\rho$, which is only reduced to $\delta\rho\rho' \mp \delta\rho'''r = 0$ in the case when $\rho'' = \mp\frac{\rho'd\rho}{dr}$.

Secondly, let us suppose that $d(\delta\rho\rho') - \delta\rho'd\rho = \pm\delta\rho'' = \mp d(\delta r\rho''')$. And we will have $\frac{dZ}{dz} \pm Z'' = \pm\frac{dZ'}{dz}$; which gives yet another different equation of $d(\delta\rho\rho' \mp \delta r\rho''') = 0$.

Scholium VII

168. Moreover it should be clearly noted that in the previous scholium it is assumed that the fluid was made up of an infinite number of layers whose densities increase or decrease by insensible degrees, or rather infinitely small ones; so that two infinitely close layers of this fluid do not differ more than infinitely little in density.

Let us suppose now that the fluid is composed of several differently dense layers and whose density difference is finite. I say that the fluid can be still in equilibrium, although the surfaces that separate these different layers were not of level. Indeed, let $AFEB$ (Fig. 10.6) a vessel in which is enclosed a stagnant homogeneous liquid whose density is δ and whose parts are driven by the natural gravity g. Having drawn any oblique line DC, let us imagine that the part $ADCB$ becomes of the

Fig. 10.6

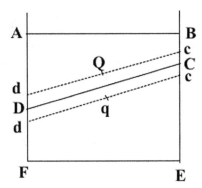

density δh, and the force that drives every particle of this part becomes g/h; it is obvious that the fluid will remain in equilibrium, and however the surface DC which separates the part $ADCB$, $DCEF$ whose densities are between them as h is to 1, is not of level. In this hypothesis, if two lines dc, dc parallel to DC, are drawn, one above and the other below, the weight of the two channels $DCcd$ will be the same, although the fluid is of a different density in the one and the other, because the weight is inverse ratio to the density.

So when a fluid is made up of layers of different densities, it is not necessary for the equilibrium that the layers are of level. Perhaps we will object that is to admit a law too unnatural and too weird, which is to assume that the weight of a fluid particle could be in inverse ratio to its density; since in this case two points infinitely close Q, q, would be driven by forces whose difference would be finite. My answer to this is: 1st, that this pretended inconvenience does not occur in the case when the infinitely close layers differ infinitely little in density, since (Scholium 6) the forces acting on a point P, can then be regulated by a function only of its distance from C and the angle ACP, function in which the density does not enter. 2nd, it does not seem to me more absurd to suppose that two points infinitely close Q and q are driven by accelerating forces whose difference is finite, than to assume that the densities of these points Q, q, have between them a finite rapport; now then, the latter assumption has never surprised anyone. 3rd, moreover it is quite true that if weight depends only on the position of the particles, the points infinitely close Q, q, will be driven by forces that only differ infinitely little from each other. Why limit ourselves to assuming that impelling forces depend only on the particle positions? If two contiguous fluids and with different density are in motion, can it not make the accelerating forces driving them unequal? Now that posed, as the force of gravity is the same for the two fluids, it follows that the forces that must be destroyed will be different.

Finally, it must be observed, that for each layer $EADF$ (Fig. 10.4) to be roughly a circle, it is required that Z, Z', Z'' are functions of the sine of the angle ACP, so that the value of these quantities remains the same when the angle z increases, either by the circumference or several times the circumference.

Scholium VIII

169. On this occasion I will remark that it seems to me that the problem of the figure of the Earth has not yet been solved in a rather general way, with the hypothesis that the attraction is in inverse ratio to the square of the distance and that the Earth is made of an amassment of fluids of different densities. Indeed, let $CA = r$, (Fig. 10.4) $CP = r + r\rho z^2$ (naming z the sine of the angle ACP, and taking the layer $EADF$ as an ellipse), R density of this ellipse, c [2π] the circumference of radius 1, φ the ratio of the centrifugal force to the weight at the equator, [ρ the ratio of the difference between the axis and the equator to the axis],[5] A which becomes $\int Rr^2 dr$, and F which becomes $\int Rd\rho$ when $2r$ becomes the axis of the earth. It will be found:

1st, that force at point P along $P\pi$ is $2z\sqrt{1-z^2}\left[\frac{2c\rho \int Rr^2 dr}{r^2} - \frac{2c \int Rd(r^5\rho)}{5r^4} - \frac{2crF}{5}\right.$
$\left.+\frac{2cr \int Rd\rho}{5} - crA\varphi\right]$; this force must be multiplied by $P\pi = \frac{rdz}{\sqrt{1-z^2}}$ and by the density R, in order to obtain the force along $P\pi$; therefore the difference between the forces along pp' and PP' is $2zdzd\left(\frac{2cR\rho r}{r^2}\int Rr^2 dr - \frac{2cRr}{5r^4}\int Rd(r^5\rho) - \frac{2cRr^2F}{5}\right.$
$\left.+\frac{2cRr^2 \int Rd\rho}{5} - Rcr^2 A\varphi\right)$.

2nd. Now, in order to obtain the equation of the spheroid this quantity must be equated to the weight of $p'P'$ minus of the pP, which is[6] $2zdzdr\left(\frac{Rd(r\rho)}{dr} - \frac{2c\int Rr^2 dr}{r^2}\right.$
$\left.-\frac{4c\rho R}{r^2}\int Rr^2 dr + \frac{6cR\int Rd(r^5\rho)}{5r^4} - \frac{4cRr}{5}(F - \int Rd\rho) - 2cRA\varphi r\right)$.

Then let $\frac{2c\rho \int Rr^2 dr}{r^2} - \frac{2c\int Rd(r^5\rho)}{5r^4} - \frac{2crF}{5} + \frac{2cr\int Rd\rho}{5} - crA\varphi = 2Kc$, K expressing an indefinite variable. We will have: 1st, multiplying this last equation by $5r^4$ and differentiating twice the following equation: $d^2\rho - \rho dr^2\left(\frac{6}{r^2} - \frac{2Rr}{\int Rr^2 dr}\right)$
$+\frac{2Rr^2 dr}{\int Rr^2 dr}d\rho = \frac{r^2}{\int Rr^2 dr}d\left(\frac{d(Kr^4)}{r^4}\right)$. 2nd, moreover the equilibrium equation will give $2zdzd(RrK)2c = 2zdzd\left[\frac{d(r\rho)}{dr} - \frac{2cR\int Rr^2}{r^2} - \frac{8cR\rho\int Rr^2 dr}{r^2} + \frac{2cR\int Rd(r^5\rho)}{r^4} + 4cRK\right]$, which divided by $2cR$ and multiplied by r^4, and then differentiated, it gives $d^2\rho - \rho dr^2\left(\frac{6}{r^2} - \frac{2Rr}{\int Rr^2 dr}\right) + \frac{2Rr^2 drd\rho}{\int Rr^2 dr} = \frac{2}{r^3 \int Rr^2 dr}\left[d\left(\frac{r^4 d(KRr)}{R}\right) - 2drd(Kr^4)\right]$. Comparing this differential equation of second degree with the previous one, and removing what is destroyed, results in $\frac{1}{r^3 \int Rr^2 dr}d\left(\frac{r^3 KdR}{R}\right) = 0$, or $\frac{r^3 KdR}{R} = Mdr$,

---

[5] Taken from the Clairaut, who calls it "ellipticity". Part II, Chap. II, §.XXIX.
[6] All these formulas are in the work of M. *Clairaut* about the shape of the Earth; these can be achieved by different methods. (*Original note*).

being M any constant. So the general equation is $d^2\rho = \rho dr^2 \left(\frac{6}{r^2} - \frac{2Rr}{\int Rr^2 dr}\right)$
$-\frac{2Rr^2}{\int Rr^2 dr} + \frac{r^2}{\int Rr^2 dr} d\left[d\left(\frac{MRdr}{r}\right)\frac{1}{r^4}\right]$.

Scholium IX

170. If $M = 0$, then we have $dR = 0$ or $K = 0$, that to say, the density is constant or the force along $P'\pi$ is null, and hence whatever layer $EADF$ of level. The spheroid equation is then $d^2\rho = \rho dr^2 \left(\frac{6}{r^2} - \frac{2Rr}{\int Rr^2 dr}\right) - \frac{2Rr^2 dr d\rho}{\int Rr^2 dr}$, which is the only one that has been found so far, but which is not as general as the previous one.

It will give again $d^2\rho = \rho dr^2 \left(\frac{6}{r^2} - \frac{2Rr}{\int Rr^2 dr}\right) - \frac{2Rr^2 dr d\rho}{\int Rr^2 dr}$, 1st, when $\frac{Rdr}{rdR}$ is equal to a constant, that is to say, when $R = Ar^n$, being A and n being whatever constants. 2nd, when $d\left(\frac{Rdr}{rdR}\right)$ will be equal to $Br^4 dr$, being B constant, that is to say when $\frac{dR}{R}$ will be equal to $\frac{dr}{r(Cr^5+G)}$ being C and G constants. In all other cases the equation of the spheroid will be more complicated than what has been assigned to it so far by wise geometricians who have dealt with this matter.

When $r = 1$, it is necessary that $K = 0$, because the first layer must be necessarily of level. Therefore, the value of R cannot be assumed as $\frac{Rdr}{r^5 dR} = 0$, or what is the same, $\frac{dR}{dr} = \infty$, when $r = 1$; which can be made by an infinite number of ways. In general it is only necessary to assume the equation between R and r represented by a curve whose tangent coincides with the ordinate R when the abscissa $r = 1$, the ordinate R being finite.

Scholium X

171. By the method we propose here to determine the figure of the elliptical Earth, it is easy to see that once the layers are of level, the weight of pP and $p'P'$ are equal. Indeed, when the layers are of level, we have $K = 0$, $d^2\rho = \rho dr^2 \left(\frac{6}{r^2} - \frac{2Rr}{\int Rr^2 dr}\right)$ $-\frac{2Rr^2 dr d\rho}{\int Rr^2 dr}$; now then, in the same case we have $M = 0$, and the general equation is reduced to $d^2\rho = \rho dr^2 \left(\frac{6}{r^2} - \frac{2Rr}{\int Rr^2 dr}\right) - \frac{2Rr^2 dr d\rho}{\int Rr^2 dr}$. So the principle of level layers, and that of equality of force between the columns and pP and $p'P'$ give the same eq. I will discuss this subject elsewhere further.

THE END

Part II
Introduction to the *Essay*

Julián Simón Calero

Chapter 11
General Considerations

11.1 The Manuscript and the Essay[1]

Late November 1749. Jean Le Rond d'Alembert had just completed the last corrections of a manuscript and was ready to send it to the Academy of Sciences of Berlin in order to compete for the prize on the resistance of fluids for 1750.[2] Leonhard Euler had proposed the prize to the Academy the year before. D'Alembert had finished the manuscript in the last days of October, and now he added an appendix and some minor details. He thought he had opened up a new way of understanding the phenomena of fluid motion by introducing the capacities of mathematical analysis in the unruly world of Fluid Mechanics. Until then only the most outstanding geometricians had dared to penetrate it, and almost always with quite meagre results. His new ideas would extend not only the borders of Fluid Mechanics, but would also provide new tools. D'Alembert was very proud of this work.

Early January 1750. The five competing works received at the Academy were sent to the Jury, composed by Euler, Johann Kies and Augustin Nathanaël Grischow.[3] In May the same year the Academy made public the decision to delay the award of the prize for 2 years, because none of the works submitted fulfilled the proposed conditions. Therefore the former participants, or any new other ones, were asked to provide experimental evidence for their calculations.

---

[1]These Commentaries incorporate the suggestions and comments done by Larrie Ferreiro and Manuel Sellés.

[2]Most of the information given here has been taken from Grimberg [1998].

[3]Both were German. Grischow was a mathematician and became member of the Academy of Berlin in 1749 and professor of Optics in 1950. In 1751 he gave up this position for the Academy of San Petersburg. Kies was mathematician and astronomer and professor of Mathematics and astronomer of its observatory.

D'Alembert was very discouraged by this decision, and matters worsened further when at the end of year some news reached Paris giving rise to suspicions about the impartiality of the jury.[4] It seems that he had already decided to withdraw his work then and to publish it separately. As a consequence his relations with Euler cooled, and their correspondence was cut short for more than 10 years. As regards the prize, it was awarded in June 1752 to Jakob Adami.[5]

What was the true reason the prize was not awarded to d'Alembert? This is an open question that we cannot answer. Clifford Truesdell, one of the most renowned historians of Mechanics in the last century, made some conjectures about it, which are not very favourable to d'Alembert.[6] He thinks that it seems very improbable for Euler to have expected that the experiments could provide some results in accordance with the theories. This is very likely true. The available experiments were quite few in number, inconclusive, sometimes contradictory, and carried out in a very limited range of options as d'Alembert himself explains. Therefore, according Truesdell, "the reason given out was only a pretext, offered in place of the truth,... that d'Alembert's reasoning was inaccurate, tortuous, incomprehensible, and that in illustration of his equations he had not succeeded in exhibiting a single flow". We might agree with the difficulty in reading the text, and also that there are matters that bear little or no relation with the fluids; but at the same time there is matter in the *Essay* that deserves a more favourable judgment. Besides, if it were true that Euler did not believe that practice could confirm the theory, what did he and the Jury expect for the new submission? What is more, if the veiled intention was to eliminate d'Alembert, then the winning work in the postponed contest would need to be of rather better quality, which, according to the prestige and capabilities of d'Alembert did not seem easy to expect. In the end, the almost unknown winner contributed nothing to Fluid Mechanics while d'Alembert undoubtedly did, although following quite a tortuous path.

At the end of December 1751, the Academy of Paris approved the *Essai d'une nouvelle Théorie de la résistance des Fluides* for printing. This was the French version of the Latin manuscript, which had the long title *Theoria resistentiae quam patitur corpus in fluido motum, ex principiis omnino novis et simplissimis deducta, habita ratione tum velocitatis, figurae, et massae corporis moti, tum densitatis & compresionis partium fluidi*.[7] Both are basically the same although with some

---

[4] According to Robert Bradley (p. 257), Grischow was summarily dismissed from the Academy of Berlin when he furtively contracted an engagement with the Academy of San Petersburg. This made him feel humiliated and he tried to make trouble revealing to d'Alembert and others his version of the events related to the prize delay.

[5] For the work *Specimen Hydrodynamicum de Resistentia Corporum in Fluidis Motorum*. See Annex III for more details.

[6] Truesdell [1954], pp. LVII–LVIII.

[7] *Theory of the resistance undergone by a body moving in a fluid, deduced only from new and very simple principles, both in the given relation with the velocity, shape and mass of the moving body, and the density and pressure of the parts of the fluid.*

minor corrections, a certain reorganization and the addition of a few new articles.[8] D'Alembert justifies the title of *Essay* as being an intent to open up a new route in the subject, expecting others to follow and further extend the limits he had discovered [Intro. VII].

In October 1752,[9] the *Essay* comes to light.[10]

In the Introduction of the *Essay*, he bitterly addresses the decision of the Berlin Academy. He states that he only pretended to agree with the best known experimental issues of the resistance, and he also remarks that this requirement had not been included in the bases of the prize. He also hints at "other reasons, in whose detail it is useless to enter" [Intro. VII], pointing out the aforementioned suspicion of lack of impartiality. To manifest his dissatisfaction and in order to retain the priority of the findings of the *Essay*, he tried to publish it prior to the judgment of the Jury. This haste was somewhat inopportune for the work, since we think that many of the criticisms made about the obscurity and reading difficulties could have be avoided with a more relaxed editing. To finish his comments about this matter, he adds with some irony "that the Judges appointed by this illustrious Company, which without any doubt have not proposed this question without assuring themselves if that solution was possible, would find something to entirely be satisfy themselves in the works that will be sent to them for the contest" [Intro. VII].

Another rather important question is to what extent this work served Euler for his own subsequent developments, specifically for the three Fluid Memoirs of 1755. Gérard Grimberg would have liked to know the opinion of Truesdell, due to his familiarity with the texts of this period of both authors.[11] However, Truesdell adds nothing; on the contrary he says that Fig. 4.5 of the *Essay*, which depicted the streamlines around a body, is very similar to Euler's in the translation of Benjamin Robins' *Gunnery*.[12] We think that this interpretation is inexact; Euler's figure showed the body with a single layer contouring it; d'Alembert's is rather more complicated. Regarding the fluid flow, Euler assumed that normal and tangent forces would exist at each point of the surface, both being functions of the local angle and the layer thickness; but there was still no rule or formula for this thickness. A qualitative jump exists between both cases.

Before publishing the three Memoirs, Euler read the "Principia motus fluidorum" in the Berlin Academy in 1752.[13] In this work Euler's ideas on how to deal with fluid motions appear for the first time with clarity, and although the scope was limited to non-compressible fluid and two-dimensional motions, it was the root of his later works. There are at least two points in which the "Principia"

---

[8]Grimberg [1998] presents a table with the correspondence between the points of the Manuscript and the articles of the *Essay* (p. 364), and the list of the new articles added (p. 14). In The Annex II we have completed this correspondence.

[9]October was probably the month when the *Essay* came to light. See Annex II.

[10]We will use the names of *Essay* and *Manuscript*, or abbreviated to *Mss*.

[11]Grimberg [1998], p. 11.

[12]Truesdell [1954], p. LII. Euler [1745], Fig. 4.3.

[13]Published later in 1761.

follows the *Essay*. One is the constancy of the volume enclosed by a surface evolving in the motion. In the *Essay* this volume is a parallelepiped, defined by eight particles in each vertex, that moves and changes shape, but with the quantity of enclosed fluid remaining the same [Fig. 4.6 and §.48]. In the "Principia" it is a triangle, whose surface will be constant when each vertex moves.[14] The second one refers to the measure of the pressure by means of a column of fluid. We recall that Daniel Bernoulli had introduced the water height manometer in his experiments, and even Newton gives inklings of this idea. However, in the *Essay* this idea acquires the value of a measurement, what it means to know a quantity, to record and compare it with others [§.33]. In the "Principia" this idea is defined in a similar way, the difference being that Euler considers the pressure as an internal action among neighbouring particles.[15] A third point could be added: d'Alembert had showed the existence of mathematical relations between the velocities in a fluid, thus opening up a vast new field for Fluid Dynamics. In these circumstances it was almost inevitable that Euler threw himself into this new world. The huge capacity he had for absorbing any new idea is well known. However, did Euler have these ideas in mind before? We cannot dismiss the probability, but no evidence exists, at least as far as we know.

Regrettably for d'Alembert, the *Essay* was only in the limelight for few short years, as Euler's *Memoirs* eclipsed it almost completely. The brief time between both works and the rivalry between the authors has inclined some modern scholar consider them as if they were in a contest. Scientists have always competed to be the first to make any discovery and to receive public recognition for it; but this has been a competition without winners or losers, they have strived to move science a step forward by aiming at a far distant goal. In our present world full of contests, not only in the cultural life but in the omnipresent world of sports, full of winners for anything under the sun, full of ranks and lists, we have become too much aggressive and competitive. In our humble opinion, we would better avoid translating our present environment to those more relaxed times.

René Dugas, after recognizing that the Euler Memoirs can be taught today almost without changing a line, points out: "Suddenly, the *Essay* found itself dethroned, but it would be ungrateful of us to forget it, because the light that we owe to the mathematical genius of Euler is obviously derived from the cutting of the rough diamond extracted by d'Alembert".[16]

---

[14]First Part, §.14-ss, Fig. 2.1.
[15]Ibid, §.43.
[16]Dugas [1952], p. 12.

11.2 Fluid Mechanics in the Eighteenth Century

Let us start with a general view of the scope and evolution of the Fluid Mechanics in the eighteenth century, which will serve to obtain a better understanding of d'Alembert's contribution.[17]

Fluid Mechanics was a part of General Mechanics, maybe the hardest to grasp. In its genesis two major milestones can be identified. The first was the Book II of Newton's *Principia*, in 1674, where we find the beginning of Fluid Mechanics as a modern scientific discipline by establishing the impact theory as an explicative model of fluid behaviour. This theory assumed that fluids were constituted by individual particles, which impacted upon a body in motion. It is remarkable that a theory which in the end turned to be unrealistic provided the foundation stone of the theorisation of fluids.

The second milestone was the publication of the three aforementioned *Memoires* by Leonhard Euler in 1755, where the equations of hydrodynamics, today known as Euler equations, were presented in such a way that has practically not aged, so that any modern reader could understand them. The fluid is no longer considered as a particle amount but as a continuous media. We must underline that the fluids were still considered physically as being constituted by particles, and because of this the impact theory made sense, but the new models had to assume the particles were continuous as the only way to fulfil the apparent reality. A consequence of this was that the flow field had a complex internal structure defined by the differential calculus, which contrasted with the simplicity of the former impact model.

This huge leap, from a fluid taken as an aggregate of particles impacting discretely on a body to a continuum media flowing along streamlines regulated by differential equations, was the outcome of the efforts of the most brilliant geometricians in the 68 years that transpired between the *Principia* and the *Memoirs*. It was a "great theorization" that yielded a set of equations that shone like a beacon upon a summit and remained there unsurpassed for many years, even for centuries. Although they continued to shine with almost ethereal beauty, in practice they had practically no application due to the lack of appropriate mathematical resources. What is more, one of the few available derivations showed that the fluid resistance of a moving body ought to be zero, a fact that contradicted the most obvious real data. In some way it provoked the need for experimentation as an alternative. This was very noticeable in two technical areas where the fluid resistance was demanded: the ballistic and the nautical.

Though we have attributed the paternity of fluids science to Newton, we cannot forget two precursors: Christiaan Huygens and Edmé Mariotte. Both interpreted the action of a fluid upon a solid as shocks of particles against bodies and conducted experiments on this subject in the 1670s. Huygens studied the effect of a jet on a plate and Mariotte the force that a current of water, which was the Seine River, exerted on plate immersed in it. The values obtained by Mariotte were very accurate

---

[17]We follow Simón Calero [2008].

and close to present day figures. In addition, Huygens considered the resistance as proportional to the square of the velocity for the first time.

Newton thought that both air and water were composed of separate particles able to move independently. In the air they were spatially separated and they moved impelled by centrifugal forces among them. By contrast, in water the particles were in contact with each other, so that the motion of any individual one was conditioned by its neighbours. In any case, a solid body moving in a fluid would strike the particles thus transferring momentum, which would be the resistance. The calculation was feasible in the air with some simplifying assumptions, but practically impossible in a liquid. For this reason Newton followed two different methods for each case.

Newton supposes that an elastic fluid is formed by particles at rest, repelling each other by centrifugal forces inversely proportional to the distances between them. However, he simplifies this model removing the centrifugal forces, leaving the particles at rest and equally spaced. He called the resulting fluid "a rare medium" but he argued that its action upon a moving body would be almost identical to the real fluid. For the individual impacts of the particles there were two possibilities: one that they rebound elastically as in a mirror, the other that they remain motionless after the impact in an inelastic shock. The difference was that in the first case the momentum transferred was double that in the second. This comprised the impact theory; we note that the local effect in any point depended only on the geometry of this point relative to the velocity. Another relevant issue was that there would only be impacts in the front part of the body, leaving a shadow zone behind it. Therefore, for a body with a given geometric shape it was mathematically possible to calculate the reflection angle at any point of its surface. Knowing the individual mass of each particle and the impact rate, that is to say the fluid density and the body velocity, the resistance was easily obtained. In the particular case of a flat plate moving perpendicular the resulting force was $C_D = 4$ for the elastic hypothesis and $C_D = 2$ for the inelastic; for a sphere both values were reduced to a half.[18]

We have to point out that once the hypothesis of the rare medium is accepted, the impact theory is fully backed by Newtonian mechanics. That is the reason we think of it as the first appearance of the fluid as a scientific discipline. However, this was not applicable in any way to the resistance of liquids. For these Newton strived for an alternative way to overcome the unmanageable aggregate of particles. The idea was to divide the problem in two parts. The first was to produce a jet stream, like the discharge of a vessel through a hole in the bottom, and to obtain the exit velocity depending on the depth of the fluid. The second was to place a sphere inside that jet, or another body, and assume that the force exerted upon it was equal to the weight of the column of fluid over the body up to the water level, that is to say the depth of the vessel. The mathematical solution would be equivalent to a problem with the

---

[18]We use the coefficient of resistance, C_D, instead of the real force in order to make the results more understandable.

depth as a parameter, which once eliminated, would give the resistance as a function of the velocity. This method had two weaknesses: one, the procedure for calculating the jet velocity as a function of the vessel depth; the other, the assumption that the force on the sphere is equal to the imaginary column of liquid above it. For the first one Torricelli's theorem could be used. At this point, let us recall that this law establishes that the efflux velocity of the liquid is that which a heavy body would acquire when falling from a height equal to the depth. Torricelli enunciated this law in 1644 based on experimental evidence and there were many attempts to confirm it, but without conclusive results. However, Newton did not follow it, but made an ad hoc reasoning that led to a velocity equivalent to a fall from half of the liquid depth. With the subsequent calculations, the resistance coefficient turned to be $C_D = 2$, irrespectively of the body shape. This was almost double that of $C_D = 1.2$ measured by Mariotte in the river Seine for a plate.

Newton would change his views on liquids somewhat in the second edition of the *Principia*, 1713, but in the meantime a group of geometricians entered the stage. Their interest swung between pure applications to more practical aspects such as ships or hydraulic machines. They all accepted the impact theory, without distinction whether it was liquid or air, using $C_D = 1$ as the proportionality coefficient; probably because they thought that nature expresses itself in whole numbers, therefore the measurement $C_D = 1.2$ would be an approximation of the reality. Jakob Bernoulli stood out among them for the differential analysis he carried out on two dimensional body shapes, sail curvature, ships velocities and leeward, etc. His brother Johann also studied the sails and a more complete theory of ship manoeuvring. Finally, Philippe de La Hire and Antoine Parent presented several works in the field of hydraulic machines.

In the second edition of the *Principia* Newton presented an exhaustive review of this part, but now maintaining the discharge with the name of "cataract". Briefly, just before the discharge started, Newton assumed that there was a funnel of ice inside the vessel, from the broad upper side to the small hole in the bottom. The water would run down sliding on the ice going out at the exit hole. Then, if the ice were to melt instantly, Newton assumed the descending liquid would continue as before without mixing with the recently melted water. Furthermore the exit jet would contract, reducing its section to a half, which he justified for experimental measurements. The result was that now the efflux velocity coincided with Torricelli's predictions. With respect to the force upon the object placed in the stream, the equivalent water cylinder would be now replaced by something like a pinnacle, also initially made of ice, which would melt just as the funnel did. The shape of the pinnacle would be between a semi-spheroid and a cone, and its weight the arithmetic average of them. Summarizing, he used a set of *ad hoc* constructions and tricks in order to reach $C_D = 0.5$, which was the figure he was looking for since it was the result of the experiments he had made dropping balls in a tank of water. In these experiments Newton, who was an excellent experimenter, measured the time that a sphere took to fall in a tank of a given depth. With the corresponding mathematical formulas, which linked all the experiment parameters and in addition taking the assumption of $C_D = 0.5$ the agreement of the theory and experimental results was very good. In addition he carried out a similar experiment dropping spheres from the dome of Saint Paul's church in London, which also resulted in a very good agreement. As a conclusion, the resistance of a sphere turned out to be $C_D = 0.5$, in both water and air. However, it is remarkable that he had found that the resistance in the air of such a sphere was $C_D = 2$ or $C_D = 1$, as we have explained before.[19]

The hypothesis of the cataract was criticized by many writers of the time, one of them being d'Alembert.[20] However the impact theory was taken as reference for many years, although successive authors added corrections, resulting in the mixed and hybrid versions of impact theory. We think that its survival was simply due to the lack of an alternative.

The impact theory continued as a reference in the decade of the 1720s. Daniel Bernoulli, Johann's son, tried to confirm its results by experiment in 1727. He proposed an imaginary experiment of a plate inside a fluid flow; the force generation was similar to the impacts mechanism, and obviously the conclusion was the same. The novelty was that he accompanied the reasoning with an experiment for measuring the force that the jet, produced by the discharge of a vessel through a hole in the bottom, would exert upon a plate. Similar experiments had been carried out by Huygens and Mariotte who had obtained a force equivalent to $C_D = 4$; however Bernoulli found a $C_D = 1$, because he placed a small cone on the plate. If a sphere had taken the place of the cone, he estimated that this value would be reduced to $C_D = 0.5$. The difference between theory and practice is remarkable.

At the same time, Bernoulli began to explore the fluid motions in channels, where the fluid was now regarded as a continuum instead an aggregate of particles. Several articles were dedicated to this subject so that in the next decade this hypothesis was the dominant one. The key work of those years was the *Hydrodynamica*, completed in 1733 but published in 1738. Now the regulating principles in the fluid motion in vessels, discharges, pipes and channels were the continuity, the plane section hypothesis and the conservation of live forces. For the first time a relation between velocities and pressures came to light.

In 1735, Daniel Bernoulli presented a rather important article about the impact of a jet upon a plate. Until then, with the impact theory, the effect of a jet was equivalent to the resistance of a body in a submerged motion: they all were impacts. But now he pointed out the inexactness of that assumption, and he indicated that the jet was deflected progressively when it approached the plate, so that it finished in a sliding motion parallel to the surface. With this new model the streamlines entered into Fluid Mechanics.

---

[19]The resistance of a sphere is function of a non-dimensional parameter call Reynolds number. At low Reynolds the flow regime around the sphere is laminar, and the resistance is about $C_D = 0.5$. At high Reynolds the regime becomes turbulent and $C_D = 0.2$. The transition between both regimes happens at a Reynolds about $5 \cdot 10^5$. The experiments of Newton were at a Reynolds less than $7 \cdot 10^4$, which is a laminar regime. Definitely, Newton was lucky.

[20]*Essay*, Intro-II.

11 General Considerations

The *Hydraulica*, by Johann Bernoulli, continued in the wake of the *Hydrodynamica* of his son Daniel. He insisted that it had been written in 1732 although it was published in 1742. There was a murky story of the father's jealousy of his son, with some plagiarism included. Nevertheless, there were many and significant new ideas in the *Hydraulica*. He introduced the concept of internal force, which allowed the separation of the fluid by imaginary surfaces. Simon Stevin had done something similar with the principle of solidification for a fluid at rest, but now we are dealing with a fluid in motion, which is rather different. One more contribution was the use of Newtonian mechanics, although Johann also employed the live forces. Johann's work was a worthy continuation of Daniel's, and we think that the known theorem of Bernoulli should be renamed as Bernoullis'.

Many and diverse works came to light in the 1740s decade including two of the most important naval treaties of the entire century; one due to Pierre Bouguer in 1746; another by Euler, in 1749. Obviously to calculate the forces on the hull and sails, a fluid theory has to be included in any naval treaty. Bouguer used the impact theory aided with a practical value which coincided with the Mariotte experiment; he also proposed a new theory for the flow in the stern, which can be interpreted as a fluid irruption. Euler proposed a construction based on the fluid that a body dragged into motion, this being a variant of the impact theory also. For the calculation of the resistance he supposed that the body transferred either momentum or live force, this yielded two results: $C_D = 2$ or $C_D = 1$ respectively. Euler chose the second based on the experimental results.

In 1742, Benjamin Robins published the *New Principles of Gunnery*, which, as its name suggests, was devoted to artillery, and a fundamental part of this was the resistance of projectiles. Robins called into question the Newtonian ideas for the motions in air and liquids. He said that when a body moves in a liquid at any velocity or in air at low velocity, the fluid completely surrounded the body; but, in the motion in air at high velocity, a vacuum would be produced behind it, because there would not be time for the air to fill the space left by the body. Obviously there would be intermediate situations when this space was only partially filled. Robins' estimations for the resistance tripled the values given by Newton. To confirm his theory he carried out experiments consisting of firing small balls with a canyon and measuring the velocity at various distances with a ballistic pendulum, a device of his invention. He found a good agreement with the theoretical predictions. Given the interest of the matter, the Prussian authorities commissioned Euler to write a translation of the book into German which was published in 1745. Euler completed the translation with very extensive comments, taking Robins ideas much further, both in the theoretical aspects and in the analysis of the experimental data.

Euler assumed that the fluid surrounded the body in a thin layer where physical impacts completely disappeared. He analysed the evolution inside this layer according to the geometrical properties of the surface, relating the pressure with the velocity, but he did not provide any means of calculating the velocity. However, in accordance with the geometry of the body contour, he obtained that the total force upon the body could be nil in some types of bodies. For the motions in air at high and medium velocity, he introduced a hypothesis about how much air would enter in the space left behind and how much resistance was induced in the process, which had to be added to the frontal resistance. The result was that the total resistance was proportional to the velocity instead of its square, which was somewhat surprising. Finally, in the reduction of the experimental firings he found values around $C_D = 1.2$–1.3 according the firing rounds. We note that these experiments were supersonic, and the C_D found was a good figure for the present day according to the Mach number.

In 1744, d'Alembert in his *Traité* still used the impacts as the basic mechanism for producing resistance, interpreting it as a change of the body momentum, which was nothing new. However, he also tried to extend the impacts to an elastic fluid. In the same book he also studied the motion in tubes by the conservation of the live forces, complemented with plane section motion, which he called the "principle of experience".

Now Alexis Clairaut entered the scene. His *Théorie de la figure de la Terre*, 1743, opened up the mathematization of the fluids. The book responded to the shape that the Earth would take assuming it to be a rotating fluid. This can be traced back to the polemic the scientific community engaged in as to whether our planet was flattened at the poles, as Newton expounded, or elongated, as a consequence of the Cartesian vortices. The controversies caused rivers of ink to flow, not only in the fields of science, but also in politics attaining even nationalist overtones. In this atmosphere the Academy of Paris sponsored two expeditions, one to Lapland, another to Viceroyalty of Peru, in present day Ecuador, with the aim of measuring a meridian degree to solve the question, in what may be understood as a crucial experiment.

For the shape of the Earth, Newton took as basic criterion the immobility of two channels coming from the pole and equator and meeting at the centre. On the other hand Huygens, taking gravity as derived from the Cartesian vortices, established that any channel in the surface should be in equilibrium. Both conditions were necessary, but Pierre Bouguer showed that they were not equivalent. Clairaut also tried to solve the problem using channels. However, the solution that he found was that the equilibrium depended only on the nature of the field forces which acted upon the fluid, irrespective of the fluid itself. Furthermore, the problem was reduced to a partial differential equation between the components of that field of forces. The fluidic problem was reduced to a mathematical one.

As we already know, d'Alembert presented his *Memoir* about the resistance to the Academy of Berlin in 1749, however it ended up published as the *Essay* in 1752 As we will show later on this work was a major step in the mathematization of the fluids, in addition we underline that for the first time the body and the fluid were considered as a system in which the body induces a velocity field on the entire fluid, not only over its surface.[21]

In the same year, Euler read in the Academy of Saint Petersburg the paper "Principia motus fluidorum", dedicated to two-dimensional and incompressible

---

[21]Grimberg [1998], p. 37.

fluid motion, which would be the seed of the new approach and the doorway to the three Memoirs of 1755. In these, Euler established the equations of hydrodynamics. He took as physical principles the continuity, the existence of the pressure as an internal force and any other mass force. He applied all of them to a differential element of fluid which would move according to Newtonian mechanics. All this was expressed with clear concepts and a clean and powerful application of differential calculus. The results were the continuity and momentum equation in partial derivatives. He also glimpsed what we call today the state equation, which would relate pressure and fluid density and even temperature. This equation together with the former ones would enclose the entire theory of fluid motion. These Memoirs represented the summit of Fluid Mechanics for the whole century and it would take many years to add something meaningful to them. However, alongside this brilliance there was the lack of mathematical resources to solve the equations that appear in them.

There is an additional theoretical work to mention in the century, the Memoir by Joseph Louis Lagrange in1781, in which he introduced the discontinuity surfaces and the small perturbations theory.

Some modern authors think that the inability of the fluid theory to explain the resistance of a body motivated the increase in experimental works. Our opinion is that these works ran in a parallel way; let us remember Robins. In any case, there were very good experimenters, some dedicated to determining the resistance of single geometrical forms, such as spheres, wedges, plates; others dealing with ships in towing basins. Let us mention Jean-Charles Borda, in 1763 and 1767; Charles Bossut, in 1777–1778, Frederik Henrik Chapman, 1768. There were others who experimented with machines, such as John Smeaton, 1759.

To finish, apart from the two aforementioned naval treaties by Bouguer and Euler, there were three more relevant ones: one by Jorge Juan y Santacilia, 1771, another by the above-mentioned Chapman, 1775, and yet another by Euler, 1773. The fluid theory used by Jorge Juan was very peculiar as he assumed that the resistance was depending on the depth and proportional to the velocity. The other two treaties employed the impact theory; it is surprising that this theory was still in use, although all the authors knew that it was false. However, as there was no alternative they were forced to use it.

Chapter 12
D'Alembert's Dynamic Conceptions

D'Alembert published his *Traité de Dynamique* in 1743, when he was 26 years old. It was here where he expounded his ideas about dynamics, including what is known as his "general principle". This is enunciated in the second part of the *Traité*, but the general ideas are in the first one. "The general principle has since become the object of considerable celebration and misunderstanding in the history of mechanics".[1]

We will give here a brief summary of this book according to our lights. For a better understanding we recommend the work on this matter given by Firode,[2] who, apart from the deep and extensive analyses of the *Essay*, tries also to understand d'Alembert's thoughts using his contributions to the *Encyclopédie* as a complement.

In the *Essay*, d'Alembert reflects his pride in the novelty and importance of his contributions, the high quality of his demonstrations and their superiority over those of his predecessors. What is more, he had no qualms in dismissing them in a way that sometimes borders on insolence. He wanted not only to deduce the mechanical principles from the clearest notions, but to show the uselessness of those proposed so far [Pref. iv].[3]

He based the dynamics on three laws: the force of inertia, the composite motion and the equilibrium, which he considers as being necessary consequences of the principle of sufficient reason.[4] In his own words "at least I hope to make clear from this *Treaty* that all this science can be deduced from these three principles" [§1].

---

[1] Fraser [1985].
[2] *La Dynamique de d'Alembert*, 2001.
[3] This quote and the followings are from the *Traité de Dynamique*, 1743.
[4] Cf. Darrigol [2005], p. 12-ss.

The force of inertia is the property of the bodies to remain always in the same state they were. Any change from this state will require the action of an external cause [§3], which he calls "power" or "driving cause" [§5].[5] Among all these causes, which can be "occasional" or "immediate",[6] only the impulsion ones allow us to determine the effects through knowledge of the cause. In all the other cases theses causes are entirely unknown, which implies that we have to infer them from the acceleration or retardation of the motion. Thus if the body receives continually, from the "power that accelerates", a velocity increment du in constant time intervals dt, the equation $\varphi dt = du$, which relates both du and dt, is given by hypothesis [§19]. However as he explains, most of the geometricians had a different view on this matter. For Daniel Bernoulli the equation was a contingent truth, while for Euler it was a necessary one. D'Alembert insists on considering it only as a definition, understanding by the term "accelerative force" something to which the velocity increments are proportional [§19]. We are not going to enter in such discussion; but we have to agree that the ontological concept of force was difficult to explain, and d'Alembert was sensitive to it as a philosopher as well as a mathematician.[7] Besides, our understanding is that d'Alembert's position, taking $\varphi dt = du$ as hypothesis or definition, is nearer to Euler than to Bernoulli.[8]

The second law is the composite motion, which means that when a body is subjected to two whatever powers, the resultant motion will follow the diagonal of the parallelogram produced for the powers, which is also known as the Varignon rule [§21]. This vector composition can be also applied inversely to decompose a velocity in two components. Thus, when a body strikes a "plane immobile and impenetrable", the impact velocity can be divided in two components, one perpendicular and another tangent to the plane. The first will be destroyed in the impact,[9] but not the other which will remain unchanged; therefore the body will change its direction [§30]. Another case of composition of motion occurs when a body runs along a curved line, because the continuous change of direction will require a "central force", which has the same nature as an accelerative force. This latter "force" will be perpendicular to the curve and normal to the tangent; that is to say, pointing to the centre of the curvature whereby it receives the name of "central" [§26].

---

[5] He says that a body is set in motion by a "power" or "driving cause" (*puissance* or *cause motrice*) [§5] and that a body will remain at rest unless an "external cause" (*cause étrangère*) moves it [§3]. Also, a body moving uniformly will continue in this state unless an "external cause", different to the "driving cause" acts upon it [§6].

[6] For the following, see [§19]

[7] Ru [1994b] compares d'Alembert with a sort of two-faced Janus, because in his thinking there is a strong dependence between science and philosophy.

[8] See the analysis of Veronica Ru [1994b] for the meaning of the accelerative force.

[9] About the nature of these impacts we recommend to see Firode, "The notion of the body" (p. 76), where he states "He [d'Alembert] resolutely adopts the idea of a world made of hard [*dur*] atoms where the phenomenon of shock necessarily entails a loss of movement and a failure of the principle of continuity" (p. 81).

The third law of equilibrium states that "if two bodies whose velocities are in inverse proportion to their masses and which move in opposite directions in such a way that one cannot move without displacing the other, an equilibrium will exist between these two bodies" [§39]. That is, equilibrium will exist when the sum of the total quantity of motion of the impacting bodies is zero, which can be extended to any number of bodies.

These principles show, as he himself manifests [Pref. xxiij] that D'Alembert pursues a mechanics of effects, instead of a mechanics of causes, thus trying to avoid the *causa agens*, especially if it is referred to as force. He declares clearly and brilliantly "I have completely banned forces inherent in a moving body, obscure and metaphysical beings which are only capable of spreading shadows on a science that is clear in itself" [Pref. xvi]. These "forces inherent in a moving body" seem to point to the "motive force impressed" of Newton's Second Law. However, the word force is widely used in the *Traité*, but as he affirms in one of his final remarks "I must warn that in order to avoid circumlocution, I often used the obscure term *force* ... but I have never pretended to attach to these terms ideas other than those resulting from the principles I have established" [Pref. xxxiv]. In more practical terms "we do not have a precise and distinct idea about the word *force* other than restricting this term to express an effect" [Pref- xxj]. Furthermore about the uncertainty of the causes: "we will never take the ratio of two forces other than for its effects, without examining if the effect is really like its cause or it is a function of this cause; a completely useless examination, because the effect is always given independently of the cause, experiment or hypothesis" [§19].

He classifies the power or driving causes in two kinds; one, those that are manifested at the same time as the effect produced by them; the other, all those that are only known by the effects although we completely ignore their nature. Examples of the first class are the motion with mutual actions and collision among bodies; and an example of the second one is the general gravitation [Pref. x]. The accelerative force, as given by the equation $\varphi = du/dt$, is the tool he introduced as the effect of those unknown causes. It has the dimension of acceleration; a word almost unused in the entire *Treaty* and *Essay*, although it is not really that. The acceleration is a kinematic concept that can be thought as being inside a time-geometrical system. On the contrary, the accelerative force implies a dynamical behaviour, that is to say causes and motions acting jointly.

We think that all these remarks require a short reflection. First of all, let us remember that the word "force" in Enlightenment Mechanics was not a clear concept and was subjected to many interpretations by each author.[10] In a general view the force was the equivalent of the Latin *vis*, used in the Mechanics of the XVII and XVIII centuries. This *vis* had a broad scope, which could be summarized as a cause able to produce a physical effect.[11] It occurred as *vis inertiæ*, *vis viva*, *vis*

---

[10]Cf. Firode, Chapter 1, "The criticism of forces".

[11]In the *Oxford Latin Dictionary*, the word *vis* has 28 entries. We find relevant (1) "Physical strength exerted on an object, force, violence". (6) "Forceful or vigorous action or movement". (15) "Power to produce some physical effect, potency, virtue".

gyratoria, vis acceleratio, etc. It was Newton who produced a precise definition of the force as physical agent in the *Principia* by the well-known equation $F = d(mv)/dt$.[12] D'Alembert thought that the *vis* or force was a rather vague meaning, even while accepting its existence, and consequently he directed his gaze to the effects that are clearly discernible. Therefore the equation $\varphi = du/dt$, which can be understood as a mirror reflection of Newton's, would be the way towards reducing these causes to physical entities able to be measured.

As we have seen, the power or force appears naturally in any dynamic system. However, there are also powers in the static systems, and these powers or causes only interact among themselves through the bodies that they attempt to move [§43]. That means that the statics must be considered as a particular case of the dynamics. Therefore the analysis would be taken by as the virtual motion of the bodies,[13] which implies that the equilibrium is one instant in a dynamic process. The action of powers must then be understood as the product of a body multiplied by its velocity or accelerative force. The single product $m_i u_i = m_i \delta x_i / \delta t$ would correspond to the virtual displacement of all the bodies, restricted by the constraints of the system, and the application of the third law. The other case, with the accelerative force, would point to $m_i du_i/dt$, clearly the Newtonian expression for a force. D'Alembert could argue that this is a consequence of a hypothesis, not a law, but in practical terms both arrived at the same point. Despite all the attempts to avoid considering the forces as main actors, they end up taking the stage.

His basic and famous principle is stated in the second part of the *Traité*. The principle is a combination of the composed motion and equilibrium laws. Firstly he posed the following problem:

> Given a system of bodies, arranged in relation to the others in any manner whatever; let us suppose that a particular motion is impressed on each one of these bodies, which the system cannot follow because of the action of the other bodies. Find the motion that each body should take. [§50]

First, let a velocity $\vec{u}$ be impressed on each body m_i; according to the constraints among bodies and boundaries, the system will respond with a velocity $\vec{v}_i$ at each body. Now let $\vec{u}_i = \vec{v}_i + \vec{w}_i$. The principle states that if each body were impressed with the velocity $\vec{v}_i$, they all would retain the motion without harming each other, and if they were impressed with $\vec{w}_i$ the system would remain at rest. We can make a free interpretation assuming that $\vec{w}_i$ is the consequence of the constraint effects, both internal and external. Therefore if the applied velocity was $\vec{u}_i = \vec{v}_i$, there is no

---

[12] In fact, Newton gave two definitions for this law (Law II and Def. VIII), neither being equal to the equation shown that is due to Johann Bernoulli. What is more, according to Westfall (p. 471), both definitions are incompatible between themselves. However, this assertion is debatable considering the nature of the two successive concepts of "moment" that Sellés [2006] found in Newton's mechanics. See also Hankins [1967].

[13] The virtual work, as we understand it today, was introduced for the first time by Johann Bernoulli.

place for such constraints and the bodies would move freely. On the other hand, if $\vec{u}_i = \vec{w}_i$, the constraints would not allow any motion.

In the *Essay* [§.1] he presents a new version of this principle. Each body, m_i is animated at any instant by a velocity $\vec{v}_i$ and impelled by a force $\vec{\varphi}_i$. After an instant dt, due to the effect of the force $\vec{\varphi}_i dt$ jointly with the internal actions of the rest of the bodies, the body m_i will move with a velocity $\vec{v}_i'$, so that $\vec{v}_i'' = \vec{v}_i - \vec{v}_i'$. Now, if each body was impelled by the velocities $\vec{v}_i''$ and $\vec{\varphi}_i dt$, there would be no motion in the system, that is to say, it would remain at rest or in equilibrium. At rest if the bodies were separated without any link among them, because $\vec{v}_i'' = \vec{\varphi}_i dt$ would occur. In equilibrium, when the bodies were contiguous and interacting among themselves, because all the $m_i \vec{v}_i''$ and $m_i \vec{\varphi}_i dt$ would be destroyed, and the entire system would end up in equilibrium, that is to say $\sum \left(m_i \vec{v}_i'' - m_i \vec{\varphi}_i dt \right) = 0$. Obviously if there were no forces at all, the solution would be $\sum m_i \vec{v}_i'' = 0$.

We can appreciate that in this version the body masses and the time come into play, although the latter as a differential. Something that is a little surprising is the coexistence of $\vec{v}_i''$ which is a finite quantity with $\vec{\varphi}_i dt$ which is a differential one. We have to interpret $\vec{v}_i''$ either as the result of a collision or a shock whose effect is instantaneous, or as a differential quantity that could be assumed as $d\vec{v}_i''$.[14] We could jump to Newtonian mechanics identifying the acting forces as $\vec{F}_i = m_i \vec{\varphi}_i$, which would arrive at the expression $\sum (\vec{F}_i - m_i \vec{\varphi}_i) = 0$, where $\vec{\varphi}_i$ would be the acceleration and $m_i \vec{\varphi}_i$ a fictitious force, called d'Alembert's force. This is how his principle has been incorporated to ordinary mechanics, "stripped of his philosophy of motion".[15] It has the peculiarity of sidestepping the intervention of internal forces, which in many cases can be advantageous.

In a more general way, nowadays the d'Alembert principle is understood as the condition $\sum (\vec{p}_i - \vec{F}_i) \delta \vec{r}_i = 0$, which a dynamic system meets; $\vec{p}$ being the momentum, $\vec{F}_i$ the external forces and $\delta \vec{r}_i$ any virtual displacement consistent with the system constrains. It is a generalization of Newton's second law and it is equivalent to the Euler-Lagrange equations of motion.

---

[14]We advance that in the *Essay*, after declaring explicitly that everything always changes by insensible degrees, he makes a similar change in the velocity, expressed as α to α'', which gives an accelerative force α''/dt [§.36-4th].

[15]Darrigol, p. 13. This treatment is applied to the Atwood machine as an example. This machine consists in two masses connected by a string over a pulley. When the masses are equal there is no motion, but when they are different the motion has constant acceleration.

Chapter 13
Forces and Fluids in the *Essay*

13.1 Forces and Pressures

The term force is one of the most widely employed in the *Essay*, more than two hundred and fifty times, only surpassed by the word pressure. Considering d'Alembert's warning that he would only use the word force in order to avoid circumlocution, this repetition seems somewhat surprising.

The forces in the *Essay* can be classified in two classes: one derived from a potential field, another coming from dynamic effects. For the first one, the most representative force is the gravity, which acts vertically in any point of the fluid and produces an equal pressure in any direction, like he says. He extends this idea to other fields of forces, with vertical and horizontal components not constants but able to produce the same type of local pressure. This generalization of the gravity is clear in the analysis of the fluid at rest [§.21–26 and Appendix], where he identifies the forces along a channel as a weight or the total force upon a body with the weight. These forces come always from an external cause and they affect the fluid either in motion or at rest.

Obviously, he has a clear concept of what gravity means. He had stated in the *Encyclopédie*,[1] that gravity is the force in virtue of which the bodies fall to the ground, and weight is the effect of this force upon a particular body. In the *Essay* the term weight is used with the same meaning as weight. However, sometimes the gravity is mentioned as natural gravity, which could indicate the possibility of a non-natural gravity. In fact, this does happen in the case of an elastic fluid [§.109].

The other class of forces are related the motion of the fluid. They are the accelerative forces. That means that when a fluid particle moves changing its velocity an

---

[1]Under the entry *gravité*. We have tried to use the entries written by d'Alembert in the *Encyclopédie* for clarification purposes. However, most of the relevant matters of the *Treaty* and the *Essay* had been incorporated in the *Encyclopédie*. As an example, the Intro. I–IV is almost totally in the entry of *Fluide*.

© Springer International Publishing AG 2018
J. Simón Calero (ed.), *Jean Le Rond D'Alembert: A New Theory of the Resistance of Fluids*, Studies in History and Philosophy of Science 47,
https://doi.org/10.1007/978-3-319-68000-2_13

accelerative force must exit upon it "which must be destroyed" by the forces acting on such particle. Although d'Alembert uses the word particle, it should be understood more properly like a mass differential element. The accelerative forces necessarily imply, or lead to, the internal forces in the fluid, even though he does not manifest it explicitly. However, he carries out integrations of these forces along narrow pipes [§.27-2nd], and in some way he accepts the same principle along a streamline [§.45-4th]. In our opinion, the concept of internal forces is in accordance with his basic principles. For him a fluid is an aggregate of corpuscles, or minuscule bodies, in contact among themselves, and the forces are transmitted through them by direct contact. Therefore in a fluid in motion, the external forces, coming from a field of forces or for the effect of any wall will be conveyed through the corpuscles to any internal point where they will manifest their effect as a change in velocity.

The term power is used a few times, mostly as equivalent to a force. Also this is used alternatively in the oscillation of bodies in water, referring to the buoyancy [§.119], and in the action of a jet against a plate [§.145] as well. Finally, in the elastic fluids [§. 80, 85, 109, 115], the force is associated with the compression; which can be understood as if the fluid were inside a flexible externally compressed vessel.

D'Alembert mentions the word pressure more than three hundred times, more than half of them followed by the qualifier "along". However the pressure is not defined anywhere, as Truesdell notes.[2] In the *Encyclopédie*, "the pressure is properly the action of a body which makes an effort to move another; such is the action of a heavy body supported on a horizontal table". Compare with the definition for a driving force "the cause that moves a body", given also in the *Encyclopédie*, the difference seems quite subtle: one is a cause, the other the action. In the entire *Essay* the pressure is shown as an action of the fluid along a direction. This is clear from the very beginning, when he notes the philosophers have reduced the laws of the fluid equilibrium to a single principle of experiment "the equality of the pressure in all directions... and to which it was necessary to refers all others" [Intro. IV].[3] That is, at any point inside a fluid, either at rest or in motion, there will be same "pressure along" any direction; this will be applicable to a channel, a streamline or the surface of a body. In many cases it alternates and plays the same role as a force, especially when the fluid is moving around a body and the pressures destroy accelerative forces.

The term pressure for d'Alembert differs of our present understanding. For us the pressure is represented by a magnitude p whose effect upon any differential surface $d\vec{\sigma}$ inside a fluid would be a force as $d\vec{F} = pd\vec{\sigma}$.[4] For d'Alembert the term pressure means exactly this $d\vec{F}$, that is a force and a direction, which he refers to as "pressure along". For the total force on a body, which would be $\vec{F} = \int pd\vec{\sigma}$, he maintains the same wording: pressure upon the body.

---

[2]Truesdell [1954], p. LII.

[3]In the *Encyclopédie*, under the entry Fluide (III), he says that it had been established by Pascal in his *Traité de l'équilibre des liqueurs*.

[4]In fact the pressure is an isotropic tensor $\bar{\bar{p}} = p\delta_{ij}$, which in a simplified way can be represented by a scalar magnitude p.

The d'Alembert's intention of building a fluid mechanics without the recourse of internal pressures has been discussed for several authors. Truesdell states that, he "managed somehow to avoid [the pressure], since it is now extremely difficult to conceive hydrodynamics without the pressure as protagonist".[5] Darrigol thinks that despite obtaining the equations without the recourse of pressure, d'Alembert uses this concept in the derivation of the Bernoulli's equation, and that "Plausibly, he favoured a derivation that was based on his own principle of dynamics and thus avoided obscure internal forces".[6] Guilbaud dedicates a chapter to this matter, analysing the forces from the *Traité de l'équilibre* up to the *Opuscules mathématiques*.[7]

We think that his understanding of pressure is an internal force directed to any direction and with a magnitude which is constant for every point of the fluid. This magnitude would be the pressure in our conception. He avoids this concept although uses its consequences, which are forces, naming them pressures. This is very clear in a quiescent fluid impelled by a field of forces. In the case of motion, when the fluid runs through a very narrow pipe of variable section, he finds that the pressure along the pipe's centre line depends on the velocity. In fact what he really finds is the difference of forces between two physical sections obtained by the destruction of the accelerative forces generated by the change of velocity. In the *Essay* he combines both gravity and motion whose effects are equivalent. Furthermore, what is an important issue, he states that the pressure at a point inside a fluid is the same that the weight of a column of a certain height of the same fluid at rest; this height will be taken as its measure [§. 33]. This height is a scalar magnitude.

The last term to deal with is the resistance, the main object of the book and mentioned some as one hundred an eighty times. He states in a general way that "the resistance that a body undergoes when it shocks with another is just, strictly speaking, the quantity of motion it loses" [Intro. IV]. Therefore, the resistances implies a change in the velocity, and this change du/dt is an accelerative force that applies to the body will give a total force $R = m_B du/dt$, that is the resistance [§.89–90]. In the *Essay* the resistance is always considered as a loss of momentum and not related to any force, with the exception of the two mentioned articles. We notice that for us the resistance is the force that undergoes a body moving in a fluid either at constant or variable velocity, while for him the term is only to the second possibility.

Thus, the effect of moving fluid upon a stationary body and the effect of a quiescent fluid upon a body at motion have a different physical origin. The first are the pressures, in turn related to the gravity, and the second the momentum loss. He will prove that both are equals.

Summarizing and making a kind of transliteration of the former concepts. We could consider the "accelerative forces" equivalent to "internal forces", the

---

[5] Truesdell [1954], p. CXVIII.
[6] Darrigol [2005], p. 23.
[7] Guilbaud, Chap. VIII.

"pressure along" to "forces due to the pressure" and the "resistance" to "loss of momentum". In addition, we can take the rule, or law, that the "accelerative forces must be destroyed" equivalent to "the change in momentum must be equal to the forces". Having this in mind, the d'Alembert dynamics can be converted in the Newtonian one.

13.2 Conception of *Fluids*

As we have seen, throughout the eighteenth century the idea of a fluid evolved from an aggregate of particles to a continuous media. This evolution occurred in the rest of Mechanics,[8] but with the fluid it implied changing the model of resistance generation.

Two statements in the *Essay* show what d'Alembert understands by a fluid:

- "I assume only what no one can deny, that a fluid is a body composed of very small particles, separate and able to move freely" [Intro. IV]
- "In all fluids which are known to us the particles are immediately adjacent by some of their points [of contact], or at least they act upon each other almost as if they were" [Intro. II].

The first assumption is established when he seeks for a non-arbitrary foundation to support the entire fluid theory. The second, made when commenting Johann Bernoulli's work, is a complement to the first. That is to say a fluid is an aggregate of independent particles, but each of them is limited in motion by its neighbours. This was a common idea in those times.

Each particle is subjected to the laws of motion, and consequently the fluid, taken as a whole, will be subjected the same laws which he had defined previously. "Let us suppose in fact that we had the advantage ... of knowing the figure and the mutual arrangement of the particles that make up the fluid: the laws of their resistance and their action will surely be reduced to the known laws of motion; because the research of the motion communicated from a body to any number of surrounding particles is only a dynamic problem, for whose solution there are all the mechanical principles that could be desired." [Intro. IV] Obviously, this possibility could be accepted as an asymptotic truth but it is practically impossible, because "as larger the number of particles is, it becomes more difficult to apply the calculus to the particles in a simple and convenient way,...but we are rather far from having all the *data* needed to be able to use this method". Furthermore, he believes there is a deep ignorance of the entrails of the system, because "we ignore not only the figure and the arrangement of the parts of the fluids; we yet ignore how these parts are pushed by the body and how these parts move between them." As a consequence "such a method would be scarcely practicable in the search for the resistance of

---

[8]See Maugin, *Continuum Mechanics through the Eighteenth and Nineteenth Centuries*, Chap. 2.

fluids". Summarizing, the laws of motion previously established are useless with the fluids.

Therefore, faced with this situation he is obliged to find another way to solve the problem.[9] The solution is to handle the fluids as a continuous media, as occurs in the hydrostatics. He said "There is anyway such a big difference between a fluid and an aggregate of solid particles, in so much as the laws of pressure and equilibrium of the fluids are very different from the laws of pressure and equilibrium of the solid".[10] The conception of a fluid as an aggregate of particles is an heir of the solid bodies, and he wants to deal with the fluids as a different class of matter. That is why he jumps to hydrostatics. In the *Encyclopédie* he will explain this process in a more straightforward way, ending by "that [for the fluids] the laws of their equilibrium and their motion are a problem ... for whose solution we are obliged to recourse to new principles.[11] Once we enter the realm of hydrostatics, we have almost to forget the former theories, because: "Experience alone has taught us in detail with the laws of hydrostatics, what the most subtle theory has never made us suspect", which he underlines as "and nowadays as well as the experiment has made known these laws, no one has been even able to find a satisfactory hypothesis for explaining and reducing them to the known principles of the statics" [Intro. IV].

In the *Essay*, the fluid taken as a set of particles is confined to the first chapter, in the all the rest of the book the fluids are treated as continuous. However, despite what we have explained previously, there is no sharp separation between both concepts because some of the theorems derived for the fluid-particle conception will be introduced later in the fluid-continuous concept. The most relevant of these is the one in which the resistance that the fluid exerts on a moving body is found to be proportional to the square of the velocity [§.8-3rd]. That is to say links exist between both conceptions. Probably, d'Alembert never completely renounced the idea of the particles as the main constituents of the fluid, and the use of the continuum was rather a tool. This can be asserted because throughout the *Essay* the wording suggests that he was always thinking of particles. However, as the brilliant mathematician he was, he knew that in a continuous media there are no

---

[9] Somehow this reminds us of the attitude of Newton searching for a way to determine the resistance in the liquids.

[10] These ideas are expressed even more roundly in the Introduction of the *Traité de l'équilibre*: "The Mechanics of the solid bodies, being supported on metaphysical principles independent of the experiment, allows the exact determination of the principles that must be the foundation of others. The fluid theory, on the contrary, must necessarily have the experiment as base, from which we only receive light well defined". [vi].

[11] Under the voice of "Fluide" (15th meaning). Reflections on the equilibrium and motion of fluids. "If the figure and mutual arrangement of the particles that make up the fluids were known, it would not need principles other than those of ordinary mechanics to determine the laws of their equilibrium and their motion; because this is always an identified problem, in order to find the mutual action of several bodies that are linked among themselves and whose figure and respective arrangement is known. But as we ignore the form and the disposition of the fluid particles, the determination of the laws of their equilibrium and their motion is a problem, which considered as purely geometric, does not contain enough data and for whose solution it is obliged to recourse to new principles".

particles at all, but differential elements. That is why we believe that for him the fluids had a dual nature: particles and continuous.

All the above is applicable for non-compressible fluids, but nothing is explained about the internal composition of the compressible or elastic ones; only that the single fluid of this type known is the air [§.114]. In the *Encyclopédie*,[12] it seems that pure air is assumed to be composed by small spherical bubbles whose size would change with compression; therefore these bubbles would take the role of the particles, all the aforesaid being applicable to the elastic fluids.

These explanations refer to ideal fluids, because the real ones are also affected by other forces, which he calls friction and viscosity [§.92–94]. Friction results from the effect of the unevenness of the surface over which the fluid runs thus generating an additional resistance proportional to the velocity. Viscosity is a force among the particles themselves producing another force that is constant; and which could be assimilated with the superficial tension.

13.3 Experience and Experiments

The word *experience* is used in the *Essay* more than one hundred times. Under this term we can understand two meanings: one the experience in a general sense, and the other the experiments. This one has a very precise sense in physics as a question asked of Nature in order to discover its laws and mechanisms, and it has been a basic stone in the Physics for several centuries. The definition given in the *Encyclopédie* for *experience* with the sense of experiment is almost identical to our present understanding.[13] However, in a general or philosophical sense its meaning is quite close to knowledge accumulated over time, close to our present word *experience*.[14] Therefore, the experience applied to the technical view has to be understood as the accumulation of data coming from observation and experiments. We want to point out that sometimes these observations not need to be supported by precise measures but they can be simple facts; for example, the resistance of a body moving in water. However, both categories, common sense observations and the experiments, are rather different. The problem is that the line between them is very fuzzy for d'Alembert.

---

[12] Under the lengthy entry "Air"; properties, (III) elasticity.

[13] In the *Encyclopédie*: "Experiment is the test of the effects arising from the mutual application or motion of natural bodies, in order to discover certain phenomena and their causes". In the *Oxford* dictionary: "A scientific procedure undertaken to make a discovery, test a hypothesis, or demonstrate a known fact".

[14] In the *Encyclopédie*: "Experience commonly means the knowledge acquired through life, jointly with the reflections that have been made on what we have seen, ...". In the *Oxford* dictionary: "The knowledge or skill acquired by a period of practical experience of something, especially that gained in a particular profession".

His most relevant to invocation to the experience is *the equality of the pressure in all directions* [Intro. IV]. For him, the single principle must be considered as the fundamental property of fluids. The second call to experience is when he finds the nullity of the pressure upon a body in a fluid stream [§.70], which also moves him to ask for more experiments.

He also introduces the observed fact that the perturbation produced in a fluid by a body vanishes with the distance as a hypothesis [§.36-1st], which could be in conflict with the theory itself. Besides, he assumes that these perturbations are limited only to very near the body surface [§.71–72], an argument which is not very rigorous because is not supported but any experiment. We also add that it is unnecessary.

The last surprising call to experience is when he wants to show that the resulting force in an impulsive action upon a body is null [§.55]. He proposes an imaginary measuring apparatus, or experiment, and he imagines the result as well.

Apart from these relevant cases, most of the mentions in the *Essay* correspond to experiments. D'Alembert was a pure mathematician, not an experimenter. However, he proposes experiments based on pendulums to measure the resistance whose bob is inside a water flow [§.75]. For very slow motions he presents an elaborate mathematical treatment [§.95–98].

Finally, he clearly comments the other authors' experiments in three places of the *Essay*. One is after the study of the motion of a fluid around the body, with Mariotte, Newton, Daniel Bernoulli and 's Gravesande [§.72–79]. The second is when he explores the non-inertial resistance, with 's Gravesande [§.93] and Daniel Bernoulli [§.99]. The third is for the non-elastic fluids, with Robins [§.114–115].

Chapter 14
Brief Analysis of the Contents of the *Essay*

Our main intention is to highlight the genuine contribution of d'Alembert to Fluid Mechanics. Therefore, instead of running through the *Essay* article by article, we have preferred to extract the core contribution and to bring it clearly to the light. Nevertheless this does not mean that we have renounced analysing the entire book. Some of the criticisms that the *Essay* has received are of its being lengthy and tortuous; we will try to mitigate these opinions by breaking up the lengthiness in small pieces and simplifying the tortuousness by reorganizing the articles.

Strictly speaking, we are following the d'Alembert own intentions. In the Introduction [Intro. V], he successively displays his master plan detailing the object of the work and the items to be addressed. Our interpretation coincides partially with this list.

Furthermore, a glance at the Table of Contents shows that the *Essay* is divided into nine chapters plus an appendix, and is preceded by an introduction. Each one contains a certain number of articles numbering 171, which are the basic building blocks of the book. There are two chapters rather larger than the rest, the IV and V, and both of these are also divided into sections and cover more than a half of the total number of pages and articles as well, which clearly indicates the importance of the matters treated therein. The theory of fluid motion around a body is developed in these chapters, although part of them dealt either with applications or collateral matters of the main theory. We understand that what d'Alembert does is to make the "theory of fluid motion around a body", instead of the "resistance of the fluid" because this is a consequence of the motion.

With respect to the rest of the chapters, the first three ones are introductory. The VI is dedicated to the oscillation of bodies in a fluid; we think that this has little if anything to do with the resistance. The last three chapters are somewhat related to the subject, but not too much. Finally, the appendix revolves around the hydrostatics pointing to the problem of the shape of the Earth, which is a little away from the main topic as well.

After these initial and general ideas, let us go into greater detail what we consider the core of his contribution. This is divided in two cases: a body at rest in a moving fluid and the contrary, a body in motion and the fluid at rest. There is an intermediate step between both cases: the impulsive motion of a fluid against a body at rest. Strictly speaking this intermediate case could be understood as a special application of the first one, but we think it is convenient to analyse separately. All this is preceded by some preliminaries.

14.1 The Preliminaries

The preliminaries, or basic tools, are covered in the three first chapters. In the first one he presents his dynamic principle applied to a system of bodies subjected to forces as we have already explained. There are two points we want to highlight. One [§.5], assuming the system is at rest, if a velocity is applied to a single body, this one will induce the rest of the bodies to move, each one following its own trajectory. Now, all these trajectories will be always the same, irrespectively of the magnitude of the initial velocity given to the body. Obviously they will move faster or slower, but each of them along the same spatial trajectory. This is a glimpse of the uniqueness of the velocity field, which d'Alembert considers as fundamental.

The second point refers to a body submerged in a fluid and moving with an initial velocity [§.8]. The fluid is constructed by a set of individual tiny bodies that, jointly with the main body, will constitute a system. The former uniqueness of the motion is applicable here, and accelerative forces will appear in the evolution with time. Consequently, a resistance will be generated upon the main body, which this must be proportional to the square of the velocity applied to this body. We note that the law of the square is demonstrated with a system of individual bodies.

In the second chapter the fluid impelled by external forces is considered as a continuum of variable density in equilibrium. This means that whichever point of the fluid will be equally pressed in any direction [§.13]. Consequently, the fluid inside any closed channel must be at equilibrium. The general equation of hydrostatics is obtained founded on this premise [§.19]. This equation is an extension of Clairaut's, although d'Alembert's found it by a different method.

The third chapter is twofold. The first part deals with a body placed inside a fluid with both the fluid and body at rest. The proposed problem is firstly to determine the force that the body undergoes at any point of its surface, which he calls pressure, and finally to find the total one on the body [§.21]. The second part refers to a fluid moving along a very thin tube of variable section; that is to say, one-dimensional motion. The problem is to find the relation between the pressure and the fluid velocity at any point. The solution is similar to the one that was obtained by the Bernoullis, father and son. It is here where the pressure is considered as equivalent to the weight of a column of the same fluid, which can be used for measuring and comparing pressures [§.33].[1]

14.2 Body in a Fluid Stream

Let us take a body, assumed axisymmetric, placed motionless in a stream of fluid moving with a uniform velocity and parallel to the body axis. It is clear that the streamlines, which are parallel to each other upstream, will curve in order to surround the body, and consequently the particles moving along the streamlines will change both direction and velocity [§.36]. The first key point that d'Alembert establishes and proves is that the pattern of this streamlines field is always the same, irrespectively of the magnitude of the upstream velocity [§.38]. The velocity at any point, obviously tangential to the local streamline, can be broken down in two components: one along the main axis of the body and other perpendicular to it. Each of them will be proportional to functions which are independent of the velocity. These two functions, known as velocity functions, depend only on the body shape and they are an invariant of the system.

The second key point that he proves is that the two mentioned velocity functions are linked through another two partial derivative equations obtained following the principles of fluid dynamics [§.45]. Nowadays these two equations are called continuity and irrotationality. Therefore the fluid problem was converted to a mathematical one, which is to find two functions that meet those equations and are consistent with the boundary condition on the body contour. These two functions turn out to be the velocity functions, which will be the solution of the problem.

Once these functions are known, and known the magnitude of the upstream velocity, the two local velocities at any point are obtained, and the pressure is derived using the Bernoulli equation [§.66]. Then the total force upon the body will be the result of the integration along its surface of the components of the differential pressure forces. This force, which he calls total pressure, turns out to be proportional to the fluid density and to the square of the upstream velocity.

However, this mathematical problem was almost impossible to solve at that time. D'Alembert tries to find a way of transforming the velocity functions in real and imaginary values of a complex variable [§.57]. This would be a very powerful tool, but it usefulness had to wait many years, until the possibility of conformal transformations among different solutions appeared. This method must be considered as an outstanding contribution, although it refers more to the mathematical side than to the fluids themselves. D'Alembert even tries to find an approximate solution based on a polynomial of power in the complex field [§.60].

---

[1]The precedent of this idea was the apparatus used by Daniel Bernoulli in his experiments. Cf. *Comm. Acad. Petrop.*, Vol. IV, 1729 (1735).

Two more remarks. One, he did not apply the mathematical model to the entire fluid, because he understood that in some parts of the body such as the two apex, there would be a conflict with experience. For this reason he introduces some artefacts in order to match theory and experiment [§.36]. Second, even when the solution of the problem was impossible in a general way, for geometrical considerations in some cases, he finds that the total force on the body turned out to be null, which is clearly against experience [§.70]. Facing this shocking result, he interpreted it as a consequence of the aforementioned difficulties of matching theory and experience. This would be named later as d'Alembert's paradox.

Finally, in any case, we have to note that the force upon the body is directly related to the upstream pressure, this is probably why he named this force fluid pressure. As we have said previously, the pressure is measured as the weight of a column of fluid, therefore, at the end the force upon the body is backed up by the gravity.

14.3 Fluid Impulsive Velocity

Before entering into the opposite problem, that is to say a moving body and fluid at rest, he introduces the case of the impulsive velocity [§.51], which could be considered as a variant of the former case. Initially both body and fluid are at rest and suddenly the fluid is given a velocity that will take the system to the same configuration as the previous case. The fluid pattern instantaneously generated will be represented by the same two velocity functions, which also will meet the same differential equations.

The pressure will be also produced in a similar way as a function of the local impulsive velocities applying the equilibrium of closed channels. The calculation is made by means of Bernoulli's equation, but using the non-steady terms. The total impulsive pressure upon the body, called pressure at the first instant, is the integration of the local pressures over the body surface. The result is the product of three factors: the density, the velocity jump and a mathematical expression that is function of the geometry of the body [§.54]. This was a very interesting finding; however d'Alembert said that for an experiment it can be easily proved that this first instant pressure must be zero [§.55]. The reasoning is based on an imaginary mechanism with a weight for retaining the body against the effect of the first impulse. He assumed that this weight would be in equilibrium; therefore, for the virtual work principle, it would be equivalent to a finite mass with an infinitely small velocity, or on the contrary, to a finite velocity and an infinitely small mass. As the density is finite, this small infinitely small mass implies a null force. We find several mistaken steps in this reasoning, as we will show later.

For us this is the biggest mistake made by d'Alembert in the entire *Essay*, and we do not understand how he could make it. The impulsive motion was a brilliant idea and the formulas he obtained are according with the values of what nowadays is known as apparent or added mass.

14.4 Body Moving in Fluid at Rest

A body moving in a fluid at rest communicates a velocity to each particle of this fluid, which implies that the body momentum is reduced in a quantity equal to the one transferred to the particles. This change in momentum is the resistance that the body undergoes or fluid resistance [Intro. IV]. Comparing this case with that of moving fluid, we can see that the force upon the body has a different source in each case; in the first it comes from the pressure, in the second from the momentum lost. Additionally, in the first case, the fluid is maintained at constant velocity with time, while in the other the body loses velocity continuously.

Until then the authors had assumed that both motions were equivalent, what is sometimes known as the reciprocity principle. This can be justified because the laws of motion used were the same either when the system is at rest or has uniform velocity. Nevertheless, d'Alembert thought that this hypothesis had to be proved [§.90].

Firstly he finds that the streamline field measured with respect to the axis fixed to a body moving at a variable velocity is also invariable, and regulated by the same equations as when the fluid was moving and the body at rest [§.86]. Now, and what is a key point, if upon this system constituted by the body and fluid, a velocity was suddenly applied equal to that the body has, but in the opposite direction, the situation would be equal to a fixed body and a moving fluid, which equivalent to the first case. However, this would be true only instantaneously, because the body was losing momentum and consequently decelerating itself. According to the terms already expressed for the impulsive velocity the streamline configuration will not be affected [§.88]. However, as the body velocity is not constant, to keep it motionless a continuous set of differential impulsions will be required, which also follow the impulsion rules. Therefore after the first finite impulse the body will receive two forces continuously: one due to fluid pressure as stated in the first case, which is proportional to the density and square of the velocity. The second, due to the successive differential impulses necessary to maintain the velocity constant, turns out to be proportional to the density, acceleration and a function that depends on the geometry of the body. This function has the dimension of a volume, therefore we call it virtual volume, and that he considers it to be null [§.89]. This has two consequences: the equality of the pressure and the resistance, even in the case of non-uniform motion, and the proportionality of the resistance with the square of the velocity. As he underlines, both things had been taken as true until then, but they had never been rigorously proven [§.90].

We wish to note that for the definition of the resistance of the body d'Alembert was obliged, we would even say he was trapped, to assume a non-uniform motion because he understood the force as a result of the change of momentum, and consequently the velocity must also change. In order to link the two cases he has to use the impulsive velocity, although regrettably we insist, he makes the mistake of taking the term of impulsive effect as null. However he includes this term with as a virtual volume in all the formulas, but warns us that it is null. The resistance ends by coinciding with the Newtonian mechanics: the product of the body mass times the acceleration.

Looking back to the *Manuscript* we can appreciate an evolution in d'Alembert's thoughts. In it the impulsive motion came inside the second case as recourse to solve the problem, while in the *Essay* it takes a separate analysis, making the approach clearer and better structured.

At this point we close his basic and sound contribution to Fluid Dynamics, which takes up no more than a third of the *Essay*. This was a real breakthrough at that time as Truesdell stated, "Nevertheless, despite its many defects, the *Essay* is a turning point in mathematical physics. For the first time, a theory is put (however obscurely) in terms of a field satisfying partial differential equations".[2] This merit is only tarnished by the above mentioned mistake, and some other minor ones.

---

[2] Truesdell [1954], p. LVII.

Part III
Analysis of the *Essay*

Julián Simón Calero

Chapter 15
The *Essay's* Introduction

We have divided the Introduction in seven clearly differentiated parts.[1] D'Alembert begins the *Essay* with a tribute to the Ancients presented by a chain of statements [Intro. I]. The Physics of the Ancients is not so limited and unreasonable as some modern philosophers think. The difference between them and us lies in our knowledge of differential and integral calculus. The Ancients had a wiser method of philosophizing than we commonly imagine; but modern geometricians have more resources, not because we are superior, but only because we have come later on the scene.

D'Alembert thought that the internal mechanism of the fluids should be a particular object of admiration for the philosophers. Only by means of modern resources is it possible to approach a rather complex matter. Notwithstanding this help, the resistance of fluids still encloses such considerable difficulties that even the greatest men have been able to give us only a slight sketch.

He points out that the main cause of the slight progress made so far was the wrong election of the true principles upon which calculus must be applied. He makes clear these two concepts: principles and calculus, and warns about the mistake of substituting principles for calculations, which will end in results contrary to the reality of Nature. Moreover, he points out that modern geometricians could incur in this fault. After all the previous kind words, this is the first subject he takes in depth: the principles and calculus; which nowadays we understand as being the theories for modelling a phenomenon and the mathematical development for representing its real evolution. He will come back to this subject at the end of the Introduction [Intro. VII]. In any case he declares that he will find out the principles first and apply the calculus after. However before presenting his principles he will review the work done previously by other authors.

---

[1]The parts II–V correspond to the *Manuscript* §1–6 although enlarged. Also parts I–IV have passed almost entirely to the *Encyclopédie* under the entry of *Fluide*.

Newton deserves a major place [Intro. II]. D'Alembert shows Newton's hypothesis about the fluid composition, the theoretical results obtained and the lack of agreement with experiments carried out by other authors. It is surprising that he does not refer to Newton's experiments, he only makes a brief mention later [§.79]. D'Alembert makes a very detailed analysis with comments and criticisms, particularly of the cataract. He finishes recognizing that "he dared for first time to clear a way to solve a problem that no one before him had ever attempted. Thus this solution, though not very exact, shines throughout with this genial inventor, this mind fertile in resources that nobody has possessed in a higher degree than him."

Newton's mistakes move him to make some reflections about how the errors of great men are useful for those who came after [Intro. III]. He affirms that most of the authors that had criticized Newton had been less fortunate than him; with the exception of Daniel Bernoulli of whom he mentions the studies on the resistance, and also the effect of a stream against a plate, making note the differences between theory and experiment in both.

After these presentations, he declares that the fluid resistance must be addressed by a completely new method without owing anything to the predecessors [Intro. IV]. The first step is that "a fluid is a body composed of very small particles, separated and able to move freely", assumption "that no one can deny". Two more propositions are added to the former: one, that the resistance is the quantity of motion lost by the moving body, and other that the dynamics can be reduced to the equilibrium, as he established in the *Traité de Dynamique*.

In a fluid of this kind a moving body can be taken as another particle, obviously not small but conceptually as if it were. If we knew the entire particle disposition and the mutual interaction among them, the problem would be converted to a static one and it would be resolved. However this is impossible, because "we are rather far from having all the *data* needed to be able to use this method". At this juncture, the fluid of particles is transformed into a continuous fluid supported by the hydrostatics. Furthermore, if the dynamics of solid bodies can be reduce to the statics, the dynamic of fluids or hydrodynamics, will likewise be reduced to the hydrostatics.

However our knowledge of the hydrostatics does not derives from any hypothesis, but from "a single principle of experience, *the equality of the pressure in all directions*; principle that they have considered (lacking a better one) as the fundamental property of fluids". To this fundamental statement, and trying to justify somewhat the lack of theoretical support, he adds a line of reasoning explaining that if we are condemned to ignore "the first properties and internal contexture of the bodies, the only resource left to our sagacity is to try at least to capture in each subject the analogy of the phenomena and to recall all of them in a small number of primitive and basic facts" [Intro. IV]. The example given is Newton and gravitation. We quote d'Alembert's beautiful words: "Nature is an immense machine whose main springs are hidden to us; we do not see this machine except through a veil which conceals from us the interplay of the most delicate parts. Among the most striking, and maybe if we dare say it, the coarsest parts that this veil allows us to glimpse or discover, there are several that the same spring sets in motion, and that's mainly what we must seek to unravel". [Intro. IV],

15 The *Essay's* Introduction

Once the plan of the work is expounded [Intro. V], he addresses the question of the experimental verification of the resistance [Intro. VI]. His main objection refers to the paucity of agreement existing among the results obtained in the experiments carried out by physicists, no fact being perfectly founded in them. He recognizes that the experiments were so delicate that some skilled persons abandoned the task. There are also difficulties in separating the different effects acting on the particles, so that it could occur that "the experiments made in small [scale], have almost no analogy with experiments made in large [scale], and sometimes even contradict them" [Intro. VI]; this indicates that he considered the scale effect to be very significant. From this he deduces the necessity of dedicated experiments for each particular case, concluding that a general result would be faulty and imperfect. But after saying that, there is a quite surprising change of viewpoint. If the experiment leaves us with very clear and exact formulas, and the lack of agreement still persists, the problem would be in the hypothesis taken for the fluid, so it would have "to renounce fully all theories of fluid resistance and to consider them as one of those questions on which the calculation can have no bearing" [Intro. VI].

The Introduction ends with bitter references to the Berlin Academy [Intro. VII], in the terms we have already mentioned. In the last paragraph he adds some more reflections about his work and basic principles used, insisting again on the differences between theories and their application.

Chapter 16
The Preliminaries

16.1 Principles of Dynamics

As we have already commented, d'Alembert presents the general principle of dynamics in a somewhat different way from the *Traité de Dynamique* [§.1]. Let us assume a system of bodies, each of them moving with a velocity $\vec{v}_i$ and impelled by a force $\vec{\varphi}_i$, which we understand as accelerative. If this force did not exist, each body would maintain the velocity $\vec{v}_i$, but due to the force $\vec{\varphi}_i$ and the mutual links among all the bodies, the velocities will change to $\vec{v}'_i$, so that $\vec{v}_i = \vec{v}'_i + \vec{v}''_i$. Then, i-body will tend to move with the sum of the velocities $\vec{\varphi}_i dt$, $\vec{v}'_i$ and $\vec{v}''_i$, but the final one will be only $\vec{v}'_i$; therefore, if each i-body was forced to move with the two velocities $\vec{\varphi}_i dt$ and $\vec{v}''_i$, the system will remain in equilibrium.

In the case that all the bodies are separated and loose without any action among them, the above condition would be $\vec{\varphi}_i dt + \vec{v}''_i = 0$. That is to say each body would move as if it were alone, and the system would remain at rest. However, in a more general case when mutual actions exist, either by contiguity or links, the components of the set $\{m_i \vec{\varphi}_i dt + m_i \vec{v}''_i\}$ must destroy themselves,[1] that is $\sum (m_i \vec{\varphi}_i dt + m_i \vec{v}''_i) = 0$. Then the solution of the problem is reduced to finding the set $\{\vec{v}''_i\}$ that meets the above condition. Once this is done, the velocities $\{\vec{v}'_i\}$ after the time dt are $\{\vec{v}'_i\} = \{\vec{v}_i - \vec{v}''_i\}$, which will be the new velocities $\{\vec{v}_i\}$ for the new instant. Therefore we will have $\{\vec{v}_i(t)\}$ and consequently the bodies position $\{\vec{r}_i(t)\}$ and the trajectories of each body $\{C_i(\vec{r})\}$. So the problem will be solved.

---

[1] We note that this paragraph was not in the *Mss*.7.

The general solution is also valid if the forces $\vec{\varphi}_i$ were zero. As before there will be a solution $\{\vec{v}''_i\}$ for $\sum m_i \vec{v}''_i = 0$. Furthermore, if the bodies tended to move with $\{k\vec{v}''_i\}$, being k any number, the equilibrium would persist [§.2].

As a consequence in a system without forces, or if we simply ignoring them [§.3], the velocities at any moment $\vec{v}_i$ will evolve to $\vec{v}'_i$. Now, if for any cause the first velocity changed to $k\vec{v}_i$, the second one would become $k\vec{v}'_i$. The proof he presents is based on the distances traversed by any particle in the time dt, which would be $\vec{v}_i dt$; then, with the velocity k-times greater, these spaces would be crossed by the bodies in a time dt/k and consequently the resultant $\vec{v}'_i$ would change to $k\vec{v}'_i$. This reasoning implies that $\vec{v}''_i$ has also to change to $k\vec{v}''_i$. This occurs in a system integrated by solid bodies where the motion is transmitted by collisions that transfer momentum.

The above reasoning is also valid if some of the velocities are null [§.4]. This leads to the limit case when all bodies are at rest except one, which would have a velocity $\vec{v}_0$ [§.5]. This case is considered by d'Alembert as fundamental. It is clear that the only moving body will induce motion in the entire system, and each body will follow a trajectory C_i. All these trajectories will always be the same, irrespective of the magnitude of $\vec{v}_0$, obviously maintaining their direction. The difference for each $\vec{v}_0$ will be the time inverted in traversing each curve. Thus calling the modulus of the given velocity u_0, if the time taken to traverse the space x of C_0 is t, with ku_0 it would be t/k. It is clear that the velocity u at the point x is a function of this space as $u = u_0 z(x)$ [§. 6]. It is easy to prove that $du/u = z'(x) dx/z(x) = -\xi(x) dx$.[2] This reasoning can be considered as a glimpse of the uniqueness of the velocity field.

Before he goes further, he explains that all these corollaries deduced with zero external forces, would be also valid with velocities proportional to the forces [§.7]. This is clear because if $\vec{\varphi}_i = k_\varphi \vec{v}_i$, the velocity $\vec{\varphi}_i dt$ can be considered as a part of $\vec{v}_i$, although under this circumstance $\vec{v}_i$ would be variable with time.

Bearing in mind that a fluid was considered to be an aggregate of very small particles that are separate and able to move freely [Intro. IV], then in the case of motion, these particles transfer the motion to the contiguous ones by thrusts and collisions and always maintain the momentum. Therefore the motion of a solid body submerged in a still and non-elastic fluid can be analyzed as a system of bodies in which the solid body takes the place of a single large body, and the rest of the fluid is composed of innumerable tiny particles [§.8]. As we have seen, the trajectory followed by the body will always be the same, regardless the initial velocity it was given. Consequently, the same principle will applicable to rest of the particles. Besides, if the resistance of the fluid was assumed as depending on the velocity, it has to be proportional to the square. We note that he has defined the resistance in the Introduction [Intro. IV][3] as the quantity of motion lost by a moving body, which also implies a velocity lost, expressed as $-mdu$, which will be related to

---

[2] D'Alembert uses the sign minus, probably expressing that the velocity is decreasing.

[3] This paragraph does not have an antecedent in the *Manuscript*.

an accelerative force $m\varphi$ or simply φ. To prove this, let us take the velocity u at any point as $u = u_0 z(x)$, thus for the "general principle of the accelerative forces" $\varphi(u) dt = -du$, or $\varphi(u)dx = -udu$. Combinimg both $\dfrac{zdz}{dx} = -\dfrac{\varphi(u_0 z)}{u_0^2}$ is reached; the first member depends only on the distance x, while the seconds depends on the initial conditions. Based on this fact, d'Alembert says that this can only be met if $\varphi(u_0 z) = z^2 u_0^2$. This is not completely correct, because the condition of independence of the second member respect to u_0 is expressed as $\dfrac{\partial}{\partial u_0}\left[\dfrac{\varphi(u_0 z)}{u_0^2}\right] = 0$, whose solution is $\varphi(u_0 z) = K(x)z^2 u_0^2$, that is $\varphi(u) = K(x)u^2$.

Now, for the resistance R [§.9], which is the momentum lost, we will have $Rdt = -mdu$, or $Rdx = -mudu$. Introducing the relation $du/u = -\xi(x)dx$, obtained previously, we will have $R = m\xi(x)u^2$.

"We will demonstrate later that the fluid resistance (excluding gravity, friction and elasticity) is actually proportional to the square of the velocity, so that the function ξ of the space traversed is reduced to a constant" [§.10]. He claims that it is the first time that the proposition of the proportionality of the resistance with the square of the velocity is demonstrated. Also, he argues that the proofs given by previous authors were unsatisfactory; because they were based on the argument of the double action of the particles. That means for the first action, that the motion communicated and received by the body from any other particle is increased proportionally to the velocity; and for the second action, that the number of particles attained by the body is increased at the same rate as well. This reasoning could only be fully justified by a motion similar to the Newtonian *rare medium*, and seems to him rather vague, an opinion we agree with.

There is another consequence applicable to a completely different subject: light moving in a fluid [§.11]. The body in motion will be a light corpuscle that passes obliquely from one fluid to another through a separating surface, which, we add, could be between air and water. The physical phenomenon is called light refraction and it is well known that the refraction angle depends on the colour of the light. According to Newton's theory, each colour corpuscle has a different velocity, which is the reason of the colours separation in the refraction. However, according to d'Alembert the curve described by a moving particle does not depend on the velocity, therefore, all colour refraction angles must be equal, and it cannot be assumed that the difference in refraction of the rays comes from the difference in their velocities.

The Chapter I finishes with the statement that the laws of the resistance of fluids depend heavily on their equilibrium laws [§.12]. We have seen that he related the resistance with the accelerative force [§.8-3rd], and this latter with the equilibrium [§.1]. Therefore, we have to go on to Hydrostatics.

One additional remark, so far he considers the fluids as corpuscular entities, in the next, and further chapters, they become continuous media. The leap between both conceptions could be understood as a pass to the limit, although the discrete and continuous media are completely different from a mathematical point of view, and consequently give rise to a different type of solution. However, d'Alembert introduces the theorem $R = m\xi(x)u^2$, obtained for the discrete media, and used it in the continuous as a bridge between both.

16.2 General Principles of the Fluid Equilibrium

The foundation stone of the continuous fluid is "a single principle of experience, *the equality of the pressure in all directions*; principle that has been taken (lacking a better one) as the fundamental property of fluids, and as to which it was necessary to refer all others" [Intro. IV]. This principle is still considered to be a fundamental property of fluids.[4]

In order to find out the fundamental law of the equilibrium, d'Alembert follows the path previously travelled by others authors in their studies of the problem of the shape of the Earth,[5] and especially Alexis Clairaut for the solution expounded in the *Théorie de la figure de la Terre, tirée des Principes de l'Hydrostatique*. However

---

[4] A classic text such as the *Hydrodynamics* of Horace Lamb states: "The fundamental property of a fluid is that it cannot be in equilibrium in a state of stress such that the mutual action between two adjacent parts is oblique to the common surface". The first edition was published in 1879. We quote the reprinting of the sixth edition in the Cambridge University press, 1945. Cf. p. 1.

[5] The question was either the Earth was flattened or elongated at the poles. Newton opted for the first solution as a result of his theory of gravitation. As he stated in the *Principia*, assuming our planet as a fluid in rotation, if two channels were driven to the center, one from one pole and the other from the equator, both must be in equilibrium but the latter will be alleviated due to the centrifugal force, therefore equator channel must be longer than the pole one. As a consequence he estimated the flattening in 230/231. On the other hand was the Cartesian theory of gravitation, in which external vortices were responsible for the attraction of the bodies. These vortices dragged the bodies towards the center of the earth, an action which Descartes denominated "conatus". With this theory the magnitude of the flattening should be less than the Newtonian one, in a value of 576/577.

In counterpoint to these values stood the geodesic measurements of the meridian arcs made in France at the end of the eighteenth century by such distinguished geometricians as Picard and Cassini. From those it was concluded that the Earth had an oblong shape, stretched towards the poles, with a difference in diameters of 1/262. This opposition between the theoretical derivations and experimental measurements, together with the difficulty of accepting the theses of Newtonian mechanics, radicalized the scientists' positions, and divided the Academy of Sciences of Paris, and even came to have national and theological implications. In words of Lafuente "Theory vs. experiment, Newtonianism vs. Cartesianism, laicism vs. scholasticism, *savant* vs. Academic, England vs. France, all were, in the end, powerful alternatives to stir up controversy and kindle all the passions" (p. 48).

The controversies caused rivers of ink to run, but their positive side was that the Academy of Sciences of Paris sponsored two important scientific expeditions: one to Lapland and the other to the Spanish Viceroyalty of Peru, in present-day Ecuador. The aim of both was to measure the meridian arcs in two very separate latitudes. With these measurements, together with those made in France, the intention was to obtain the longitude of a meridian degree at various points of the Earth, and thus to clarify the value of the Earth's flattening. It was a crucial experiment about whose was the true theory: whether it was that of Newton or that of Descartes.

16 The Preliminaries

Fig. 16.1 Channels in a fluid

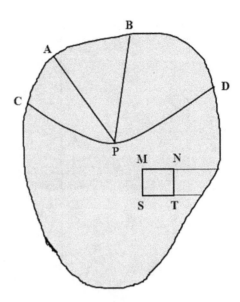

d'Alembert does not quote this author, instead he cites Colin MacLaurin, who, he says, was the first to make use of this principle in his research in the *Treatise of fluxions* [art. 639] and *On the Cause of the ebb and flow of the seas*, [§.14].[6]

D'Alembert assumes a mass of fluid whose particles are subjected to forces so that the fluid remains in equilibrium [§.13]. There is no indication about the nature of these forces; they could be understood as virtual accelerative forces, as we have explained. However, we think that even when this fluid mass can be understood in a very general manner, he probably has in mind the Earth considered as a fluid mass. Consequently these forces would be due to the action of gravity plus rotation; that is to say, the weight. Later this will seem clearer [§.21].

In any case, an internal point as *P* (Fig. 16.1) is pressed equally in all directions. This is also applicable to the particles of the channels *PA* and *PB*, which both reach to the surface, therefore the fluid in the rectilinear siphon *APB* is in equilibrium as well [§.13]. Next, by adding and subtracting channels is easy to prove that any closed one, rectilinear or curvilinear, must be in equilibrium [§.15–18].

Now, let $R(x,y)$ and $Q(x,y)$ be the forces acting along each axis upon any particle and *SMNT* (Fig. 16.2) a small rectangular channel [§.19]. Due to the equilibrium, the force at the vertex *N* will be related to the *S* via either the columns *ST* plus *TN* or *SM* plus *MN*. Omitting the channel width for simplicity, we have:

$$dp_{STN} = \rho R dx + \left(\rho + \frac{\partial \rho}{\partial x}dx\right)\left(Q + \frac{\partial Q}{\partial x}dx\right)dy \qquad (16.1\text{a})$$

---

[6]In the *Manuscript*, it is only quoted the second work as Paris 1740, p. 210, prop. 1, art. 3.

Fig. 16.2 Closed rectangular channel

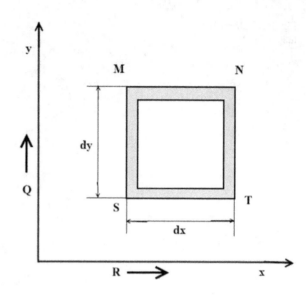

$$dp_{SMN} = \rho Q dy + \left(\rho + \frac{\partial \rho}{\partial y}dy\right)\left(R + \frac{\partial R}{\partial y}dy\right)dx \qquad (16.1b)$$

Being $\rho(x, y)$ the fluid density. Equating and eliminating the second order terms the resulting expression is:

$$\frac{\partial(\rho Q)}{\partial x} = \frac{\partial(\rho R)}{\partial y} \qquad (16.2)$$

D'Alembert says: "proportion that was already known but that nobody, it seems to me, had yet demonstrated by a method as simple as we just have done" [§.20].

Certainly, a similar proposition, but with constant density, had been established by Clairaut applying also the equilibrium of the internal channels. However, Clairaut considers that the expression $Rdx + Qdy$ must be an exact differential, because the total force between two points L and H, expressed as $\oint_L^H \rho(Rdx + Qdy)$, is always the same independently of the path taken from L to H. Then, according to a theorem obtained by him previously, a sufficient condition for the functions R and Q is $\frac{\partial Q}{\partial x} = \frac{\partial R}{\partial y}$.[7] Comparing both approaches, apart from the inclusion of the density, D'Alembert's seems to us more physical while Clairaut's is more mathematical.

---

[7] Cf. *Traité*, §XVI–XVII.

16.3 Pressure on Submerged Bodies

The next step is to place a body inside a fluid and to calculate the effects of the fluid forces on it. We encountered some difficulties in understanding of the text. On the one hand, almost all the terms relatives to the forces (force, pressure, power, gravitation and weight) are used in quite a few pages; on the other, it seems that the French version from the *Manuscript* was rather careless.[8]

Let us assume a weightless fluid, which can be either indefinite or finite, with a body G placed inside the fluid and surrounded by a surface Σ, as shown in Fig. 16.3 [§.21]. Inside Σ, both body and fluid are subjected to forces in such a way that they are in equilibrium, while outside Σ there are no forces by hypothesis. The problem is to find the pressure at any point of the body depending on the forces on the fluid.

In fact, the surface Σ "can be considered as the external surface of a fluid in equilibrium", because the forces upon it "are either absolutely zero or at least perpendicular to Σ". The hypothesis of the absence of forces outside Σ is not absolutely necessary; in the *Manuscript* it is said that those particles "are impelled by many forces",[9] which is also partially true, because the condition must be the perpendicularity of forces upon Σ, or in others words, that Σ has to be a level curve.

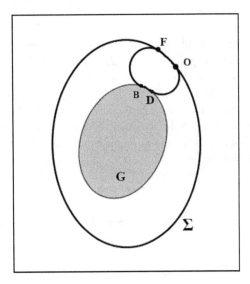

Fig. 16.3 Body in fluid

---

[8]Truesdell wrote that this chapter contained many propositions and corollaries which were incomprehensible for him [p. LII]. We will try to solve some of these incomprehensibilities and uncertainnness scrutinizing the *Manuscript* in order to untangle what d'Alembert's original thoughts could have been.

[9]"Nam siquidem particulæ fluidi extra spatium [Σ] positæ, multis (hyp.) viribus agitantur" *Mss*.25. "Because the fluid particles placed outside Σ are impelled by many forces [hyp.]."

Fig. 16.4 Body in fluid under vertical forces

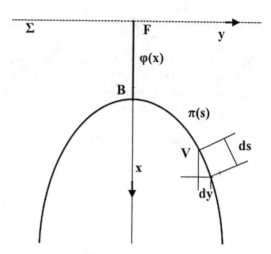

Now, the channel *FBDO*, formed by the segment *FO* over Σ and *BD* over the body plus any other two lines *FB* and *DO*, must be at equilibrium and consequently the "weights" of channels *OD* and *FBD* are equal. "Therefore the pressure in point *D* will be the same as if this point were pressed perpendicular to the *BDC* surface by a force equal to the weight of the *FBD* channel". It is obvious that we have to understand the term "weight" in a different way as the effect of the gravity forces, such as $p_D = \rho \oint_{FBD} \vec{f} \cdot d\vec{s} = \rho \oint_{OD} \vec{f} \cdot d\vec{s}$, being $\vec{f}$ the forces, which could coincide with the gravity in some cases, but not in general.

The application of the above to a more conventional body [§.22] does not follow this procedure strictly. The body, Fig. 16.4, is submerged in a limited fluid and subjected only to vertical forces. The above mentioned Σ will be the fluid horizontal surface crossing at *F*, and the forces will be only a function of the depth, called *x*, without any lateral component in accordance with Eq. 16.2.

The pressure at the point *V* on the surface is the result of two addends, one vertically from *F* to *B*, plus a second one from *B* to *V* along the surface. In the first the "weight" is expressed by a function $\varphi(x)$ and in the second by $\pi(s)$, *s* being the distance along the surface. The result, in which we have included the density, will be:

$$p_V = \rho \int_F^B \varphi dx + \rho \oint_B^V \pi ds \qquad (16.3)$$

It seems clear, as d'Alembert will show in next article [§.23], that $\pi ds = \varphi dx$, and according to the previous procedure, the solution would be the integration vertically from Σ to *V*.

The resulting pressure along the axis *X* will be $dF_x = p_V dy$, that, neglecting the first addend if it is small, leads to:

Fig. 16.5 Body submerged

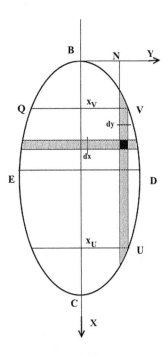

$$dF_x = dy \cdot \rho \oint_B^V \pi ds \qquad (16.4)$$

Two points to highlight. We have included the density in order to maintain the coherence of the formulas; this is not used in the *Essay*, although in the *Mss*.26 there was a paragraph saying that it had been omitted as it was considered as being one, but it should be included. The second point is that d'Alembert employs "pressure along" instead of force, this is a constant fact throughout the entire book, as we have explained before.

If the previous force was gravity [§.23], designated as ψ, it would be $\pi = \psi \frac{dx}{ds}$, so $\int \pi ds = \int \psi dx$. Therefore the pressure on point V will equal to the weight of the column VN; that is $dF_x = -\rho g(x_U - x_V)dy$, with $\psi = g$, which is the ancient principle of Archimedes.[10]

If ψ is not constant $\psi(x)$ the total "weight" of the body could be calculated by a double integrating of dF_x from E to D (Fig. 16.5). However, an alternative method is the sum of horizontal layers:

---

[10]In the *Essay* says "the variable force ψ" which we think is a lapse for "the force ψ", as expressed in the *Manuscript* [*Mss*.27].

Fig. 16.6 Body surrounded by a channel

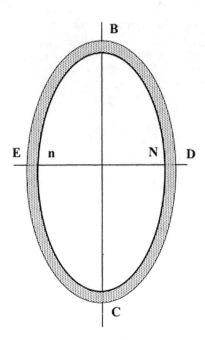

$$W = \rho \int_{x_B}^{x_C} c(x)\psi(x)dx \qquad (16.5)$$

This is easy to prove because $\iint \psi(x)dxdy = \int_{x_B}^{x_C} dx \int_0^{c(x)} \psi(x)dy = \int_{x_B}^{x_C} c(x)\psi(x)dx$, with $c(x)$ being the contour of the body.

There are two applications for a body surrounded by a channel (Fig. 16.6). The wording of the first one [§.24] would be more comprehensible with the insertion of three sentences picked out from the Mss. 28. The body, whose maximum width at nN is b, is surrounded by the channel BDCE and subjected to any force in the upper part that is EBD; while there is no force at all in the lower part, ECD. The result of the upper forces will be a pressure at E or D, said $p_E = p_D$, that will be transmitted to the entire lower channel, because of the absence of forces. Therefore it results an upwards force $F = \rho p_E b$.

The next one [§.25] is based on the same configuration, but superimposing two force fields. One, ψ, constant and vertical, acts upon the entire channel; whose results is $\rho\psi \int yds$, being $\int yds$ the surface of the body. The other, a variable tangential force π acts only upon the upper part, resulting in a force on the body of $\rho \int dy \int \pi ds$, and also a pressure as has been said in the previous case at D and E of $\rho \int \pi ds$, which generates an upward force of $\rho b \int \pi ds$. The final result will be

$$F_x = \rho\psi \int yds + \rho \int dy \int \pi ds - \rho b \int \pi ds \qquad (16.6)$$

The *Mss.* 29 says that this corollary will be very useful later for defining the resistance of fluids.

Finally, everything said so far for a symmetrical and plane body can be transformed for an axisymmetric one with some minor mathematical changes [§.26].

16.4 Motion in Tubes

One more general principle is required to complete the fundamentals of the pressure in fluids: it consists of the change in pressure along a streamline, i.e. the Bernoulli theorem or an equivalent. This subject had been already addressed by D'Alembert in the *Traité de l'équilibre*.

For the basic approach [§.27], let us take a very narrow pipe (Fig. 16.7) whose part *FABG* has a constant section that increases from A to D, or at least changes. Through the pipe a homogeneous and weightless fluid flows with the constant velocity in any slice. This occurs because *PM* is assumed to be very small and is also due to the adherence of the particles in virtue of their viscosity. In other words, it is a unidimensional motion.

The fluid velocity when passing from the point P to p, separated by dx, will change from v to $v + dv$. Therefore applying the fundamental principle of motion, the slice *PMmp* would be in equilibrium if it was impelled by the accelerative force $\varphi = du/dt$. If the pressure at *PM* is p, the force acting on the slice will be Sdp, being $S(x)$ the section of the tube, and its mass $\rho S dx$. Therefore, the equilibrium equation will be:

$$Sdp + \rho S dx \frac{du}{dt} = 0 \qquad (16.7)$$

Taking the section *AB*, where the change of section starts, as reference for continuity we have $Su = S_0 u_0$, and the solution of former equation will be:

$$p - p_0 = \frac{1}{2}\rho(u_0^2 - u^2) = \frac{1}{2}\rho u_0^2\left(1 - \frac{S_0^2}{S^2}\right) \qquad (16.8)$$

We must notice that d'Alembert introduces neither the section nor the density, assuming them as unity; and also that $p - p_0$ is indicated by the single variable P, which means that $p_0 = 0$ at *AB*.

Fig. 16.7 Fluid motion in narrow tube

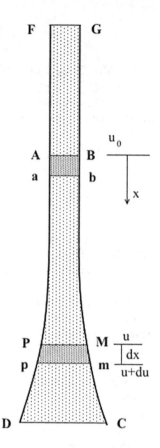

From the above formula it is easy to see that the pressure increases when the area does. Then, if the section of the tube decreased (Fig. 16.8),[11] consequently the pressure would do so likewise, which would lead to negative values. To avoid that, in a rather confused explanation [§.30], d'Alembert takes the pressure as null at *QN* and maximum at *AN*, using Eq. 16.8 to find the values at any intermediate point. However this leads to another problem, because if there is any pressure force on *AB* upwards along *AF*, there should be another one from *F* to *A*; but as the fluid moves uniformly from *F* to *A*, no force can exist along *FA*.

To solve the question he imagines that the length of the tube *AFBG* is infinity, and that the pressure at *AB* is supported by the continuous impact of the fluid mass above it. If such a length, called l, was finite, this mass would be $\rho S_0 l$ and its momentum $\rho S_0 l u_0$; in the impact with the mass *AQNB* the former velocity u_0 will change to v, so that $\rho S_0 l(v - u_0) = \rho v \int \frac{S_0}{S} dx$. Therefore $v = \frac{S_0 l}{l + \int \frac{S_0}{S} dx} u_0$, and $v = u_0$

---

[11]The narrative description given in the *Essay* for this figure does not correspond at all with the depicted as Fig. 3.6 in the *Essay* [§.30]. The one we show here is similar to that given in the *Manuscript* as Fig. 12.

Fig. 16.8 Tube narrowing

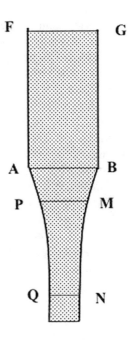

would takes place only if $l=\infty$. Obviously this a misleading reasoning, because a power is always required to maintain the flow, as pressures, velocities and powers are physically related.

Now, d'Alembert develops three variations of Eq. 16.8. In the first, the assumption of the constancy of u_0 is eliminated, which leads to a non-stationary motion [§.28]. In this case, du is obtained from the continuity equation as $d(uS)=d(u_0S_0)$, giving $du = \frac{S_0}{S} du_0 - \frac{u_0 S_0}{S^2} dS$, which introduced in Eq. 16.7, and after making some operations, leads to:

$$\left(p+\frac{1}{2}\rho u^2\right) - \left(p_0 + \frac{1}{2}\rho u_0^2\right) = -\rho S_0 \frac{du_0}{dt} \int_{x_0}^{x} \frac{dx}{S} \quad (16.9)$$

The second is the introduction of the gravity, both when u_0 is constant or variable [§.29]. The equation will be now:

$$\left(p+\frac{1}{2}\rho u^2 + \rho g(x+l)\right) - \left(p_0 + \frac{1}{2}\rho u_0^2 + \rho g l\right)$$
$$= -\left(S_0 \int_{x_0}^{x} \frac{dx}{S} + l\right)\rho\frac{du_0}{dt} \quad (16.10)$$

Where the length *l* of the part *FA* is included. We have jointed the terms in brackets to show the similitude with the Bernoulli equation.[12] If the tube is inclined an angle θ with the horizontal, g must be changed to $g\cos\theta$ [§.31].

Taking the velocity as $u = \kappa u_0$ the terms involving pressure and velocity can be expressed as $p - p_0 = \frac{1}{2}\rho u^2(1-\kappa^2)$ [§.32–33]. Besides, the velocity can be related to the height h by $u^2 = 2gh$, therefore the pressure can be written as $p - p_0 = gh(1-\kappa^2)$. Then, "the pressure would be the same as that of a column of stagnant fluid of gravity g and height $h(1-\kappa^2)$. By this it is seen that the formula found here for the amount of pressure can be used for registering and comparing easily known pressures" [§.33]. This is a method for measuring the pressures, irrespectively of how they have been produced.[13]

The third is assuming that the density is variable [§.34]. There are only a few lines dedicated to it, but it deserves some attention. In the tube, both section and density are related to their values at the section *AB* by means of two non-dimensional parameters $\kappa = \frac{S_0}{S}$ and $\sigma = \frac{\rho_0}{\rho}$. Consequently $\kappa\sigma = \frac{u}{u_0}$ and $du = u_0 d(\kappa\sigma) + \kappa\sigma du_0$, which introduced in Eq. 16.7 leads to the equation:

$$p - p_0 = -\rho_0 u_0^2 \int \kappa d(\kappa\sigma) - \rho_0 \frac{du_0}{dt} \int \kappa dx \qquad (16.11)$$

It is clear that $\kappa(x)$ is given by the geometry of the pipe, but this does not happen with the density $\sigma(x)$, which will be an additional unknown.

---

[12] See Annex I.

[13] As we have mention previously, Guilbaud analyzes the antecedents of this formula in the *Traité de l'équilibre*.

Chapter 17
Resistance of a Body Moving in a Fluid

This is the core of the *Essay* and Chaps 4 and 5 deal with it. We prefer to take this matter as a whole, leaving for a posterior analysis the parts of these chapters not clearly connected to the subject. This manner of dealing with it tries focus on the more essential items, which in our opinion would fully justify the *Essay*. As we have explained before, d'Alembert attacks the problem in three steps: moving fluid, impulsive fluid and moving body. We consider see each step in detail.

17.1 Fluid in Motion

The model used is depicted in Fig. 17.1. An axisymmetric solid body *ADCE* is placed in a fluid stream flowing uniformly from *Q* to *H* [§.36]. The fluid, which is homogenous and weightless, unlimited or confined in a vessel, will exert an action upon the body, which is maintained at rest by any external means, for example a power pushing upstream. The goal is to find out the fluid pressure upon the body.

But before going into the matter, and in order to understand the subsequent propositions, d'Alembert presents five necessary observations.

1st. Particle trajectories. Upstream all fluid particles move following parallel lines, shown as T, P_1, P_2, P_z in Fig. 17.1. However the presence of the body makes them change their direction gradually at some distance from the body, F, K_1, K_2, etc., and makes them describe the curves FD, K_1S_1, K_2S_2, etc. The curvature of these lines is greater the closer they are to the body, and at some distance the lines become straight as if there were no body, as in line P_zS_z. This means that the effect of the body is assumed to be confined to a finite dominion of the fluid.

2nd. Steadiness of the velocity field. Excluding any accelerative forces upon the fluid, the motion is steady state, so that the curve pattern is always the same. According to our present terminology we call these curves streamlines or

Fig. 17.1 Body in a fluid in motion

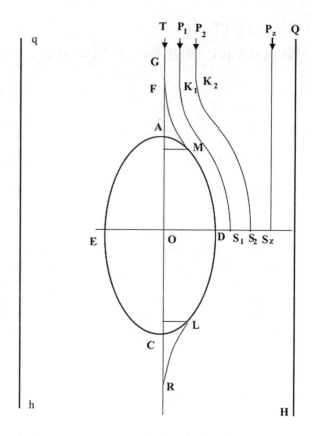

trajectories. Although both are conceptually different, they coincide when the motion is in a steady state.[1]

3rd. Stagnation zone. " Each body in motion that changes direction only changes by imperceptible degrees". This is the ancient and well known aphorism *nature non facit saltus* [*nature does not make leaps*] that was raised and named as principle of continuity by Leibniz, and which took on real meaning within differential calculus. Then the particles in the axis *TF* should must change their direction abruptly by 90° when arriving at the apex *A*, contradicting the former statement. Therefore in order to avoid this kind of turn, they leave the axis at a point *F* and touch the body at *M*; from here they will slide over the body surface until point *L*, where they will separate in an inverse process to reach the axis again at *R*. Therefore two particular

---

[1] A trajectory is the path traversed for a particle along the time, which is represented by the equation $\vec{x} = \vec{x}(t)$ depending on the initial conditions. The results are $x(t)$ and $y(t)$, which represents the trajectory in a parametric form. A streamline is the geometrical place of the tangent to the velocity in an instant. Its equation is $y = y(x, x_0)$ is determined by $y' = v_x/v_y$, also depending on an initial parameter for each one. In steady state condition at any point the trajectory is always tangent to the velocity, therefore it will coincide with the streamline.

spaces will be formed, the *FAM* in the front and the *LCR* in the back, "where the fluid is necessarily stagnant".

4th. Pressure generation. In the motion the velocity of fluid particles changes either the direction or the magnitude, or both at the same time. The difference between the velocity at any instant and the next one will be the accelerative force. Then "the fluid pressure would be the same as if it were stagnant and its parts were subjected to the [former] accelerative force"; and consequently the body can only undergo the pressure that comes from these forces.

5th. Pressure calculation. The force upon the body, which is called pressure as we have seen, depends on the accelerative force along the channel *TFMD*, therefore the problem is reduced to finding both the curvature of this channel and the forces in it.

17.1.1 Stagnation Zone

The existence of the stagnation zone brings some additional problems. Let us look at this area in more detail. It is limited by a streamline which leaves the axis tangentially at the point F (Fig. 17.2) and reaches the body also tangentially at M, hence it has to be curved. To prove "that no pressure can result from the particles

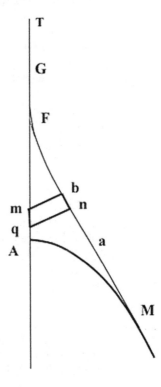

Fig. 17.2 Stagnation zone

contained in the part *FM*" [§.36 last by one], he assumes firstly that no accelerative force acts along *FM*; or what comes to the same thing, the velocity must be constant. If it was not, in the channel *bnqm*, there would be some pressure from *b* to *n*, which should be compensated by pressures on the others sides, "but it has been proved that the fluid is stagnant in the space *FAM*" and obviously in the channel as well. We must note that the only proof shown was a supposition in order to avoid a finite turn at *A*. With respect to the normal forces, whose effect is to curve the trajectory, it is true that they will not produce any pressure along *FM*, as they will be "perpendicular to the channel walls"; however, they will affect the sides *bm* and *nq*, altering the hypothetical equilibrium. Furthermore, the line *FM*, along which the particles are moving, is contiguous with a quiescent fluid, which implies a discontinuity in the velocities; hence the stillness of the inside fluid seems very difficult to understand. In fact this construction reminds us rather too much of Newton's cataract, so much criticized by many authors and even by Newton himself in the Introduction [Intro. II].

Nevertheless after this previous reasoning, d'Alembert surprisingly presents two alternatives for the velocity along *FM* [§.36 last]. One, where it is constant and finite as stated above; the other, where it can be variable, moreover in this case the velocity must be infinitely small, so that it can be taken as null. "We will make clear later that it is the second case which takes place here", which means that he discards all the previous arguments leaving them with a sense of inconsistency. The aforementioned clarification will appear later in [§.52]. Following his reasoning, if the velocity along *FaM* is null, this means that the particles undergo a deceleration along the axis from their original value upstream at *T* up to zero at *F*. The non-answer question is which accelerative force is the responsible for this.

Looking at the *Manuscript*, we found that the *Essay* [§.36] corresponds almost equally to the *Mss*.39, with the exception of the paragraph referring to the smallness of the velocity. In the *Manuscript* the hypothesis of the constancy of the velocity along *TFM* is fully developed. "We have supposed for more generality that the velocity, though uniform in channel *FaM*, is not necessarily uniform in the *TF* part, or necessarily equal to the uniform fluid velocities at various points in the *TG* line".[2] According to this (Fig. 17.4), the velocity v_T at *T* starts changing at *G* to reach v_F at *F*, which is maintained until *N*. Nevertheless, the paragraph ends by affirming that $v_T = v_F$, which will be proved in *Mss*.52. This proof is really given in the previous *Mss*.51. We must note that between *Mss*.39 and *Mss*.51 the development of the velocity equations and their relations takes place. This means that the true nature of this zone will become clear later, and d'Alembert presents an advance of it in *Mss*.39. In *Mss*.51, he makes use of the streamlines structure of Fig. 17.4 in order to show that the point *G* must coincide with *F* and *B* with *K*, based on that in the channel *FNMK* the force in the side *FK* is null. He starts by assuming that the

---

[2]"Supposuimus autem majoris generalitatis causa velocitatem in canali FaM, licet uniformen, non tamen necessario uniformem esse in parte TF, seu necessario æqualem velocitati uniformi fluidi in variis punctis lineæ Tg" [*Mss*. 39-5th].

velocity at G, besides being equal to the one at T, reduces its magnitude, but maintains its direction until F, where it turns in the angle ε. There are three different reasons to prove that ε is an infinitely small angle of the second degree. Firstly, a finite angle would imply that the curvature should change by an instantaneous jump (*per saltum*). Secondly, if the angle would be infinitely small but of first order, i.e. a tangent point, a pressure along the channel FK would be required, but $TPKF$ is in equilibrium. Thirdly, that any force that were to exist on FK, would be governed by the equation $\frac{\partial p}{\partial x} = \frac{\partial q}{\partial z} + \lambda$ instead of $\frac{\partial p}{\partial x} = \frac{\partial q}{\partial z}$, but this is not possible as operating the equations $\lambda = 0$ is obtained. Besides, the points G and F must coincide, because the width of $TPFK$ must be constant and the velocity as well [*Mss*.52]. We found all these arguments rather weak, and probably this is why d'Alembert did not include these articles in the *Essay*. He only included the final results and also recovered the arguments dedicated to the former λ case, but in a different context [§.49].

Coming back to the *Essay* [§.36-last], it seems that d'Alembert was satisfied with the conclusion of all the previous arguments that lead to $v_T = v_F$. According to this, all fluid particles will have this velocity, except those inside the stagnation zone; consequently the curve FaM (Fig. 17.2) would be a discontinuity line between u and zero. Probably, later on when he analyses the impulsive motion, he would find the need of reducing the size of this area, taking it to be almost null [§.52–53]. Consequently he came back and added the new clarification. We advance his new concerns: "Now it would be shocking that while the particles contained in the *FAM* space stop suddenly, the particles that are on the curve FaM, which is the limit of this space, have a velocity not infinitely small, since nothing is done in nature by *leaps*, but by insensible degrees" [§.52]. Obviously, the solution was this reduction. To modern eyes, this is closer to our present understanding of the phenomena. This solution was not introduced in the rest of the *Essay*, which maintains the large stagnation zones following the *Manuscript*. The size of the stagnation zone has no effect on the derivation of the velocity equations, but it is relevant in the calculations of the total force upon the body.

17.1.2 Streamline Field Invariance

D'Alembert clearly establishes that whatever the upstream velocity and fluid density are, the set of curves followed by the particles are always the same [§.39]. In other words, the streamline configuration or field is an invariant; or in broader terms this means the uniqueness of the solution. As we have explained, as the motion is a steady state, the streamlines are also trajectories.

This will be proven in two premises. First, "that it can be assumed that each of these curves is always the same"; second, "that they must be necessarily supposed as such". That is to say, if the uniqueness of the field can be a solution, this must be it.

For the first premise two proofs are presented. One is based on the flow continuity [§.39-I-1st], as the fluid is supposed to run between a streamline and the neighbouring one, changing the velocity according to the separation between them. In in general way let express each curve in parametric coordinates as $x = x(s, \zeta)$ and $z = z(s, \zeta)$, ζ being a parameter for each of them and s the distance traversed; for example, taking the plane at T as reference (Fig. 17.1), ζ could be the ordinate for each streamline. The separation between the curves ζ and $\zeta + d\zeta$ would be an expression like $dw = w(s, \zeta)d\zeta$ and consequently the relation between the velocity at T and at any other point will be $w(s_T, \zeta)v_T = w(s, \zeta)v$, so at M it would be $v_M = W(s_M, \zeta)v_T$. Next, let us take another velocity at T as $v_{TA} = kv_T$ which would generate a new set of curves $x_A(s, \zeta)$ and $z_A(s, \zeta)$, which would lead to $v = W_A(s, \zeta)v_{TA}$. Therefore, the velocity at point M it would be $v_{MA} = W_A(s_M, \zeta)kv_T$. Now let us also assume that $v_{MA} = kv_M$ which would lead to $W_A(s, \zeta) = W(s, \zeta)$: this is the same set in both cases, which means that d'Alembert's first premise is legitimate.

The second proof concerns the accelerative forces [§.39-I-2nd] that act upon all particles and "that must be destroyed" by the forces π to maintain the equilibrium among them. This means $\vec{\pi} = \rho d\vec{v}/dt$, according to the wording of the text even though this formula is not written. The acceleration $\vec{\gamma} = d\vec{v}/dt$ can also be expressed in terms of the geometry of the streamlines. This is a bit more complicated because the normal and tangential components exist for both acceleration and force, as he had just recognized in the previous articles [§.37–38]. At any point the normal acceleration would be $\gamma_n = \dfrac{v^2}{R_C(s, \zeta)} = \dfrac{W^2(s, \zeta)}{R_C(s, \zeta)}v_T^2$ and the tangential $\dfrac{dv}{dt} = \dfrac{dv}{ds}\dfrac{ds}{dt} = W(s, \zeta)\dfrac{dW(s, \zeta)}{ds}v_T^2$; therefore in general we would have $\vec{\pi} = \rho v_T^2 \vec{\Gamma}(s, \zeta)$. Now, working as in the first step, on changing both density and velocity to $\rho_A = h\rho$ and $v_A = kv$, the configuration $\vec{\Gamma}_A$ would be equal to the $\vec{\Gamma}$ if $\vec{\pi}_A$ is assumed equal to $hk^2 \vec{\pi}$. We warn that in the text g is used instead of k^2, which would not make any difference, but to say that for a velocity gv the accelerative force becomes $g\gamma$ is not correct, it must be $g^2\gamma$ as he will find later in [§.43]. We only add that he never mentioned that the forces needed to move the fluid or to maintain the body at rest, must be also multiplied by the factor g or k^2.

The second premise was that "since the particles of the fluid *can* always describe the same curves in the two cases,... Therefore they really *must* describe them" [§39-II]. However, the only argument in favour of this is that the "reasoning is completely analogous" to the motion in the vacuum under a Newtonian gravitational force that leads to a unique conic section, ellipse, parabola or hyperbola; a reasoning "which is accepted by all geometricians". Nevertheless, there is a rather big difference between this and the gravitation case. This is backed by a theory substantiated in mathematical formulations that support the uniqueness of the solutions; something which is lacking, at least until now, in the fluid motions. We think that if anything, he could have applied as a justification here the propositions of the uniqueness of trajectories obtained for the body motions in systems [§.3–6].

Finally, the set of curves "neither depend on the fluid density nor on the body mass, but only on the figure and volume of the body". Also the rate v/v_T at any point

17 Resistance of a Body Moving in a Fluid

will be always the same; then the two components v_x and v_z along the axis X and Z can be expressed by two functions q and p as:

$$v_x(x,z) = v_T q(x,z); \quad v_z(x,z) = v_T p(x,z) \tag{17.1}$$

This two non-dimensional functions q and p will completely define the problem, which will depend only on the body shape [§.40–42].

The problem would be reduced therefore to a search for these two functions. However q and p are not completely independent of each other, but they are linked through some relations. D'Alembert's next task will be to find these relations. It is here where his talent shines to most advantage, and where his major contribution to fluid mechanics lies, without forgetting the establishment of the invariance of the fluid field and the separation of the velocity into its two components. Nevertheless, the way in which he approaches the question is sometimes not easy to follow. Although the idea is brilliant the explanations are somewhat confused, with frequent changes in the meaning of the symbols, lengthy and tortuous paths, and everything aggravated by misprints. We will try to present his reasoning as clearly as possible, introducing modern terminology in order to make easier reading.[3]

17.1.3 Velocity Field Equations

The first thing that d'Alembert carries out is the kinetic analysis of the motion of a fluid particle along a streamline. The particle comes from the point F in the instant $t - dt$, to reach N in t, and afterwards m in $t + dt$ (Fig. 17.3).

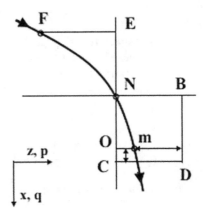

Fig. 17.3 Particle trajectory

---

[3]We have to remember that the typographical recourses available then were limited, which obliged him to repeat the symbols frequently. Years later the aids of the sub index would make the tasks clearer and easier. An example is given in the next note.

The variations of both q and p can be expressed as [§.43][4]:

$$dq = \frac{\partial q}{\partial x}dx + \frac{\partial q}{\partial z}dz; \quad dp = \frac{\partial p}{\partial x}dx + \frac{\partial p}{\partial z}dz \qquad (17.2a, b)$$

Therefore, the accelerations will be:

$$\gamma_x = \frac{dv_x}{dt} = v_T\left(\frac{\partial q\,dx}{\partial x\,dt} + \frac{\partial q\,dz}{\partial z\,dt}\right) = v_T^2\left(q\frac{\partial q}{\partial x} + p\frac{\partial q}{\partial z}\right) \qquad (17.3a)$$

$$\gamma_z = \frac{dv_z}{dt} = v_T\left(\frac{\partial p\,dx}{\partial x\,dt} + \frac{\partial p\,dz}{\partial z\,dt}\right) = v_T^2\left(q\frac{\partial p}{\partial x} + p\frac{\partial p}{\partial z}\right) \qquad (17.3b)$$

In the *Essay* they both have negative signs, because they are taken as accelerative forces, or forces that must be destroyed. This analytical treatment of the fluid along a streamline is carried out here for the first time, and it allows the finding of the subsequent relations in partial derivatives.[5]

These equations are obtained in the *Essay* by a geometrical approach based on Fig. 17.3 [§.43–44]. The forces are $(FE - Om)/dt^2$ and $(NE - NO)/dt^2$ along the axis OZ and OX respectively. The velocities at N are v_Fq and v_Fp; and at F they were the result of subtracting from the former the differential part due to the segments FE and EN; applying Eq. 17.2a, b and considering that the slope at N is q/p.

Now we come to the axisymmetric body, Fig. 17.4, and the fluid layer infinitely near the surface that surrounds it. This layer starts upstream as a circle of radius w_T and becomes a conical ring around the body, whose width at the point N is $w \cos \gamma$. Assuming that w is very small respect to the radius y, the continuity condition leads to the equation [§.45-1st-2nd]:

$$\pi w_T^2 v_T = 2\pi z w v \cos \gamma = 2\pi z v_T q w \qquad (17.4)$$

Now the streamline slope at M is obtained by two different ways: one uses the velocities, the other the variation of the channel width. The calculation is made mixing kinematic and geometry and it is a bit cumbersome [§.45-3rd].

In the first one, $v_T q$ and $v_T p$ being the components of the velocity at N, the values at M are:

$$v_x = v_T\left(q + \frac{\partial q}{\partial z}w\right); \quad v_z = v_T\left(p + \frac{\partial p}{\partial z}w\right) \qquad (17.5)$$

Introducing the value of $w = \dfrac{w_T^2}{2qz}$ from Eq. 17.4, the slope will be

---

[4]In the *Essay* these equations are expressed as $dq = Adx + Bdz$ and $dp = A'dx + B'dz$, that is $\frac{\partial q}{\partial x} = A$, $\frac{\partial q}{\partial z} = B$, $\frac{\partial p}{\partial x} = A'$ and $\frac{\partial p}{\partial z} = B'$.

[5]Rouse, *History of Hydraulics*, p. 102 and Grimberg [1998], p. 42–43.

17 Resistance of a Body Moving in a Fluid

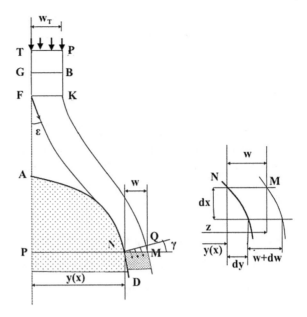

Fig. 17.4 Flow around an axisymmetric body

$$\Gamma_M = p + \frac{\frac{w_T^2}{2qz}\frac{\partial p}{\partial z}}{q + \frac{w_T^2}{2qz}\frac{\partial q}{\partial z}} \tag{17.6}$$

For the second, the ordinate of point M is $z = y + w$, being $y = y(x)$ the contour of the body. Then the slope will be $\Gamma_M = \frac{dy}{dx} + \frac{dw}{dx}$, but $\frac{dy}{dx} = \frac{p}{q}$, then:

$$\begin{aligned}\Gamma_M &= \frac{p}{q} + \frac{d}{dx}\left(\frac{w_T^2}{2zq}\right) \\ &= \frac{p}{q} - \frac{w_T^2}{2}\left(\frac{1}{z^2q}\frac{dz}{dx} + \frac{1}{zq^2}\frac{dq}{dx}\right) \\ &= \frac{p}{q} - \frac{w_T^2}{2zq^2}\frac{\partial q}{\partial x} - \frac{w_T^2}{2}\left(\frac{1}{z^2q}\frac{\partial q}{\partial z} + \frac{1}{zq^2}\right)\frac{p}{q}\end{aligned} \tag{17.7}$$

Equating both, reducing and neglecting the terms in w_T^4, we arrive at:

$$-\frac{\partial p}{\partial z} = \frac{\partial q}{\partial x} + \frac{p}{z} \tag{17.8}$$

This is the first velocity equation.

If the former equation is backed by the kinematics, the second one will be backed by the forces. The pressures at points N and M can be deduced by means of the equation Eq. 16.8, or Bernoulli, applied to the channels *TFN* and *PKM*, knowing that the conditions for the upstream T and P are the same [§.45-4th]. So, we have:

$$p_M = \frac{1}{2}\rho(v_T^2 - v_M^2); \quad p_N = \frac{1}{2}\rho(v_T^2 - v_N^2) \tag{17.9}$$

Therefore:

$$\Delta p = p_N - p_M = \frac{1}{2}\rho(v_M^2 - v_N^2) \tag{17.10a}$$

At N the total velocity is defined by the components q and p as $v_N^2 = (q^2 + p^2)v_T^2$ and for M the q and p from Eq. 17.5. After the corresponding operations, and neglecting the w^2 terms, the above formula becomes:

$$\Delta p = \rho v_T^2 \left(p\frac{\partial p}{\partial z} + q\frac{\partial q}{\partial z}\right)w \tag{17.10b}$$

But we have found that in the channel NM exists and an accelerative force $\gamma_z = v_T^2\left(q\frac{\partial p}{\partial x} + p\frac{\partial p}{\partial z}\right)$, (Eq. 17.3b), which produces the force $\rho\gamma_z w$ that must be in equilibrium. Therefore, equating both:

$$\frac{\partial p}{\partial x} = \frac{\partial q}{\partial z} \tag{17.11}$$

This is the second velocity equation that, jointly with the Eq. 17.8, will define the relation between the velocity and space.

With the aid of Eq. 17.8 the Eq. 17.2b can be rewritten as:

$$dp = \frac{\partial q}{\partial z}dx - \frac{\partial q}{\partial x}dz - \frac{pdz}{z} \tag{17.12a}$$

$$d(pz) = z\frac{\partial q}{\partial z}dx - z\frac{\partial q}{\partial x}dz \tag{17.12b}$$

This one and the Eq. 17.2a have to be exact differentials [§.46]. We can see that in both formulas there are only the derivatives of q, as $\partial q/\partial x$ and $\partial q/\partial z$.

The equations and their exact differentials are valid for the entire fluid, which is more clearly explained in the *Mss.* 49.[6] In the *Essay* [§.47], he notes that the values for q and p meet these conditions in the upper part of FM and adjacent curves, for MD and also its neighbouring curves. However the curves FM and MD do not belong to the same equation. This is a consequence of the different nature of both

---

[6]"In demonstration præcedenti eum tantum fluidi rivulum contemplati sumus, qui superficiem corporis MD immediate tangit. Sed patet æquationes [Eq. 17.12a], et [Eq. 17.12b] eodem præcise ratiotinio inveniri posse pro quovis alio rivulo, existente generatim z distantia puncti cujus vis M' ab exe AP, sive punctum illud sit superficies MD, sive non." *Mss.*49. "In the preceding demonstration we contemplated the fluid thread that immediately touched the body surface MD. However, it is obvious that the equations [Eq. 17.12a] and [Eq. 17.12b] can be found by the same reasoning used for any other thread…whose point M' is at the MD surface or not."

Fig. 17.5 Volume fluid in motion

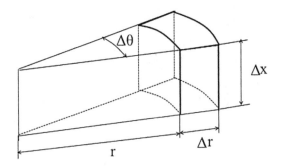

curves: *MD* is clearly defined by the body geometry, while *FD* is only the limit of the stagnation zone, conjectured but not defined. He might have applied here the hypothesis of the smallness of this zone.

Before continuing d'Alembert proposes obtaining the former equations by another method: "somewhat more general than the previous one" [§.48]. The difference is that in the previous method the calculations were made in the fluid layer nearest to the body, which involved the body as a component in the process, while now the calculations will be made in the fluid layer farthest from the body with the only condition of axisymmetric motion. The differential element taken is now in the midst of the fluid flow and isolated and separated from the body. It consists of a fraction of a rectangular ring of radius z, with a width Δz and height Δx, limited by two meridian planes forming an angle $\Delta\theta$; which can be considered almost as a parallelepiped (Fig. 17.5). This element will evolve in the fluid motion, changing its position but maintaining its volume, except for differences of high order. This idea will be used extensively by succeeding authors.

The initial volume of the differential element is $\Delta V = \Delta x \Delta z (z\Delta\theta)$ and after an interval of time dt the previous dimensions will change in factors of $1 + v_F(\partial q/\partial x)dt$, $1 + v_F(\partial p/\partial z)dt$ and $1 + v_F(q/z)dt$. By the constancy of the volume, in dt it will vary in:

$$\Delta x \Delta z (z\Delta\theta) = z\left(1 + v_T \frac{\partial q}{\partial x} dt\right)\left(1 + v_T \frac{\partial p}{\partial z} dt\right)\left(1 + v_T \frac{p}{z} dt\right)\Delta\theta \qquad (17.13)$$

Calculating and eliminating the higher order terms, the expression already known Eq. 17.8 is found again. Basically, there is a strong parallel with the former procedure, since in both the continuity is the basis of the reasoning; here as a differential element, there as layer of fluid. In both cases it is a kinetic construction.

In order to obtain the other equation, d'Alembert reduces the motion to a hydrostatic equilibrium by applying the general principle considering the accelerative forces γ_x and γ_z as a force field, because "these forces must destroy themselves" [§.48]. Then, applying the condition of hydrostatic equilibrium (Eq. 16.2) to these forces, the result would be:

$$\frac{\partial \gamma_x}{\partial z} = \frac{\partial \gamma_z}{\partial x} \qquad (17.14)$$

Considering Eq. 17.3a and 17.3b:

$$\frac{\partial}{\partial z}\left(q\frac{\partial q}{\partial x}+p\frac{\partial q}{\partial z}\right) = \frac{\partial}{\partial x}\left(q\frac{\partial p}{\partial x}+p\frac{\partial p}{\partial z}\right) \qquad (17.15)$$

Operating, we attain:

$$\frac{\partial q}{\partial z}\frac{\partial q}{\partial x}+q\frac{\partial^2 q}{\partial x \partial z}+\frac{\partial p}{\partial z}\frac{\partial q}{\partial z}+p\frac{\partial^2 q}{\partial z^2} = \frac{\partial p}{\partial z}\frac{\partial p}{\partial x}+p\frac{\partial^2 p}{\partial x \partial z}+\frac{\partial q}{\partial x}\frac{\partial p}{\partial x}$$
$$+q\frac{\partial^2 p}{\partial x^2} \qquad (17.16)$$

Now, he affirms that this equation will occur if the two velocity equations Eq. 17.8 and Eq. 17.11 are met. Taking the two expressions for the velocity Eq. 17.1 as exact differentials and applying Eq. 17.11 to them, he finds

$$\frac{\partial^2 p}{\partial x^2}=\frac{\partial^2 q}{\partial z \partial x}; \quad \frac{\partial^2 q}{\partial z^2}=\frac{\partial^2 p}{\partial x \partial z} \qquad (17.17)$$

We think that an easy way to arrive at these expression would be to derivate Eq. 17.11 directly with respect to x and z. In any case, the Eq. 17.15 is simplified to:

$$\frac{\partial q}{\partial z}\frac{\partial q}{\partial x}+\frac{\partial p}{\partial z}\frac{\partial q}{\partial z}=\frac{\partial p}{\partial z}\frac{\partial p}{\partial x}+\frac{\partial q}{\partial x}\frac{\partial p}{\partial x} \qquad (17.18)$$

But in accordance with Eq. 17.1 and Eq. 17.11 both members of this equation are equal to $-\frac{p}{z}\frac{\partial q}{\partial z}$, "therefore the two quantities $\partial \gamma_x/\partial z$ and $\partial \gamma_z/\partial x$ are actually equal". [§.48].[7]

Summarizing, the equality of the cross derivatives of the acceleration is proven, assuming the existence of the two velocity relations that are employed several times in the developments, which we add is a quite elaborate and intricate procedure. This means that these equations are a sufficient condition for the motion, although not a necessary one. That is to say, there will be possible motions that do not follow these hypotheses.

---

[7]Truesdell [1954], p. LIV, considered that this step is false because the d'Alembert principle only can be applied when the solution is known by other methods. He found that the solution given by Eq. 17.16 is more general that the former Eq. 17.11 and "to reconcile both he has to recourse to a logical fallacy"; Truesdell thinks that this shows the "unreliability of the principle". Grimberg [1998], p. 51, rebuts these arguments stating that the use of the principle is rigorous, and that the error that could be attributed to d'Alembert is to be interpreted in modern terms as saying that the conservation of the acceleration field (Eq. 17.14) is irrotational, which is not always true.

17 Resistance of a Body Moving in a Fluid

Apparently, as a sequel of the previous reasoning [§.49], he establishes a new condition for Eq. 17.14. This condition is a remnant from the *Mss*.51, where, as we have said before, it was one of the proofs for the constancy of the velocity on the *FN* curve. Here it is out of place.

Before going on to the next subject it is worth noting the modern form of d'Alembert's two equations. The first one Eq. 17.8 can be written as div $\vec{v} = 0$ or $\nabla \cdot \vec{v} = 0$,[8] which means the constancy of the fluid contained in an element $\Delta x \Delta z$ in its evolution with time, therefore the equation is also called a continuity. The second equation Eq. 17.11 would be as $rot\ \vec{v} = 0$ or $\nabla \times \vec{v} = 0$,[9] which means that the fluid element does not rotate with time,[10] that is, the motion is irrotational.

In the case of a plane figure instead of a body of revolution [§.73] the equation Eq. 17.11 remains the same but the Eq. 17.8 changes to:

$$-\frac{\partial p}{\partial z} = \frac{\partial q}{\partial x} \qquad (17.19)$$

This case is considered as if the fluid were flowing through a rectangular channel bounded by the walls *qh* and *QH* (Fig. 17.1) and the planes *qhQH*, and another plane parallel to it at some height; the body would be a cylinder whose section was the figure given and which would cover the full height.

In this point we skipped the order followed in the *Essay* jumping until what we estimate as the logical sequel: the determination of the functions *q* and *p*.

17.1.4 Determination of the Functions q and p

D'Alembert was not fully satisfied at having established the equations of *p* and *q*, so he also tried to find a method to calculate them in practical cases. He knew that this was a very difficult task, so he had the brilliant idea of drawing a bridge from here up to a class of functions of complex variables, transforming the problem from real to complex variables. These jumps from one field to another, opening up new roads, are frequent in great mathematicians. However the difficulties remained and the method of complex functions would not yield fruits until almost 150 year later.

First, he assumes the hypothesis that $dq = Mdx + Ndz$ and $dp = Ndx - Mdz$ [§.57], and he proposes as a new mathematical problem of finding *M* and *N*, being $Mdx + Ndz$ and $Ndx - Mdz$ exact differentials [§.58].

---

[8]The divergence of a vector field means the capacity of sinking or sourcing of a field element.

[9]The rotational means the rotation of one field element.

[10]It is easy to see that the side Δx of any element would turn in dt the angle $\frac{\partial v_z}{\partial x} dt$ and Δz will turn $\frac{\partial v_x}{\partial z} dt$, consequently they must compensate each other.

If the two former expressions are exact differentials; both $Mdx + N\frac{dz}{i}$ and $Nidx - Midz$ are as well.[11] Introducing $dx - idz$ and $dx + idz$ as factors,[12] that they are also exact differentials, and carrying out several linear manipulations, it is found:

$$Mdx + Ndz + i(Ndx - Mdz) = (M + iN)(dx - idz) \qquad (17.20a)$$
$$Mdx + Ndz - i(Ndx - Mdz) = (M - iN)(dx + idz) \qquad (17.20b)$$

Which means:

$$M + iN = \Phi(F + x - iz) \quad M - iN = \Psi(G + x + iz) \qquad (17.21)$$

Where both F and G are constants and Φ and Ψ are any complex functions. Next, remembering that the velocity equations for the two-dimensional case were (Eq. 17.19 and Eq. 17.11):

$$\frac{\partial p}{\partial z} = -\frac{\partial q}{\partial x}; \quad \frac{\partial p}{\partial x} = \frac{\partial q}{\partial z} \qquad (17.22a, b)$$

That means that both $qdx + pdz$ and $pdx - qdz$ are exact differentials [§.59] and therefore applying Eq. 17.21 we have:

$$q + ip = \Phi(F + x - iz) \qquad (17.23a)$$
$$q - ip = \Psi(G + x + iz) \qquad (17.23b)$$

In these expressions we find that the two components of the flow field have been converted into a single complex variable $\boldsymbol{q} = q + ip$, or into its conjugate $\bar{\boldsymbol{q}} = q - ip$, and the *xz-plane* in the complex *z-plane*, where $z = x + iz$ and $\bar{z} = x - iz$. The result is that the flow field now is expressed by a complex function $\boldsymbol{q} = \boldsymbol{q}(z)$. The equations Eq. 17.2a, b are the first occurrence of the Cauchy-Riemann equations.[13] By addition and subtraction of these and with some considerations that end by taking both F and G as zero, the values of p and q can be separated as:

$$p = -i\xi(x - iz) + \zeta(x - iz) + i\xi(x + iz) + \zeta(x + iz) \qquad (17.24a)$$
$$q = i\xi(x - iz) + i\zeta(x - iz) + \xi(x + iz) - \zeta(x + iz) \qquad (17.24b)$$

Now the problem of finding p and q has been transformed into a search for the complex functions $\xi(z)$ and $\zeta(z)$ which must depend exclusively on the geometry of the body. The solution continues to be difficult, and to cope with this d'Alembert assumes "as an example" [§.60] that both ξ and ζ are third degree polynomials with

---

[11] D'Alembert employed $\sqrt{-1}$ to express the imaginary unit, but we will prefer the symbol i. This one was used occasionally by Euler although it becomes of general application with Gauss in the nineteenth century.

[12] In the *Essay* they are written as $dx + \frac{dz}{i}$ and $dx - \frac{dz}{i}$.

[13] Truesdell [1954], p. LV.

unknown coefficients, $\xi = az + bz^2 + cz^3$ and $\zeta = e\bar{z} + f\bar{z}^2 + g\bar{z}^3$. These formulas once introduced in Eq. 17.24a and 17.24b and operating give p and q as polynomials of third degree for p and of second degree for q with five different coefficients.

At this point we would like to highlight that the introduction of complex variable functions was a very noticeable contribution.[14] The development of the complex algebra was still in its early phases and had to wait until the next century for a full understanding. The application of this method to two-dimensional aerodynamics would come at the beginning of the twentieth Century, when Jukowski presented the conformal transformation $w = z + 1/z$ which converted a circle and its already known flow field from the *z-plane* to an airfoil in the *w-plane*. This allowed the understanding of the aerodynamic lift and its relation with the velocity circulation, as established by the theorem of Kutta-Jukowski.

Coming back to the expansion in complex series, he probably found it tedious to deal with the successive development of $(x \pm iz)^n$, but taking advantage of the structure of the polynomials obtained previously, he changes to a direct expansion of powers of x and z which seemed easier for computing [§.61]. So, p would be a polynomial of degree m that must meet the well-known $dq = \frac{\partial q}{\partial x} dx + \frac{\partial q}{\partial z} dz$ and $d(pz) = z \frac{\partial q}{\partial z} dx - z \frac{\partial q}{\partial x} dz$. Therefore:

$$p = \sum_{i+j \leq m} \alpha_{ij} x^i z^j \tag{17.25}$$

In the *Essay* only a third degree polynomial is used, although the formula is ended by an "etc". Identifying $d(pz) = \frac{\partial (pz)}{\partial x} dx + \frac{\partial (pz)}{\partial z} dz$ with the former $d(pz)$ it will give $\frac{\partial q}{\partial x} = -\frac{1}{z} \frac{\partial (pz)}{\partial z}$ and $\frac{\partial q}{\partial z} = \frac{1}{z} \frac{\partial (pz)}{\partial x}$ that introduced in the definition of dq leads to:

$$dq = -\frac{\partial (pz)}{\partial z} \frac{dx}{z} + \frac{\partial (pz)}{\partial x} \frac{dz}{z} \tag{17.26}$$

That expresses dq as function de p and is also an exact differential. Introducing the p from the polynomial expansion in this expression, we reach:

$$dq = dz \sum i \alpha_{ij} x^{i-1} z^j - dx \sum (j+1) \alpha_{ij} x^i z^{j-1} \tag{17.27}$$

The condition of exact differential will be:

$$\sum i(i-1) \alpha_{ij} x^{i-2} z^j \equiv -\sum (j+1)(j-1) \alpha_{ij} x^i z^{j-2} \tag{17.28}$$

This is actually an identity, so it must be accomplished for whatever x and z. Expanding both sides we will have two polynomials, and equalizing the terms of the

---

[14]Morris Kline quotes this case as the first historical contribution of the complex variable functions [Chap. 27], although complex numbers had already been used for solving other mathematical problems.

same power a certain number of linear equations among pairs of the coefficients α_{ij} will be found. At the end we will have for the α_{ij} calculation a recurring formula such as: $(i+2)(i+1)\alpha_{i+2,j} + (j+3)(j+1)\alpha_{i,j+2} = 0$, jointly with $\alpha_{0j} = 0$ and $\alpha_{1j} = 0$. For this, the final series for p will only have a limited number of terms with an even smaller number of independent coefficients. In the case of $m=3$, there are four terms and three unknowns; written as $p = b'z + e'xz + h'x^2z - \frac{1}{4}h'z^3$. In the case of $m=4$ there are seven terms and five unknowns and if $m=5$ there are eleven and eight respectively.

There is no mention for q in the *Essay*, although in the *Mss*.58 it said that is very easy to find it. Presumably going back through Eq. 17.26 and looking for a new polynomial, such as $q = \sum \beta_{ij} x^i z^j$, whose coefficients β_{ij} will be functions of the α_{ij}.

The practical way to calculate the α_{ij} for a given body, whose contour is defined by the equation $y = f(x)$ [§.62], will be to choose a certain number of points (x_k, y_k), in accordance with the unknowns of the polynomial p. In each of them the velocity is tangent, that is $\frac{p_k}{q_k} = \left(\frac{dy}{dx}\right)_k = \eta_k$, therefore we will have a set of equations as $p_k - \eta_k q_k = 0$, each of them an homogenous polynomial in the unknowns α. The solution of this system due to its homogeneousness will only give the relative values of the α, but not the absolute ones, so an additional condition is needed [§.63]. This will come from the equality $\psi = \mu + \Omega - \pi \Gamma z_L^2 = 0$, defined in Eq. 17.42. However, in this formula the positions of the two stagnation zones, points x_M and x_L in Fig. 17.1, intervene directly and also through the calculation of Ω and Γ, which will introduce two more unknowns. Besides, in both points the velocity is zero, that is $\sqrt{p^2 + q^2} = 0$, which would give the position of M and L, and with such positions and $\psi = 0$ the last unknown would be determined. However, we think that this was only wishful thinking on d'Alembert's part, and very likely the problem has no solution by means of this kind of polynomials. What is more, there are more boundary conditions not mentioned in the *Essay*: such as that upstream and downstream the velocity is uniform, which means that when $x = \pm\infty$, then $q=1$ and $p=0$. Besides this is applicable laterally, either at $z = \pm\infty$ or at the channel walls, if these walls exist. It is difficult to believe that d'Alembert did not realise advert these incompatibilities, because he keeps insisting that this solution was correct [§.74].

Nevertheless d'Alembert persists in the searching for a method to solve the problem. The new idea is to concentrate the effort on the fluid thread adjacent to the body [§.64], leaving out the rest of the fluid. On the surface $p = \eta q$ and also $pz = \eta z q$, "z being always the same as y"; which differentiated gives $d(pz) = \eta z dq + q\eta dz + qz d\eta$; introducing $d\eta = \eta_x dx + \eta_z dz$ and $dq = \frac{\partial q}{\partial x} dx + \frac{\partial q}{\partial z} dz$ leads to:

$$d(pz) = \left(\eta z \frac{\partial q}{\partial x} + q z \eta_x\right) dx + \left(q\eta + \eta z \frac{\partial q}{\partial z} + q z \eta_z\right) dz \qquad (17.29)$$

This equation must be an identity, and he insists that identity means not only equality but the equation must be expressed with the same symbols as the already well known $d(pz) = z \frac{\partial q}{\partial z} dx - z \frac{\partial q}{\partial x} dz$. That means the equality of the terms in dx and

17 Resistance of a Body Moving in a Fluid

dz., therefore $\frac{\partial q}{\partial z} = \eta \frac{\partial q}{\partial x} + q\eta_x$ and $-z\frac{\partial q}{\partial x} = q\eta + \eta z\frac{\partial q}{\partial z} + qz\eta_z$. From these two linear equations $\partial q/\partial x$ and $\partial q/\partial z$ are obtained:

$$\frac{\partial q}{\partial x} = -\frac{q}{z}\frac{\eta + z\eta\eta_x + z\eta_z}{1+\eta^2} \tag{17.30a}$$

$$\frac{\partial q}{\partial z} = -\frac{q}{z}\frac{\eta^2 + \eta\eta_z - z\eta_x}{1+\eta^2} \tag{17.30b}$$

Introducing these equations in dq and after a chain of operations, bearing in mind that $p = \eta q$ and the value of $d\eta$, the following equation results:

$$\frac{dq}{q} = -\frac{dz}{z} - \eta_z dx \tag{17.31}$$

This formula is simple and even elegant, and "it seems at first that nothing is easier than determining q by this equation..., since η_z is calculated from η and this is given by the equation of the curve $dz/dx = \eta$" [§.65]. However, d'Alembert warns against this idea because "the ratio dz/dx can be expressed of an infinite numbers of ways,... which are not identical although equal". It is a consequence of the identity requirement, because the same curve can be expressed algebraically by different equations, and each of them will have also different formulations for η. As he points out "η cannot be taken at will". To prove it, two variants of the former Eq. 17.42 will resulted if dx is substituted using $\eta = dz/dx$, or η_z from $d\eta = \eta_x dx + \eta_z dz$. These two are:

$$\frac{dq}{q} = -\frac{dz}{z} - \frac{\eta_z dz}{\eta}; \quad \frac{dq}{q} = -\frac{dz}{z} - \frac{d\eta}{\eta} + \frac{\eta_x dx}{\eta} \tag{17.32a, b}$$

Each one give a different solution for q.

At this point let us quit a moment and look to the η_x and η_z parameters. The body contour has been defined as a function as $z = f(x)$, that can be expressed in a more general way as $C(x, z) = 0$, in which the former is contained. The set C_x, C_z, C_{xx}, C_{xz} and C_{zz} are obtained by derivation as:

$$\eta = -\frac{C_x}{C_z}; \quad \eta_x = -\frac{\eta C_{zx} + C_{xx}}{C_z}; \quad \eta_z = -\frac{\eta C_{zz} + C_{xz}}{C_z} \tag{17.33}$$

The derivatives are expressed by sub-indexes in order to make the writing easier. Obviously for the explicit case $C \equiv f(x) - z = 0$, they are $\eta = f'$, $\eta_x = f''$ and $\eta_z = 0$. The circumference is presented as an example, the function C can adopt several forms as $x^2 + z^2 - a^2 = 0$ or $\sqrt{a^2 - x^2} - z = 0$ or $\sqrt{a^2 - z^2} - x = 0$, all of them different. Consequently, the algebraic expressions of the η-derivatives will be also different for each one, which is the reason for the diversity in the q functions.

The last attempt is to equalize, not to identify, the two former expansions of $d(pz)$, to obtain:

$$\eta z \frac{\partial q}{\partial x} + q\eta_x z + q\eta^2 + \eta^2 z \frac{\partial q}{\partial z} + zq\eta\eta_z = z\frac{\partial q}{\partial z} - \eta z \frac{\partial q}{\partial x} \qquad (17.34)$$

"But as the unknown $\partial q/\partial z$ remains to be determined, this method is perhaps not very useful". That closes his attempts for a solution.

For our part, before leaving this matter, we have to note a failure in this reasoning, which lies in the assumption of the identity of $d(pz)$ and $d(\eta zq)$ functions. The phenomena is stated as limited to the adjacent layer to the body, which implies that both dx and dy are not independent, but linked by the tangent condition, the aforesaid $dz = \eta dx$. It does not make sense to think of $d\eta$ with $dx = 0$ and $dz \neq 0$; rather it should be $d\eta = (\eta_x + \eta\eta_z)dx = z'' dx$, which would affect the subsequent calculations.

17.1.5 Pressure Upon the Body

The pressures upon the body depend on the accelerative forces on its surface. At any point, like D for example, this force has two components: the tangent and the normal, named π and π' respectively. The pressure upon D is assumed as being the sum of π along TD because the action of π' on the point is infinitely small; this means that only the component $\int \pi ds$ is significant [§.37]. But as $\pi = -dv_N/dt$, the question is reduced to finding v_N along the line TD [§.38], and this velocity is related to q and p by Eq. 17.1.

Therefore, once the functions p and q are known for a given body, the next step is to calculate the pressures over its surface and the total force upon it [§.66]. From p and q the velocity at the point N is $v_N = v_T\sqrt{p^2+q^2}$ and by means of the theorem of Bernoulli (Eq. 16.8) the pressure would be: $P - P_T = \frac{1}{2}\rho v_T^2(1 - p^2 - q^2)$. We have included in this formula the density ρ and the pressure P_T upstream where $p = 0$ and $q = 1$; d'Alembert takes it as zero, which is the source of several problems. Thus in the same article he indicates that $\sqrt{p^2+q^2}$ must be less than one along the thread MDL (Fig. 17.1), otherwise negative pressures would appear; but if the calculations lead to $\sqrt{p^2+q^2} > 1$ in some points; then, it will necessary to seek where $\sqrt{p^2+q^2}$ is maximum, which will happen when $pdp + qdq = 0$, and to take this point as reference. Calling it K, and $k^2 = p_K^2 + q_K^2$ we have:

$$P - P_K = \frac{1}{2}\rho v_T^2(k^2 - p^2 - q^2) \qquad (17.35)$$

This can be interpreted either as he accepts negative pressures or that there must be some pressure at infinity. His remark that "some readers may perhaps imagine that the velocity along the thread MDL must be greater than v_T; they can truly base themselves on daily experience, seems to verify that the fluid accelerates when turning around the body" [§.67] causes us some surprise However, if the calculation indicates that $\sqrt{p^2+q^2} < 1$, the explanation is that "in this theory only the thread that immediately touches the body surface is taken; this thread escapes observation, and it may be that threads which are at very little distance from the body have much more velocity than it". It seems just a pretext, we recall that something similar was argued in the flow through a tube [§.30].

The force, which he calls total pressure, is obtained by the integration of $\frac{1}{2}\rho v_T^2 2\pi y(k^2 - p^2 - q^2)dy$ along the surface, this is the reason for using y instead of z [§.68]. This integral will be:

$$\varphi = 2\pi \int_M^L (k^2 - p^2 - q^2) y \, dy + \pi k^2 y_M^2 - \pi k^2 y_L^2$$

$$= -2\pi \int_M^L (p^2 + q^2) y \, dy \qquad (17.36)$$

Where the two stagnation areas are excluded because in them the velocity is null, that is $p = q = 0$. In the case of $\sqrt{p^2 + q^2} < 1$ the solution will be the same.

The value of φ depends only on the body shape and is independent of the velocity, because the points M and L are independent as well. Then the "total pressure" will be $\rho v_T^2 \varphi$ [§.68–69]. Let us observe the similitude with our present formulation of the resistance as $D = \frac{1}{2}\rho v_T^2 C_D$, where C_D is the drag coefficient.

Now we are faced with one of the most commented items in the history of Fluid Dynamics: the paradox of d'Alembert, although it was not known by this name until sometime later [§.70]. The paradox would embarrass scientists for more than a century and a half.[15] For a body with equatorial symmetry (Fig. 17.6), the slope at

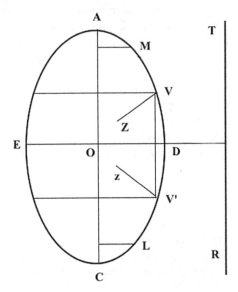

Fig. 17.6 Symmetrical body

---

[15] About this paradox cf. Simón Calero [1996] and Grimberg et al. [2008].

the points V and V' have the same value but different signs; therefore $p_V = p_{V'}$ and $q_V = -q_{V'}$, so the integral element $(p^2+q^2)y\,dy$ at V balances the corresponding element at V'. Then if the arcs LD and DM were equal, the integral "would be equal to zero, so that the body would not undergo any pressure from the fluid, which is against experience". Therefore these arcs cannot be equal, although at first glance anybody could consider both as equal. To quote him "if we stand by the theory alone, we would be moved to think ... that these arcs must be equal in effect". It is clear that d'Alembert is rather surprised by this result and the only way to escape from it is to assume some difference between these arcs. As experience also teaches $LD > DM$, otherwise the pressure would be negative. He reverts to experiment once again: "Hence it is clear how experiments are needed in the present question".

Years later, in his his *Opuscules Mathématiques*,[16] he will come back to this problem under the title of "Paradox proposed to geometricians on the resistance of fluids". There he will avoid the stagnation zones by supposing the body to have a small cone shape at both ends in order to eliminate the possibility of any arbitrary assumptions to explain the vanishing of the resistance: it becomes a true paradox.

This surprising result had appeared before in Fluid Mechanics. Euler had arrived at similar results in his commentaries to the translation of Robins' *Gunnery*. However, Euler had approached the problem using a method derived from the impact theory, while d'Alembert, as we have seen, used the fluid field equations for the first time. D'Alembert requires the equatorial symmetry of the body, while Euler needs certain geometrical conditions at the ends of the body. In spite of their differences, both tried to escape from this embarrassing conclusion, by imagining the existence of fluid zones in which the theory is not satisfied.

The paradox was a matter of discussion during many years. Quite soon after, Borda would touch the subject using the live forces to arrive at a similar solution.[17] Almost a century later, Saint-Venant pointed to the viscosity as being responsible for the resistance[18]; but the final answer came with Ludwig Prandtl, who introduced the boundary layer concept in 1904.[19]

After the mathematical development, d'Alembert ponders on his solution of which he is rather proud [§.74]. "It seems to me [it is] based on principles less vague and less arbitrary than all the ones that have been given so far. Everything is rigorously proven, and this is perhaps why it is so difficult to apply the calculation to it [the problem of resistance] and to compare it with the experiment". The last sentence is surprising, but his concerns refer to the procedure used to find q and p by approximations, a method "so long that it can discourage the most intrepid

---

[16]Vol. V, 1768 Memoire 34, I.

[17]Grimberg et al. [2008].

[18]Cf. Darrigol [2005], p. 134.

[19]With the "Über Flüssigkeitsbewegung bei sehr kleiner Reibung", (Motion of Fluids with very little Viscossity), Heidelberg, 1904. However, althought the boundary layer was a "breakthrought to bridge the gap between theory and practice", "it took years before the scope of the new theory was recognized", as Michael Eckert points out (p. 38).

calculator". He continues "However, I do not think a more direct and simple method can be found for determining the resistance and fluids pressure, and I even dare to assert that if this method does not agree with what will be found by experiment, it should almost despair of finding the resistance of fluids by theory and *by the analytical calculation*; since all the physical principles on which our analysis is supported have been demonstrated in rigor". This method is obviously the assumption of the existence of the functions q and p, although "strictly speaking that assumption can be disputed, but in this case all hope to determine the fluid pressure by calculation and therefore by theory, must be renounced".

We appreciate in these words a hint of insolence that differs from the spirit of the author of the "Preliminary Title" of the *Encyclopédie*. We are inclined to consider this article, which is without any precedent in the *Manuscript*, as an angry reaction against the Academy of Berlin.

The analysis of the pressure upon a body should finish here, but there are two articles that we have skipped [§.71–72], which deserve to be commented. In these d'Alembert tries to show that according to experiment the perturbation induced by the body is limited to the fluid layers nearest to the body. This idea is repeated several times in the *Essay*, and is the final attempt at calculating q and p [§.64–65], and in consequence the body resistance [§.85-4th] and also motions in plane sections [§.101]. We think that the first occurrence of this idea was the statement limiting the velocity field by a parallel line [§.36-1st]; which is repeated here as the line *TR* (Fig. 17.6) "that separates the parts of the fluid where the velocity and direction are not changed from those where the velocity and direction are changed" [§.71]. Outside this line $q=1$ and $p=0$, and both velocity equations Eq. 17.8 y Eq. 17.11 are met, therefore motions with this condition can exist, but it is an additional condition rather than a theoretical hypothesis. As an common example we can consider the motion in a channel whose walls were planes inside the fluid; it is clear that $q_o = 1$ and $p_o = 0$ outside the channel as before, while inside it would be $q_i \neq 1$ and $p_i \neq 0$. That is to say the separation would be a discontinuity that must be supported by a physical wall, otherwise it would disappear. Even more, moving this wall closer to the body would change the streamline pattern- in order to comply with the new conditions.

However, d'Alembert proposes a "common and simple experiment" to prove that the separation line, *TR* in Fig. 17.6, is quite close to the body. The body will be the bob of a pendulum which is placed in the middle of a channel, and which due to the effect of the resistance is displaced from the vertical in a plane parallel to the flow. Next, the pendulum is moved laterally "so that it is much closer to one wall than the other", and "[the bob] will seem to rise to the same level in a vertical plane also passing through direction of the flow". Therefore the forces around the body are equal both when the pendulum is in the lateral or middle position. "Whereby it results that the parts of fluid nearer to the body are the only ones where the motion is changed significantly by the effect of the body". This experiment could be carried out and the practical extension of the perturbed zone could be shown with a very careful measurement of the displacement angles, both in the sense of the stream and perpendicular to it. However, no such experiment was made.

In the next article [§.72] another even stranger experiment is proposed. A floating body is placed in the bottom of a vessel full of water and the time it takes to reach the surface is measured. He says that this time is always the same, independently of its position, either near the walls or in the centre. If the pendulum string deviation is not an easy task to measure, the determination of the velocity would be even more difficult.

The article finishes with a rather striking argument. In accordance to the uniqueness of the streamline field, the position of the line TR must be always at the same distance from the body. However he adds "the experiment shows that when the velocity is very small the motion and direction of the parts of the fluid are not altered until a relatively small distance from the body. So in general, it is quite close to the body, regardless of the velocity a". This is a fallacy, because he is mixing absolute and relative velocities. Besides, to propose an experiment, whatever it may be, and to invent the results based on facts as vague and ambiguous as the experiment is to act in a manner unworthy of him.

In an infinite and ideal fluid the body perturbation would spread until infinity, although in a decreasing ratio. In practical terms, with real fluids such perturbation vanished at a certain distance, as experience shows. This is probably what d'Alembert wants to introduce it in order to comply with experience. However, his concerns to restrain the body effect only to a very near part of the fluid are not clear to us. We think it is an unnecessary proposition, because the pressure on the body depends only on the fluid layer touching the surface, no matter how the rest of the velocity field is.[20]

17.2 Impulsive Motion

The values obtained for q and p correspond to any instant of the motion, now "it is good to know the values of p and q at the first instant" [§.50]. This leads to we have called impulsive motion, which here he considers as" useful to determine the fluid pressure", and in some articles later on he insists is "absolutely necessary" [§.54]. This matter underwent an important modification from the *Manuscript*, because there the impulsive motion was considered as a limit of accelerated motion and his analysis included in the moving body case, while in the *Essay* they are separated, resulting in and clearer but more elaborate view.[21]

---

[20]Just as an example, taking the sphere potential from Appendix, the velocity increment produced by the body at the diametric plane perpendicular to the fluid velocity is $v_i = v_0(a/z)^3/2$; that is, 1/2 at the surface, 1/16 at a radius and so on.

[21]The *Mss*.77–84 covers the moving body, which includes the impulsive case as part of *Mss*.83 and some of *Mss*.80. Part of them plus many additions are integrated in the impulsive case of the *Essay* [§.50–56]. The rest of these *Manuscript* articles are moved to the §.86–90 of the *Essay*, with minor modifications. See the Annex II for more details.

17 Resistance of a Body Moving in a Fluid

Both body and fluid are at rest. Suddenly the fluid receives an impulse so that it jumps as a whole to a finite velocity. The question is to find out the effect produced by the body upon the fluid parts [§.51]. This problem may seem strange at first glance; however it was not only a mechanism to start the motion, but can also be extended for the generation of non-steady motions. As far as we know, it is the first time that this type of motion is presented.[22]

Two things are clear [§51]. First, that the fluid particles contiguous to the body surface, *EAD* in Fig. 17.1, will be forced to change their direction, and the same will apply to the neighbouring ones, at least until a certain distance. Second, a stagnation zone will formed in the apex, represented as *FAM*. As a consequence the fluid particles will describe the curves $P_i S_i$, which will be the same as those formed in the case of moving fluid.

In the *Essay* two velocities are introduced, called U and u. The latter is clearly the impact or jump, which we prefer to designate as u_I, while the origin of U is not clear. "A velocity parallel to *AC* equal to Uq can be assumed in the parts of the fluid, and another perpendicular to *AC* equal to Up, U being in a given ratio with u; such that instead of Uq, it can be written uq and up instead of Up". It seems to us that U was a consequence of u through an unknown relation. We find some enlightenment in the *Mss*.83A.[23] Here the body is not fixed but moves with a velocity v; then once the fluid receives the velocity u there will be a relative fluid velocity V, which can be broken down into Vq and Vp.[24] It seems possible that in the new wording in the French version, U takes the place of the former V, even when it was unnecessary. Therefore, we will have the same $q(x,z)$ and $p(x,z)$ as before, and the velocity components $u_I q$ and $u_I p$.

D'Alembert continues "Now it is necessary (*art.* 1) that the parts of the fluid moving due to the velocities of tendency u_I and $-u_I p$, $-u_I q$ are in equilibrium. Now then, the velocity u_I being the same in all them, [therefore] they would be already in equilibrium in virtue of the single velocity of tendency u_I. Therefore they must be in equilibrium in virtue of the singles velocities $-u_I p$, $-u_I q$". We find the expression "velocity of tendency" for the first time in the *Essay*,[25] and that it is associated with the equilibrium. Firode points out that d'Alembert makes use of this notion for explaining the equilibrium of two bodies at rest which could be moved.[26] Besides, we also remember that the mentioned *art.* 1, deals with the response and equilibrium conditions of a system of bodies to applied velocities [§.1]. Therefore, if u_I induces $u_I p$ and $u_I q$, according with the *art.*1 the system would be in equilibrium

---

[22]Truesdell [1954], p. LIV, said that this type of motion was asserted by Lagrange and Cauchy years later.

[23]We have divided the *Mss*.83 in three parts; 83A up to "perpendicular et = vp"; 83B up to "Ergo A′ = B"; 83C the rest.

[24]In *Mss*.83A says that V is composed of qv and pv. Probably it is a misprint.

[25]It will appears once more in §.88.

[26]Cf. p. 32. He quotes the *Traité de dynamique* (p. xiv) "All geometricians agree that two bodies, whose directions are opposite, are in equilibrium when their masses are in an inverse ratio to the velocities which they tend to move with."

applying $-u_I p$ and $-u_I q$. We have given the expression $\sum \left(m_i \vec{v}_i'' - m_i \vec{\varphi}_i dt \right) = 0$ to explain this. In the present circumstances $\vec{v}_i'' = u_I$, which reduces the formula to $M u_I - \sum m_i \vec{\varphi}_i dt = 0$, being $M = \sum m_i$ and that in some way the last sentence justifies the quote: the equilibrium in virtue of $-u_I p$, $-u_I q$. Next he goes on to analyse the equilibrium of the channel *MNnm* (Fig. 17.8). In the *Mss*.83, he arrives at the same point but he jumps directly to it from the mention of *art*. 1.

For clarification, we can imagine a body inside an accelerated fluid as in Fig. 17.7. If both fluid and body densities were equal, they would move together without any relative velocity. As result, the acceleration $\dot{u}_F$ will produce a pressure varying linearly and decreasing to the rear of the body; therefore the body will be subjected to a force $F_V = \rho_F V_B \dot{u}_F$, similar somehow to a buoyancy force, which must be equal to the inertia force F_I. Though if the body density was greater than the fluid one, it would move with the acceleration $\dot{u}_B$, which means $\dot{u}_F - \dot{u}_B$ relative to the fluid and a new force $\rho_F \chi (\dot{u}_F - \dot{u}_B)$ would appear, where χ is the resistance or pressure coefficient for accelerated motion, which plays the same role that φ does for the constant velocity. So the new equilibrium equation would be:

$$\rho_F V_B \dot{u}_F + \rho_F \chi (\dot{u}_F - \dot{u}_B) - F_E = \rho_B V_B \dot{u}_B \quad (17.37)$$

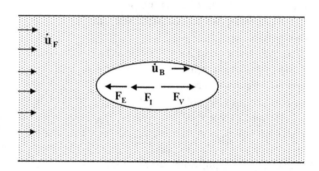

Fig. 17.7 Body in an accelerated fluid

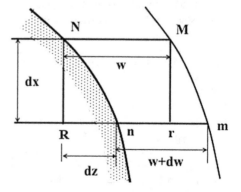

Fig. 17.8 Impulsive pressures

17 Resistance of a Body Moving in a Fluid

Where F_E represents any other external force which might exist. If the body were to stay motionless, that is to say $\dot{u}_B = 0$, the following must occur:

$$F_E = \rho_F(V_{B+}\chi)\dot{u}_F \qquad (17.38)$$

The first addend is the buoyancy produced by the acceleration in the entire fluid, and the second is due to the body itself, and more precisely to the streamline field generated. If $\dot{u}_F = 0$, then

$$F_E = -(\rho_F\chi + \rho_B V_B)\dot{u}_B \qquad (17.39)$$

According to this formula, the acceleration resistance can be interpreted as the body mass $\rho_B V_B$ increased by $\rho_F\chi$. Then this last term is often called added or virtual mass, which can be explained as the consequence of the kinetic energy transferred to the fluid. This effect is important in phenomena where the body and fluid density are comparable, such as in naval architecture or turbo-machinery and it is very relevant in motions with air bubbles.

The equation $-\frac{\partial p}{\partial z} = \frac{\partial q}{\partial x} + \frac{p}{z}$ (Eq. 17.8) is obtained here from the constancy condition as in Fig. 17.5 and Eq. 17.13. We recall that it is based on the continuity principle, whose physical conception is also applicable to the impulsions. However, it is not possible to say the same about the forces. In the moving fluid they were derived from the Bernoulli equation (Eq. 16.8) for steady state, which is not suitable here. The way that d'Alembert follows is to calculate the impulsive pressures in the closed channel $MNnm$, represented separately in Fig. 17.8, and establishing the condition that the pressure at point m from N must be equal along both Nnm or NMm [§.51]. That is: $\Delta p_{Nn} + \Delta p_{nm} = \Delta p_{NM} + \Delta p_{Mm}$; but $\Delta p_{Nn} = \Delta p_{NR} + \Delta p_{Rn}$ and $\Delta p_{Mm} = \Delta p_{Mr} + \Delta p_{rm}$, so the equation can be reorganized as:

$$(\Delta p_{NM} - \Delta p_{nm}) + (\Delta p_{Mr} - \Delta p_{NR}) + (\Delta p_{rm} - \Delta p_{Rn}) = 0 \qquad (17.40)$$

It includes only terms on segments parallel to axis X or Z, and they are also grouped conveniently.

In order to obtain the difference in impulsive pressure between two near points, we will use the Bernoulli equation now for a non-steady motion (Eq. 16.9). If these two points are separated by the distance Δx, the searched pressure will be $\Delta p_{\Delta x} = -\rho\Delta x \dot{u}$. We can consider $u_I = \dot{u}\tau$ with $\tau \to 0$, therefore $\Delta p_{\Delta x}\tau = -\rho\Delta x u_I$ in general, but if we consider the generic Δx as l_x or l_z and the corresponding velocities as $u_I q$ and $u_I p$, the jumps in pressure in the directions of X and Z become $-\rho u_I q l_x$ and $-\rho u_I p l_z$ respectively; which can be represented as $-l_x q$ and $-l_z p$ omitting the other factors that are constant, and the impulsive pressures are related to the geometry.

Coming back to Eq. 17.40 and the Fig. 17.8, the channel width NM is w, which was defined by $\pi w_T^2 = 2\pi zw$ (Eq. 17.4), and the segment Rn in found by $dz/dx = p/q$ at the point N. To obtain nm and rm. we must use dw and $d(p/q)$. With all these observations it is easy to arrive at:

$$\Delta p_{nm} - \Delta p_{NM} = d(wp) = \frac{w_T^2}{2}d\left(\frac{p}{qz}\right) \qquad (17.41a)$$

$$\Delta p_{Mr} - \Delta p_{NR} = w\frac{\partial q}{\partial z}dx = \frac{w_T^2}{2qz}\frac{\partial q}{\partial z}dx \quad (17.41b)$$

$$\Delta p_{rm} - \Delta p_{Rn} = w\frac{\partial}{\partial z}\left(\frac{p^2}{q}\right)dx = \frac{w_T^2}{2qz}\frac{\partial}{\partial z}\left(\frac{p^2}{q}\right)dx \quad (17.41c)$$

Which introduced in Eq. 17.40 gives:

$$-d\left(\frac{p}{qz}\right) + \frac{1}{qz}\frac{\partial q}{\partial z}dx + \frac{1}{qz}\frac{\partial}{\partial z}\left(\frac{p^2}{q}\right)dx = 0 \quad (17.42)$$

Operating this formula with the aim of eliminating $\partial p/\partial z$, and leaving only dx as factor, we arrive at $\frac{\partial p}{\partial x} = \frac{\partial q}{\partial z}$ already found in Eq. 17.11. Summarizing, the velocity field is equal and ruled by the same equations; in his own words "the quantities p and q are found at the first instant by the same equations as in the following instants".

To find the pressures generated in the first instant of the impulsive motion it is considered "absolutely necessary for determining of the quantities p and q" [§.54]. For the pressure along a streamline d'Alembert makes use of the Bernoulli equation in a non-steady motion Eq. 16.9, whose term $-\rho S_0 \frac{du_0}{dt}\int_{x_0}^{x}\frac{dx}{S}$ is converted to $p_I = \rho \int v ds$ taken $\frac{du_0}{dt} \to v_I$. In the Mss.80 the body was moving under acceleration, and the pressure had one term for the steady motion and another for the non-steady one which is not necessarily impulsive.[27] Expanding the formula for the impulsive component[28]:

$$p_I = \oint \rho v_I ds = \rho u_I \oint \sqrt{q^2 + p^2}ds = \rho u_I \oint qdx + pdy \quad (17.43)$$

This integral is extended along any streamline and in particular on the body contour. Taking the point M, in which the front stagnation zone APM finished (Fig. 17.9), as reference, the pressure would be: $p(x) = p_M + \rho u_I \int_{x_M}^{x} qdx + pdy$.

Therefore, the force, or pressure as he calls it, upon the body turns out to be:

$$F_B = \rho u_I \int_{PMLI} 2\pi y p(x) dy$$

$$= \rho u_I \int_{PMDLI} 2\pi y \left(p_M + \rho u_I \oint_{x_M}^{x} qdx + pdy\right) dy \quad (17.44)$$

---

[27] The pressure at a point is $-\rho\frac{du}{dt}\int\sqrt{p^2+q^2}ds + \rho\frac{u^2}{2}(1-p^2+q^2)$, Mss.80.

[28] We have: $ds^2 = dx^2 + dy^2$ and $= u_I\sqrt{q^2+p^2}dt$, $dx = u_I q dt$, $dy = u_I p dt$, and so $\sqrt{q^2+p^2}ds = qdx + pdy$.

17 Resistance of a Body Moving in a Fluid

Fig. 17.9 Body in fluid impulsion

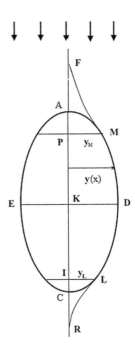

The part of the integral due to p_M is zero, as well as the part upon the front PM. For the rear IL the pressure from C to A is constant over the circle of radius y_L. The calculation in the *Essay* is a bit confused due to handling of the integral limits. This is done with the help of two functions:

$$\Gamma = \oint_{x_M}^{x_L} pdy + qdx \tag{17.45a}$$

$$\Omega = \int_{x_M}^{x_L} 2\pi r dr \oint_{x_M}^{x} pdy + qdx \tag{17.45b}$$

In accordance with Eq. 17.38 $\chi = \Omega - \pi y_L^2 \Gamma$, and including the term of the volume, the "pressure at the first instant" will be:

$$D_I = \rho u_I \left(V_B + \Omega - \pi \Gamma y_L^2 \right) = \rho u_I \psi \tag{17.46}$$

This formula can also be used when the velocity is variable, $u(t)$, because the velocity can be assumed as the successive application of du_I differential increments to match $u(t)$ [§.56]. Besides, the formula can also be applied to accelerated motions, like $D = \rho \psi \dot{u}$ [§.88] or in *Mss*.80. This fact, which the force upon an accelerated body is proportional to the acceleration and to a factor ψ, has to be recognized as a very important finding for d'Alembert. This factor has two addends:

the body volume plus $\Omega - \pi\Gamma y_L^2$, by which it can be understood as a virtual or added volume.[29]

Nevertheless, and regrettably, he will ruin the merit of this finding in the next article, in which he aims to prove that this force is zero. Here we reproduce the article integrally.

> It can be easily proved by the experiment that $\mu + \Omega - \pi\Gamma b^2 = 0$ $[V_B + \Omega - \pi\Gamma y_L^2 = 0]$. Because a weight may be found which is capable, by its own mass, of keeping the body ADCE in equilibrium from the first instant of the impulse of the fluid, and to prevent that, the body is set in motion by this impulse. Now then, the action of a weight that is in equilibrium is equivalent to a finite mass animated by an infinitely small velocity. Therefore, the force with which the weight will be in equilibrium will also be infinitely small, thus the quantity $u\delta(\mu + \Omega - \pi\Gamma b^2)$ $[\rho u_I(V_B + \Omega - \pi\Gamma y_L^2)]$ must be equivalent to a finite mass animated by an infinitely small velocity, or an infinitely small mass animated by a finite velocity. So since the velocity u is finite, then it follows that $u\delta(\mu + \Omega - \pi\Gamma b^2)$ $[\rho u_I(V_B + \Omega - \pi\Gamma y_L^2)]$ should be necessarily infinitely small; that is equal to zero. [§.55].

The first sentence is rather surprising because he invokes the experiment in order to contradict something that has been proved mathematically; and even more, the experimental proof comes in the form of an imaginary or thought experiment. The idea is to support the impulsive effect upon the body by means of a weight. In Fig. 17.10 we depict our interpretation of this experiment or measuring device; it consists in a beam balance which will transfer the momentum received by the body through a lever to a mass. The lever is assumed to have equal arms for simplicity. It

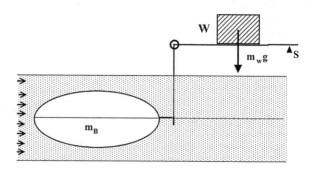

Fig. 17.10 Imaginary apparatus

---

[29] As an exercise, we can apply the above formula ψ to one sphere. The velocity potential for it is given in Annex I in spherical coordinates as $v = -\frac{R^2}{r^3}\cos\theta$; although it would be more easier to work with this type of coordinates, we have transform this formula to cylindrical coordinates. Then, $q = -\frac{R^3}{2}(x^2 + y^2)^{-\frac{3}{2}} + \frac{3R^3x^2}{2}(x^2 + y^2)^{-\frac{5}{2}}$ and $p = \frac{3R^3xz}{2}(x^2 + y^2)^{-\frac{5}{2}}$. On the contour $x^2 + y^2 = R^2$ is $xdx + ydy = 0$ and we have $[q]_C = -\frac{1}{2} + \frac{x^2}{R^2}$ and $[p]_C = \frac{3}{2}\frac{xy}{R^2}$; and $[pdy + qdx]_C = -\frac{1}{2}$. Operating we obtain for $\psi = \pi\left[\frac{4}{3}\pi R^3 - \frac{x_L^3}{6} + \frac{x_M^3}{6} - \frac{R^2}{2}(x_M - x_L)\right]$; which for the case without stagnation zones, $x_M = -R$ and $x_L = R$ gives $\psi = \frac{4}{3}\pi R^3 + \frac{2}{3}\pi R^3 = 2\pi R^3$. That means an added mass equivalent to half a sphere; this result coincides with the one given in the Annex I.

17 Resistance of a Body Moving in a Fluid

is obvious that initially the system must be maintained at rest by a kind of fixing S, otherwise the weight will fall down.

Physically, a percussion on the body will induce an instantaneous velocity v_u in the system formed by the body mass $m_B = \rho_B V_B$ and the weight mass m_w, according to the relation $\rho u_I \psi = (m_B + m_W) v_u$. More realistically the system will have no velocity in the first instant, but the weight W will start moving upwards and the body to the right. In addition they are both submitted to an acceleration $g m_W / (m_B + m_W)$ due to the gravity acting upon W, and in consequence both will keep moving during a time $t_u = \frac{\rho u_I \psi}{g m_W}$, which is the time taken to transfer the fluid impulse $\rho u_I \psi$ to potential energy. Next the motion would reverse to the initial position if we ignore the resistance in the steady state motion. We can see that the only way to prevent motion would be to use an infinity weight. This is the first error. In the case of considering that the resistance after the impulsion is $\rho \varphi u^2$, as he does later [§.88], this force would be transmitted to the right hand lever of the balance, so that the vertical force would be $m_w g - \rho \varphi u^2$, thus the evolution of the system would depend on the sign of this quantity.

Next comes the equivalence of the action of a generic weight in equilibrium to a finite mass animated by a infinitely small velocity. But what does "the action of a weight in equilibrium" mean? This reminds us of the virtual works principle used in the *Traité de Dynamique* as we have explained previously, where the action of powers was understood to be the mass times velocity. Thus, if the generic weight was W and the m the finite mass, the necessary power to keep the weight in equilibrium during the instant δt would be $\delta F = m \delta v = W \delta t$. He continues by saying that this force is infinitely small, which is correct, but it acts in an infinitely small time, thus to maintain the situation permanently an accelerative force $\delta v / \delta t$ would be needed. Translating this to our case, δF would be the force upon the body, so that $\delta F = \delta(\rho \psi u_I)$, which we understand should be $\delta(\rho \psi u_I) = \rho \psi \delta u_I$, and δu_I could be either a differential impulse as in [§.56] or $\delta u_I = \dot{u} \, \delta t$ a constant acceleration, which is the same thing; in either case the weight W would be in equilibrium. However, his assessment is different. He considers that the virtual force δF can be understood in two ways $\delta F = m \delta v$ or $\delta F = v \delta m$: one with an infinitely small velocity and a finite mass and the second one the reverse. It is clear that mathematically this is correct, but not physically as the motion allowed to the mass m must be infinitesimal. This is the second error. Then he ponders $\rho \psi u_I$ as being equivalent to one of these two possibilities, and he takes the second as being the valid one, arguing that u_I is finite. But in addition to what we have said, u_I is the velocity of the fluid, and not the virtual velocity of the body. This is the third error.

The reasoning given in the *Mss.* 83 had some differences. The body is free without any counterbalance weight. The impulse u_I produces a velocity v on the body and the pressure $\rho \psi v$, which must be equal to the quantity of motion lost by the body, so $\rho_B V_B (u_I - v) = \rho v (V_B + \Omega - \pi \Gamma y_L^2)$. "However the experiment confirms that the velocity lost in the fluid motion in the first instant is infinity small, which clearly agrees with this principle: *nothing in nature is done by leaps, all change in motion is resolved by insensitive degrees*. Therefore $u_I - v = 0$;

consequently $\rho v(V_B + \Omega - \pi \Gamma y_L^2) = 0$".[30] We see that he had accepted the transfer of momentum, only he says it is null because it goes against the continuity principle and the supposed experiment. The incongruence is obvious: following this principle it will be impossible to produce the jump u_I. Furthermore, he is inventing the result of an experiment that was not carried out.

Summarizing, we find his entire reasoning weak and tricky, belittling his previous contribution. We have to say again that in our opinion this is his biggest mistake and we are unable to justify the causes that led him to introduce these strange reasons. Even more, he likes to boast of himself, sometimes arrogantly, as the champion of rigorous reasoning, so then why does he renounce all these principles?

17.3 Body in Motion

This is the third phase of the process. The body is set in motion by an initial velocity; but this motion will be slowed down due to the effect of the fluid, and consequently will move with a non-uniform velocity. This is important, because the resistance undergone by the body will be equal to the rate of the momentum lost by it. Thus we have the initial velocity v_{B0}, the instantaneous one $v_B(t)$ and the space traversed $r(t)$. Now, the theorems obtained for a set of individual bodies [§.6–9] are applicable to this system. Basically they establish that the entire set of trajectories induced by the motion of a single body is unique, irrespective the magnitude of the initial velocity, and depending only on the space traversed by that body. This is complemented by the laws for the velocity $-dv_B/v_B = \xi(r)dr$ and for the resistance $R = \xi(r)v_B^2$. All these parameters are referred to axis fixed to the fluid; however, d'Alembert will try to prove that the flow pattern relative to the moving body is constant and equal to the one already obtained in the two previous phases [§.86]. We make a presentation of the problem with the help of Fig. 17.11.

Both fixed (XZ) and body (XY) axis are positive to the right side, while the body moves to the left, which means that the depicted velocity v_B is negative. A fluid particle, such as P, will traverse a closed trajectory $P_0P_1P_2P_3$ with a velocity $\vec{U}_P$ relative to the fixed axis, induced by the approaching body. The trajectory starts at P_0, reaching the maximum at P_2, when the body is just behind it, and comes back to P_0 again when it is far away on the left side. Moreover from the previous [§.8], $\vec{U}_P$ depends on the body position r and also on its initial velocity v_{B0}. That is to say $\vec{U}_P = \vec{U}_P(v_{B0}, r)$. Then, considering the motion relatively to the body axis, the above particle P will have a relative velocity $\vec{v}$ and will follow a trajectory like the

---

[30] *Experientia autem constat velocitatem quam in fluido motum amissit primo instant, esse infinite parvam, quod quidem huic alteri principio congruit, nihil in natura per saltum fieri, omnem, que motus mutationem per gradus insensibiles absolvi. Ergo* $u - v = 0$, proinde $V_B + \Omega - \pi \Gamma y_L^2 = 0$. Mss.83C.

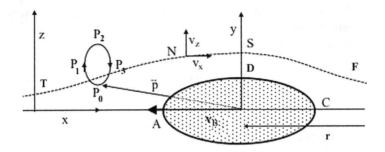

Fig. 17.11 Body moving in fluid

curve *TNSF*. Therefore the absolute velocity will be $\vec{U}_P = \vec{v} + \vec{v}_B$, whose components along both axis are $v_B + v_x$ and v_z.

Now, as the trajectories only depend on the modulus of the initial velocity v_{B0},[31] as expressed in [§.6], the rates $(v_B+v_x)/v_B$ and v_z/v_B are functions of the relative position of the particle with respect to the body and of the space traversed by it, depicted as $\vec{p}$ and r [§.86]. However, the body velocity also has to meet $-dv_B/v_B = \xi(r)dr$ [§.6], therefore both v_x/v_B and v_z/v_B will depend on the position and the initial velocity v_{B0}. In these circumstances he poses the question whether v_{B0} would disappear in v_x/v_z or not. He points out that this happens in the body contour $y(x)$, because on it $dy/dx = v_z/v_x$. "Then let us suppose at first that the quantity v_{B0} is not found in the ratio of v_x to v_B and v_z to v_B and let us see what will result from this hypothesis" [§.86].

The first thing is to define the functions q and p as $v_x = v_B q$ and $v_z = v_B p$. Next, taking a point as reference like N, a parallelepiped similar to Fig. 17.5 is placed here and the constancy of the fluid enclosed is applied in the same way as in [§.48]. We need not repeat all the formulation, already given in Eq. 17.8, and the result is $\frac{\partial p}{\partial z} = -\frac{\partial q}{\partial x} - \frac{p}{z}$.

The following step is to analyse the acceleration in a streamline passing by N, as was shown in Fig. 17.3, and applying a reasoning as in [§.43]. However, in deriving the acceleration it must be taken into account that v_B changes with time; that is $\gamma_x = \frac{dv_x}{dt} = \frac{d(v_B q)}{dt}$ and $\gamma_z = \frac{dv_z}{dt} = \frac{d(v_B p)}{dt}$, therefore the term $\frac{dv_B}{dt}$ has to be added to γ_x as[32]:

$$\gamma_x = \frac{dv_B}{dt} + q\frac{dv_B}{dt} + v_B^2 \left(q\frac{\partial q}{\partial x} + p\frac{\partial q}{\partial z} \right) \quad (17.47a)$$

$$\gamma_z = p\frac{dv_B}{dt} + v_B^2 \left(q\frac{\partial p}{\partial x} + p\frac{\partial p}{\partial z} \right) \quad (17.47b)$$

---

[31] D'Alembert understands this dependency taking the body velocity as v_{B0}/g, being g a factor.
[32] In the *Essay* §.86 the body's velocity is taken positive towards the left. This explains the difference in signs of our formulas Eq. 17.47a and 17.47b.

This will take the place of the Eq. 17.3a and 17.3b, although with additional terms. Appling the cross derivatives condition Eq. 16.2, as had been done in [§.48], we will arrive at the new equation that replaces Eq. 17.25:

$$\frac{dv_B}{dt}\frac{\partial q}{\partial z} + v_B^2 \frac{\partial}{\partial z}\left(q\frac{\partial q}{\partial x} + p\frac{\partial q}{\partial z}\right) = \frac{dv_B}{dt}\frac{\partial p}{\partial x} + v_B^2 \frac{\partial}{\partial x}\left(q\frac{\partial p}{\partial x} + p\frac{\partial p}{\partial z}\right) \quad (17.48)$$

The $\frac{dv_B}{dt}$ has no effect in this equation since it affects the entire fluid. As q and p do not depend on the velocity v_B, the former can be split in two equalities, one equal to the former Eq. 17.25 and subjected to the same reasoning, and the other:

$$\frac{dv_B}{dt}\frac{\partial q}{\partial z} = \frac{dv_B}{dt}\frac{\partial p}{\partial x} \quad (17.49)$$

This has the solution $\frac{\partial q}{\partial z} = \frac{\partial p}{\partial x}$. Consequently the set of streamlines around the body is identical whether it is the body or the fluid moving or at rest, leaving open the possibility, at least in theory, for the calculation of the velocities.

Nevertheless, before proceeding, let us go back to the assumption of $v_x = v_B q$ and $v_z = v_B p$ [§.87]. Given the above mentioned reasons of dependency of v_B on v_{B0}, he proposes an alternative formulation like $v_x = q\kappa(v_{B0})$ and $v_z = p\kappa(v_{B0})$, being κ any function. The final equations would be the same and the pressures would be proportional to $v_B^2 \kappa(v_{B0})^2$, "which should not be surprising, since in general the resistance R (*article* 9) is proportional to ξu^2". Besides, he says that $\kappa(v_{B0}) = 1$ will be proven in the subsequent articles.

Now, after all that had been said, d'Alembert heads straight on to determine the fluid resistance; the next articles [§.88–90] are very important items in his thoughts. The main idea is to apply to the whole fluid and body system an impulse velocity equal in magnitude to what the body has, but in the opposite direction. After that operation, the body will be left at rest and the fluid will strike it at the velocity v_B, "but for the primitive laws of motion, the fluid pressure on the body will not be changed" [§.88]. The situation is converted into the case of fluid in motion and body at rest, in which the fluid pressure upon the body could be calculated, at least at theory. But, to which "primitive laws of motion" does he refer? However, the body was not moving at constant velocity, but with an acceleration $\dot{v}_B$; therefore, the body is only motionless for an instant, and to keep this state an additional differential impulse must be given to this system, which will be precisely $-\dot{v}_B dt$. In the *Essay* this is expressed as "any accelerating or retarding force proportional to kdt" acts in every instant upon the parts of the fluid", which will not disturb the streamline pattern, as was said in [§.56], and it will increase or decrease the velocity proportionally to $kdt\sqrt{p^2 + q^2}$. Now, this kdt, called "velocity of tendency", will produce a pressure of $kdt(V_B + \Omega - \pi\Gamma y_L^2)$, but as he has taken $\psi = V_B + \Omega - \pi\Gamma y_L^2$ as zero [§.55] the pressure is null. Two small comments: in the next paragraphs dv_B/dt will take the place of k, and the expression "velocity of tendency" had been also used in the impulsive motion [§.51].

17 Resistance of a Body Moving in a Fluid

Summarizing, the pressure at any instant will contain two addends. One, $\rho\varphi v_B^2$ coming from the velocity; the other, $\rho\dot{v}_B\psi$ from the "variable velocity", that is $\dot{v}_B$. Both φ and ψ have been defined in Eq. 17.36 and Eq. 17.46 respectively. However, as he has taken $\psi=0$, the "fluid pressure" is only proportional to v_B^2. Additionally, $\kappa(v_{B0})=1$, as he had announced in [§.87].

He argues that all authors of hydraulics had taken "for principle that the resistance of a moving body in a fluid is equal to the pressure that this fluid, moving with the same velocity, exerts against the body assumed at rest" [§.88].[33] D'Alembert claims that they have not paid to attention the situation where the velocity is variable, and therefor there will be an additional term of resistance proportional to $\dot{v}_B$, besides another term proportional to v_B^2. Furthermore, he continues saying that these authors have only proved the proportionality with v_B^2 "in a very vague way". We think this is very important because it is the first time, as far we know, that these two components are found although he had taken the second as zero. He considers a merit for himself to have proved "that the coefficient of dv_B/dt is zero, and that the coefficient φ of v_B^2 is always the same, no matter what v_B is".

Before entering into the calculation of resistance, we wish to mention that until now, the forces or pressures upon the body have been translated to the pressure exerted by fluid in motion, which in turn are also referred to the forces of gravity. However, the resistance undergone by the body must be equal to the momentum lost by it, or, in a clear Newtonian reference, the mass times acceleration. Thus [§.89], at any instant, this force will be $\rho_B V_B dv_B/dt$, being V_B the volume of the body and ρ_B the density. This force must be balanced by the pressure $[\rho\varphi v_B^2 + \rho\psi dv_B/dt]$. That is:

$$\rho_B V_B \frac{dv_B}{dt} + \rho\psi \frac{dv_B}{dt} + \rho\varphi v_B^2 = 0 \qquad (17.50)$$

He insists on $\psi=0$, but the term is included in the equation, which to our understanding means that the main fact is the existence of the term itself. So the former equation becomes the well-known expression:

$$\rho_B V_B \frac{dv_B}{dt} + \rho\varphi v_B^2 = 0 \qquad (17.51)$$

The arguments are repeated in the next article [§.90], but it is worth noting the statement: "This did not seem easy to prove [that $\psi=0$], because of the difficulty of expressing the quantities Γ and Ω analytically; but fortunately we have arrived at the end for the consideration of the primitive velocity of the body, without needing to know these quantities". The sentence "primitive velocity of the body" looks

---

[33] We assume that among the mentioned authors of hydraulics Newton was included as he had stated that "the action of a medium upon a body is the same (by Cor. V of the Laws) whether the body moves in a quiescent medium or whether the particles of the medium impinge with the same velocity upon the quiescent body" [Book 2, Prop. XXXIV].

obscure, but it seems that he was fully convinced that the term in $\dot{v}_B$ must be null in any case.

In all the previous calculation both fluid and body were considered weightless. If the gravity is assumed as acting along the X axis, all the reasoning will be the same, but two additional terms will appear, one coming from the weight of the body and other from its buoyancy. That means that the pressure $\rho\varphi u^2$ must be incremented in $gV_B(\rho_B - \rho)$ [§.91].

Before leaving what is for us the more relevant part of the *Essay*, it is time for some reflections. We have to acknowledge the merit of d'Alembert in how he was able to determine the resistance leaving aside the forces as physical entities which he does not deny, but avoids for their "metaphysical" character. We recognize as truly brilliant how the fluid resistance, which is loss of momentum, is transferred to the pressures, which are backed by the gravity. Besides, along the way he found how the resistance was also affected by the acceleration; which proved that the reversibility fluid-body was not applicable to non-steady motions. For all this he deserves to be praised. However, adducing contrived reasons in an indirect way, he pretends to have proven that the acceleration effect was null. We regret that he misunderstood the true consequences of his former development; furthermore, we say again and for the last time, this was his major mistake in the entire *Essay*. For this he deserves to be criticized. Two faces of the same great man. We do not know the opinions of his contemporaries concerning this problem; in particular we would like to know if Euler noticed anything. Among his modern scholars, Dugas and Truesdell comment on it, but they only say that he was wrong.[34]

---

[34]Dugas [1952], p. 11 and Truesdell [1954], p. LVI.

Chapter 18
Other Resistances and Fluids

18.1 Friction and Viscosity

The resistance analysed until now comes from the transfer of momentum between fluid and body and it can be understood as being derived from the inertia of matter. However, the authors knew that there were other sources of resistance related to the properties of the fluids, as yet not well known, and named with ambiguous terms like lubricity, slipperiness, fluidity or viscosity. As one example, apart from the resistance due to the inertia, Newton spoke of two more arising from the viscosity and friction,[1] assuming the first to be constant and the second proportional to the "moment of time", that is the velocity.[2] Later he identified "the resistance arising from the want of lubricity" as proportional to the velocity gradient,[3] which is how the viscosity is understand nowadays.

D'Alembert starts assuming that "the friction of the fluid over the body may only come from the relative velocity of the fluid respect to the body" and that the friction is proportional to the velocity according to the experiments made by Musschenbroek [§.92].[4] Therefore the friction must be proportional to the velocity as μv, being μ a coefficient to be determined experimentally. Consequently the forces at any point of a body must be decreased by $-\mu v_B q$ and $-\mu v_B p$ along the respective axis; that means that. The quantities $-\mu v_B \frac{\partial q}{\partial z}$ and $-\mu v_B \frac{\partial p}{\partial x}$ must be added at the left and right sides of equation Eq. 17.48. The calculation of the pressure due to $\mu v_B q$ and $\mu v_B p$ over the body surface is mathematically equivalent to the effect of

---

[1]*Principia*, Book 2, Prop. XL, Sch., p. 366.
[2]Ibid. Prop. XIV, Sch.
[3]Ibid. Sec. IX, Hypothesis.
[4]He did not specify what these experiments were. We have browsed both the *Elementa Physicæ* and the *Essai de Physique* without success.

© Springer International Publishing AG 2018
J. Simón Calero (ed.), *Jean Le Rond D'Alembert: A New Theory of the Resistance of Fluids*, Studies in History and Philosophy of Science 47,
https://doi.org/10.1007/978-3-319-68000-2_18

the impulsive velocity already studied, so $\int_\Sigma \vec{v} d\vec{\sigma} = v_B(\Omega - \pi\Gamma y_L^2)$; with Ω and Γ as defined in Eq. 17.45a, b. And according to his supposition $\psi = V_B - \Omega + \pi\Gamma y_L^2 = 0$ it follows that $\Omega - \pi\Gamma y_L^2 = V_B$. Thus including the gravity as well, the equivalent to formula Eq. 17.51 would be:

$$\rho_B V_B \frac{dv_B}{dt} + \rho\varphi v_B^2 + gV_B(\rho_B - \rho) + \lambda\rho v_B V_B = 0 \qquad (18.1)$$

But besides this force of friction, there is another resistance, "that comes from the viscosity of the parts of fluid; and, as far as we can conjecture from all experiments" [§.93]. Proofs of its existence are the floatability of bodies heavier than water and the drop retention when they adhere to the lower surface of a body. Nowadays we know that these phenomena derive from the superficial tension, something unknown to at this time. However, it seems as if d'Alembert had some doubts about it; we think that it was easier for him to understand the resistance due to the friction, rather than the one derived from the viscosity, because "whether it comes from a compressive force or from the attraction of parties, [it] is a constant force such as gravity, though very small compared to it". In this way he raises an objection to the constancy of this force because a pendulum moving in water should stop not at the vertical, but instead close to it due to this constant force. The answer given is that the angle would be very small and not easy to observe. But in its favour he says that it is "a truth that seems consistent with reason, and which is supported by an infinity of experiments".

Next, he takes a lengthy commentary on the experiments and ideas of 's Gravesande about this matter [§.93].[5] We will comment later on these experiments, but as d'Alembert explains, that author tried to determine the relative value of the two components of the pressure of a fluid moving against a body at rest: one proportional to the single velocity v, and due to the fluid viscosity, and the other to its square v^2, due to the force of inertia, something like $k_1 v + k_2 v^2$. However, for the case of a body moving in a fluid, 's Gravesande thought that the term $k_1 v$ should be substituted by a constant one, otherwise the body would never stop. This is mathematically correct, as d'Alembert says, but he puts the question of how it can be explained that in one case the viscosity produces a force proportional to the velocity, and in the other case it is constant. He found 's Gravesande's arguments obscure if the viscosity is understood as the force that particles oppose to being divided, as he thinks. Thus, when the body moves it must separate the particles and this makes it lose velocity, but in the other case the viscosity seems more a passive force rather than an active one. After some more reflections he ends by admitting that the viscosity is equivalent to increasing the weight of the fluid.

Calling $\kappa\rho$ to this force, the former equation became [§.94]:

---

[5] *Physices Elementa Mathematica Experimentis Confiermata.*

$$\rho_B V_B \frac{dv_B}{dt} + \rho\varphi v_B^2 + gV_B(\rho_B - \rho) + \lambda\rho v_B V_B + \kappa\rho = 0 \qquad (18.2)$$

This general equation of the motion can be reduced to the type $\frac{-du}{au^2+\beta u+\gamma} = dt$, which had been study previously and whose solution is $\frac{(u+k)(g+k')}{(u+k')(g+k)} = e^{-\alpha t(k-k')}$. The problem is now reduced to the analysis, therefore he leaves it and goes on to other researches, though he notes that he came to this formula by an entirely new method.

As a final comment, neither of these sources of friction or viscosity has supported the test of time. The only one that has prevailed is the interpretation of the viscosity as proportional to the velocity gradient, which was established by Newton in the vortex motion,[6] but it was not included in the theories until many years later.

18.2 Resistance in Non-elastic and Finite Fluids

"All fluid wherein a body moves is elastic or non-elastic"; we would say compressible or uncompressible. The elastic fluids can be compressed or expanded when the container recipient is reduced or enlarged, which does not happen with the non-elastic ones [§.80].

Besides the classification, based only on the fluid itself, d'Alembert adds another depending whether or not a vacuum is produced in the fluid behind a moving body. However, this effect can always happen in the elastic ones, but only under some circumstances in the non-elastic fluids.

In the non-elastic one, if the fluid is infinite or it is confined inside a vessel closed on all sides, there will be no possibility for such a vacuum [§.81]. However, if the vessel has an open side, like a pond, and the body is moving near the free surface, as in Fig. 18.1, the upper level will be altered and the fluid rearranges itself internally and an empty space is produced behind the body [§.82]. Obviously enough velocity is needed.

For an elastic fluid, either indefinite or enclosed, there will be always a compression in the front and an expansion in the back of the body, which allows the formation of an empty space. Depending on the velocity, the vacuum part could be partially refilled by the surrounding air which will rush inside [§.83].

The ideas of the vacuum left behind the bodies in motion, and the capability of the fluid to fill it had some relevance in the epoch. Let us recall that the impact theory implied a vacuum behind the body, or more precisely in its shadow, which was clearly contrary to daily experience. The direct observation of the whirls in the stern of a boat, or another similar object, seemed to reveal that the fluid falls down into this space. Pierre Bouguer in his *Traité du Navire*[7] used this idea to explain the

---

[6]*Principia*, Book 2, Sec. IX, Hyp.
[7]Cf. Lib. V, Sec. VII.

Fig. 18.1 Motion in a finite fluid

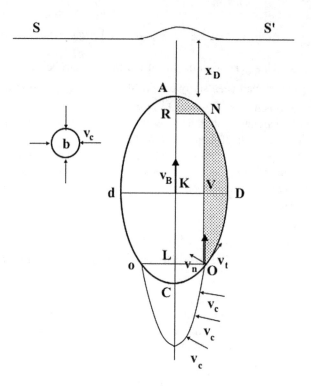

resistance in the stern of the hull of a vessel, assuming that the water entered in the space left by the motion with a fall velocity as it fell from a height equal to the depth. For the air there were studies by Daniel Bernoulli in the *Hydrodynamica*[8] about the velocity with which the air penetrated in a vessel at vacuum. Later on in the *Gunnery*, Robins studied this matter in detail as d'Alembert recognized, and as we shall see further on.

Therefore, there are three different cases: the non-elastic and infinite, which we have studied, the non-elastic and finite, and the elastic. Let us deal with the second case.

D'Alembert explains that the void is generated only when the body moves fast and is near the surface. The deeper the body is, the greater the velocity should be in order to produce a vacuum, in such a way that if it is deeply submerged a vacuum will never exist. Due to the movement of the fluid particles, he also thinks that the fluid surface must rise above the stationary level [§.106]. Also the number of particles affected in the motion differs whether a vacuum is produced or not, because of the lack of contact between body and fluid in the vacuum zone [§.107].

To deal with the vacuum he introduces a velocity due to the compression of the particles, which will be the eruption velocity of the fluid when it enters a bubble, as shown in Fig. 18.1. This velocity, which we call v_c, is the same over the entire

---

[8] Cf. Sec. X, §32-ss.

surface of the bubble [§.108]. We know that it will depend on the depth, but for now it is assumed to be constant throughout the entire fluid.

When the body moves upwards, the velocity v_B is broken down at any point of the surface in two components, one tangential and other normal, called v_t and v_n. The hypothesis of d'Alembert is that the vacuum will be produced if $v_n > v_c$; that is, the compression velocity of the surrounding particles is not fast enough to erupt and fill the empty gap left in the motion.

For the tangency condition $v_n = v_B \frac{dy}{ds}$, therefore there will be a vacuum in the surface points if $v_B \frac{dy}{ds} > v_c$; consequently, if $v_B \leq v_c$ there will be no vacuum at all. A similar idea had been used by Bouguer as mentioned before.

We note that d'Alembert superimposes these new velocity components v_t and v_n on the former ones $v_B q$ and $v_B p$, knowing that $v_B \sqrt{q^2 + p^2}$ was tangential to the body. In the Mss103 he had tried to relate $\frac{v_c}{v_B}$ with p and q, but nothing is mentioned in the $Essay$. We may think of this criterion as arbitrary, but we have to agree that the problem was too complicated and the proposal had the advantage of being easy to deal with.

The origin of v_c is the force that compress all the particles of the fluid equally [§.109], and which is designed as ψ. Although he does not explain how this force could be generated, an interpretation could be a constant pressure upon the free surface as $p_e = \rho \psi$, which would be transmitted uniformly to the entire fluid mass. Now he suggests that ψ can be assumed as the pressure of a fluid under a gravity g' and a height h_ψ such as $\psi = g' h_\psi$, and then $v_c = \sqrt{2g' h_\psi}$. This modified gravity does not exist in the Mss.104, but only the ordinary one g is considered; even though both solutions are equivalent in the end. D'Alembert, probably wanted to highlight that this pressure "comes from another cause other than gravity" [Mss.105]. We have to add, that strictly speaking, as the fluid is heavy, the depth must be also taken into account in this calculation, so that $v_c = \sqrt{2g' h_\psi + 2g(x + x_D)}$; probably the second addend is considered too small so it can be ignored.

The pressure at any point would be $p = \rho \psi + \rho g(x + x_D) = \rho g' h_\psi + \rho g(x + x_D) =$, and the total force over the body $F_x = \int_0^{y_0} \rho \pi y^2 \left[\rho g' h_\psi + g(x + x_D) \right] dy$, resulting in:

$$F_x = \rho \left[\pi y_0^2 g' h_\psi - g(V_{NDO} + V_{ANr}) \right] \quad (18.3)$$

The first term is the "compression" due to ψ, and the second the buoyancy, being V_{NDO} and V_{ANr} the volumes generated in the rotations of areas NDO and ANr. Then the equation of the movement will be:

$$-\rho_B V_B \frac{dv_B}{dt} = \rho \varphi v_B^2 - \rho g(V_{NDO} + V_{ANr}) + \rho_B g V_B + \rho g' \pi y_0^2 h_\psi \quad (18.4)$$

An Equation where the values of V_{NDO} and V_{ANr} depend on the position of the point O, which itself depends on v_B.

In the case where the pressure ψ comes only from the weight, the term $\rho g' \pi y_0^2 h_\psi$ should be removed [§.110], but nothing is said about the velocity due to the compression. In the *Mss*.105 it is added that now that velocity is $v_c = \sqrt{2g(x+x_D)}$, which will be variable.

It only remains to find the value of φ as $\varphi = \oint_I^O 2\pi y (1 - p^2 - q^2) dy$, where the position on the point I is unknown, which must correspond with the end of the stagnation zone. The suggested method is to introduce the resistance due to the acceleration in Eq. 18.4, and make it equal to zero [§.111]. No more details are given, but in the *Mss*.104 it is identified as $-\rho \frac{dv_B}{dt}(V_{LODNA} + \Omega)$; then with the assumption of its nullity $V_{LODNA} + \Omega(I) = 0$, from where the position of I can be found and after the quantity φ.

18.3 Resistance of Elastic Fluids

As we have seen, for an inelastic and undefined fluid, in other words incompressible and unlimited, a vacuum behind the body never will be produced [§.112]. However when the fluid is compressible, or non-elastic, it will always be produced, because in the motion the fluid is condensed in the front part and expanded in the rear, both with greater intensity in the measure that the velocity is higher. Besides, the highest values of both will be at the respective ends, changing gradually along the body surface until an intermediate point where neither will exist, because everything in nature happens by insensible degrees [§.113]. The rear dilatation will produce a vacuum, "because the motion of the body leaves an empty space behind the body into which the fluid rushes with more velocity in the measure that its compression is greater" [§.114].

After these general considerations, d'Alembert introduces Robins' ideas on this matter, recognizing that he was the first to assume the existence of an empty space behind the body when the velocity was great enough. In the *Gunnery*[9] Robins questioned Newton's theories, which he substituted by his own ideas. His brilliant analysis was complemented by experimental measurements of the projectile velocities using ballistic pendulums, an instrument of his invention.

Robins thought that all fluids were continuous.[10] Moreover, if they were compressed enough or were moving at slow velocity, a vacuum will never take place; this happens always in water and also in air with the body at low velocity. However,

---

[9]The *Gunnery* was translated into German by Euler as *Neue Grundsätze der Artillerie, aus dem Englischen des Herrn Benjamin Robins übersetzt und mit vielen Anmerkugen versehen* published in Berlin 1745. Euler also added very extensive commentaries to explain the total and partial vacuum generation, and he did a much better exploitation of the experimental results. It is not known if d'Alembert knew this book.

[10]"All the fluids... that their particles either lie contiguous to each other, or at least act on each other in the same manner as if they did" [*Gunnery*, Chap. II. Prop. I, p. 69].

if the fluids were not compressed enough a vacuum can be produced. Here we note that unlike d'Alembert, he never admitted the possibility of a vacuum in liquids.

Robins agreed with Newton's solutions, who give a resistance coefficient of $C_D = 0.5$ for a sphere in a liquid, and $C_D = 1$ for a slow corpuscular motion, that is two times the former. However, when the velocity went faster, the degree of compression on the frontal part was also increased giving a factor up to three times, so $C_D = 1.5$.[11] D'Alembert explains Robins' thoughts and ideas of how, in a continuous motion, the flow of particles from the front to the rear part of the body helps to reduce the resistance. However, when there is a vacuum this reflux cannot take place, and consequently the resistance is increased. D'Alembert concludes "this proposition, or, if it is preferred, this conjecture seems to be confirmed by experiments that *Robins* has made. But as he has not given another theory on this subject, I thought that it would not be useless to expose here some views on this matter".

After the comments about Robins, d'Alembert introduces some observations about the air [§.115-1st], which is the only elastic fluid known at that time. The air, in its natural state, "is compressed by a force equal to that a column of water about 32 pieds [10.4 m]".[12] And as the air is 800 times lighter than the water then "the air is compressed by a force corresponding to an air column of approximately 32x800 pieds [8,320 m]". That means that the atmosphere is assimilated to a sea of air with constant density and a depth of 25,600 pieds, a very common idea in this epoch. Then at ground level, thinking in terms of a vessel discharge, the air would erupt at 1240 pieds per second [403 m/s]. Consequently, if the body moves at a higher velocity, a vacuum must be produced.

The compression force is taken as a measure of the atmospheric pressure [§.115-2nd], like $f_a = \rho_a g h_a$, being ρ_a "the density of the air in its natural state" and $h_a = 25,600$. Now, if ρ is "the density in another state", the compression would be $h_a \frac{\rho}{\rho_a}$, and the force $f = \rho g h_a$. That means that the air will always erupt in an empty space with the same velocity $v_e = \sqrt{2gh_a}$, as he clearly points out.

In the Fig. 18.2 a body is shown moving in an elastic fluid, and passes from the position $ADCd$ to $aD'cd'$ [§.115-3rd]. It is clear that the greater compression will happen in the front, being Aa a measure of it, and it will decrease when moving backwards on the surface, as shown by the segment NV. Therefore, "it seems to me that it will not deviate far from the truth assuming that the density at A is $\rho = \rho_a \left(1 + \varepsilon \frac{v_B}{v_\varepsilon}\right)$", which means that when the velocity was v_ε the density would be the natural density multiplied by $(1 + \varepsilon)$; both v_ε and ε being known values. For another

---

[11] Ibid. p. 72.

[12] The pied of Paris was equivalent to 325 mm. According to the present values, the standard atmospheric pressure at sea level is 101.3 kPa and the air density 1.219 kg/m^3, therefore this density is 1/820 of the water. The atmosphere equivalent high is 8480 m and the eruption velocity 408 m/s.

Fig. 18.2 Motion in elastic fluid

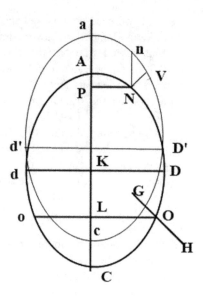

point, as N, the factor $\frac{NV}{Aa} = \frac{dy}{ds}$ must be added as $\rho = \rho_a \left(1 + \varepsilon \frac{dy}{ds} \frac{v_B}{v_e}\right)$, a formula applicable from the compression at A to an expansion at C, where $\frac{dy}{ds}$ is negative.

Finally, the criterion for determining the vacuum zone is equivalent to the one used before in the non-elastic and finite fluids, which was that the normal component of the body velocity was greater than the velocity of compression, which here is 1240 pieds/s. That is the vacuum starts at the point where $-\frac{udy}{ds} = 1240$ pieds per second [§.115-4th].

It seems clear that d'Alembert envisaged that some relation should exist between velocities and densities, because the local velocities depend on pressures and these can alter the natural densities. He was right, but the tools needed for that were in the realm of the thermodynamics, a science still far off from d'Alembert and his world.[13]

For determining the pressure he follows the same procedure as in non-elastic cases, although he includes the density [§.116]. The first step is to take a differential element of the fluid, as shown in the Fig. 17.5, and to express the constancy of the mass contained in the element, instead of its volume as he did before. The right side of the equation Eq. 17.13 must now include the factor $\left(\rho + \frac{\partial \rho}{\partial x} v_B q dt + \frac{\partial \rho}{\partial z} v_B p dt\right)$ due to the variation of the density. The new result is:

---

[13]The relation between the density at any point and the stagnation one is $= \rho_\infty \left(1 + \frac{\gamma-1}{2} M^2\right)^{\frac{1}{\gamma-1}}$, where γ is the ratio of specific heats and M the Mach number. For low velocities the formula is approximated as $\rho = \rho_\infty (1 + 4.325 \cdot 10^{-6} v^2)$, where we note v^2 instead of v. For a velocity of 35 m/s, which would be a hurricane, the ratio of densities is about 1.005, almost nothing. We wish to note that the effect of the compressibility was almost insignificant in the field of aviation until the appearance of jet motors allowed the velocity to be increased to sonic levels.

18 Other Resistances and Fluids

$$\frac{\partial(\rho q)}{\partial x} + \frac{\partial(\rho p)}{\partial z} + \frac{\rho p}{z} = 0 \tag{18.5}$$

We note that the possible variation of the density with time has not been considered.

The equation for the acceleration Eq. 17.14 needs the density as a factor $\frac{\partial(\rho a_x)}{\partial z} = \frac{\partial(\rho a_z)}{\partial x}$ resulting:

$$\frac{dv_B}{dt}\left(\frac{\partial \rho}{\partial z} - \frac{\partial(\rho q)}{\partial z} + \frac{\partial(\rho p)}{\partial x}\right)$$
$$- v_B^2 \left[\frac{\partial}{\partial z}\left(\rho q \frac{\partial q}{\partial x} + \rho p \frac{\partial q}{\partial z}\right) - \frac{\partial}{\partial x}\left(\rho q \frac{\partial p}{\partial x} + \rho p \frac{\partial p}{\partial z}\right)\right]$$
$$= 0 \tag{18.6}$$

There are only a few indications about how to use this equation [§.117]. Firstly, it is assumed that $\rho = \rho(u, x)$ so that $\frac{\partial \rho}{\partial z} = 0$, and that the remaining equations will be exactly similar to those used in the analysis of the body moving in a fluid. He also says that ρp and ρq can be found by the method of approximation by series. As the body contour is known, the density along it can be calculated by the equation $\rho = \rho_a\left(1 + \varepsilon \frac{dy}{ds} \frac{v_B}{v_e}\right)$, and finally the quantities p and q and the pressures.

All this seems no more than a weakly founded list of intentions. For the case in which a vacuum is produced, there is only a sentence to point to the case the non-elastic fluids.

In the *Mss*.117 and App. he tries to work out the solution by a series approximation. We will never know why d'Alembert omitted this reasoning in the *Essay*; probably, given the magnitude of the problem, which he was aware of, he preferred just to leave a brief outline. In this sense, the last words of [§.117] which I quote may be significant: "other hypothesis more real on the value of ρ would make the calculation even more complicated; and all this is no more than a light essay" [§.117].

Chapter 19
Experiments and Theories

19.1 Reflections on Experiments

D'Alembert quotes experimenters in several places in the *Essay* using the results of their experiments in his own reflections as we have already seen. There are also mentions in the Introduction, but apart from this, he adds a complete section [§.75–79] concerning experimenters under the title "Reflections on the experiments that have been made or can be made to determine the pressure of fluids". The title is a bit surprising, because it is not clear what he wanted to say when he refers to experiments that can be made. We imagine the shadow of the Berlin Academy behind these words and very likely the inclusion of this separate section was motivated by the criticism received in the rejection of the *Manuscript*. It contained three articles, *Mss*.73–75 + A, that were a scholium to a corollary, and were rather short.

Edmé Mariotte, Isaac Newton, Daniel Bernoulli and Willem Jacob van's Gravesande, who are all well known to us, are the experimenters who appear in this section. 'S Gravesande will be quoted again [§.93]; and also Benjamin Robins [§.114] and Krafft [§.142–143].

Before beginning our comments, we should clarify a couple of concepts about measurement of the resistance in the epoch. Firstly, the fluid velocity was commonly expressed by a height, as we have said previously. Instead of the concept of "distance traversed by unit of time", they employed "the height due to velocity", this was the height from which a body must fall in order to reach the fluid velocity, or otherwise the vertical ascension it reached if the velocity was turned upwards. We have called this magnitude "kinetic height", and its value is $h_R = v^2/2g$. This concept was widely used by all the authors, and its decline commenced with the introduction of the differential equations in the motion. Secondly, for quantifying the resistance there were two magnitudes. One, was to relate it to the force able to annul the body momentum in a certain time, this was used by Newton. The other

© Springer International Publishing AG 2018
J. Simón Calero (ed.), *Jean Le Rond D'Alembert: A New Theory of the Resistance of Fluids*, Studies in History and Philosophy of Science 47,
https://doi.org/10.1007/978-3-319-68000-2_19

was to assimilate the resistance to the weight of a cylinder of fluid whose base was the cross-section of the body and with a certain height. This last led to height that was related with the former h_R. Both are quoted in the *Essay*. In what follows we will make almost exclusive use of the resistance coefficient C_D, because this is closer to our modern understanding.

Mariotte was the first who obtained experimental values for a square plate inside a stream. The experiments were carried out in the river Seine and the apparatus was basically a plate attached to a set of rods and balanced by a weight, everything fitted in boats anchored in the river. The experiment, as related in his *Traité*,[1] consisted in a squared plate of 6 pouces (162 mm) in a current of 3 ¼ pieds/s (1.06 m/s) with a balance weight of 3 ¾ livres (18 N),[2] working with this figures the value $C_D = 1.22$ results. Mariotte repeated the experiment moving the boat closer to the river bank with less velocity and obtaining $C_D = 1.25$.

D'Alembert reports only $C_D = 1$, which corresponds to a cylinder with a height equal to the kinetic one [§.75]. We think that his interpretation follows the wishful principle that considered t that Nature loves integer numbers, therefore the decimals were only deviations from the true value. He also says that this apparatus was the simplest, which is true; and he adds a quite long list of observations and comments, that are rather superfluous and irrelevant for any experimenter, who would surely know how to do his own job. Even more, he proposes a pendulum within the same framework.[3] The Mariotte test was very simple, but it seems odd that in 70 years nobody had repeated it, or if somebody did there was nothing published.[4]

When he talks about Newton [§.77] D'Alembert quotes the value of $C_D = 0.5$ for the resistance of a sphere in a liquid,[5] obtained by the theory of the cataract. He also adds "this formula that he [Newton] says to have confirmed by a large number of experiments". Certainly Newton carried out a set of experiments[6] that consisted in letting a globe fall in a water tank and measuring the falling time. In the descent the globe is subjected to gravity and buoyancy, and the falling velocity increases asymptotically until it reaches the limit velocity. The process is regulated by a differential equation, whose solution Newton gives as a table to use in the calculations. He made twelve rounds with globes of different diameters and weights, around one inch (25 mm) and from 70 to 300 grains (4.5 to 20 g) in two tanks of 112 and 186 inches (2.8 and 4.7 m) depth.[7] However, besides those experiments, he

---

[1] *Traité de movement des eaux*, p. 214–216. This value is close to the present one.

[2] The conversion factors for the Paris measure units are: 1 pied = 324.83 mm; 1 livre = 0.4895 kg; 1 pouce = 1/12 livre.

[3] Respect of the use of pendulums for measure the resistance we only know the works of Newton, who tried to measure the damping ratio as an index of the exponent of the velocity [Book 2, Prop. 31, Sch.]. Although in theory the correlation is possible, in practice there are many problems, whose results did not satisfy him.

[4] Pierre Bouguer in his *Traité du navire* (1746) [Book. III.I.II.§.I] presents an experimental value of $C_D = 1.21$ without quoting the source, but it is very close to Mariotte's figures.

[5] *Principia* [Book 2, Prop. 39].

[6] Ibid. Prop. 40. This was included in the second edition.

[7] For conversion, 1 inch = 25.4 mm, 1 foot = 12 inch and 1 grain = 0.0648 g.

also made some more with globes falling from the apex of the dome of Saint Paul's church in London at 220 feet (67 m), now with a diameter of 5 inch (127 mm) and a weight of 510–640 gr (33–41 g). Newton checked the results calculating the space that theoretically the balls must traverse in the registered time assuming $C_D = 0.5$ and, comparing the space with the tank depth, he found a very good agreement. We have taken several cases from his data, and working back we have obtained a C_D coefficient from 0.51 to 0.57, quite close to the present values.[8]

Regarding Bernoulli [§.77], d'Alembert presents the formula given with its justification, which turns to be the same as the previous one, that is $C_D = 0.5$ and "that he has equally confirmed by experiments".[9] This experiment was the effect a jet produced in a vessel discharge through a hole in the bottom upon a plate as shown in Fig. 19.1.[10] We are sure that the experiment was carried out, but Bernoulli does not give any practical data, although he says that type of experiment had been done many times before.[11] However, it is worth explaining how Bernoulli reached his conclusions. Previously, in his "Disertatio", he had tried to find the resistance by an imaginary experiment based in inelastic impacts that yields $C_D = 4$, the same as Newton, which is obvious because he had used the impact principles. In these circumstances he moved to the aforementioned experiment but included a small cone in the zone where the fluid impacted, which reduced the effect to $C_D = 1$, as can be seen in the left hand picture of Fig. 19.2. The next step was to assume that this cone was an sphere, and to apply the rule of the square of the sinus for the impact angle, which made an additional reduction to $C_D = 0.5$.

D'Alembert summarizes the situation [§.78]. For Newton both the globe and the cylinder in a liquid have the same resistance, the aforesaid $C_D = 0.5$, although he "does not seem to have sufficiently demonstrated the pretended equality". For Bernoulli the pressure on the cylinder is double that of the globe, but he proved this based on Newton's theory for corpuscular fluids, which would not be applicable to a continuum. He also recalls that the value for the cylinder agrees with Mariotte's results. However, he quoted that "*Daniel Bernoulli* assures that he has tried several experiments on the resistance cylinders, and that they agree with his theory; disregarding the viscosity of fluids, which often contributes to increasing the resistance, especially in the cylindrical bodies". And he continues "That is why, waiting for new experiments on this subject, we take $\pi \delta g h_k$ and $\pi \delta g h_k/2$ for the pressures of the cylinder and the globe", which in our words are $C_D = 1$ and $C_D = 0.5$ respectively.

---

[8]We have to note that in the flow surround a sphere there are two regimes: laminar and turbulent. In the first the resistance is about 0.52, but in the turbulent regime it comes down to 0.2. The transition in governed by a value of the Reynolds number of about 5×10^5. Fortunately for Newton, all the cases were laminar.

[9]"Disertatio de actione fluidorum in corpora solida et motu solidorum in fluidis", *Comm. Acad. Petrop.* Vol. II, 1727.

[10]Ibidem.

[11]Experiments of this kind had been made by Huygens and Mariotte.

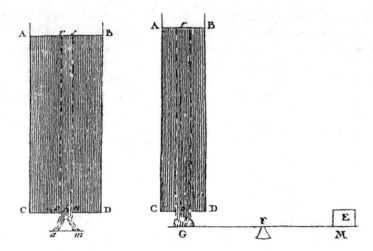

Fig. 19.1 Daniel Bernoulli experiment

Fig. 19.2 Gravesandre apparatus

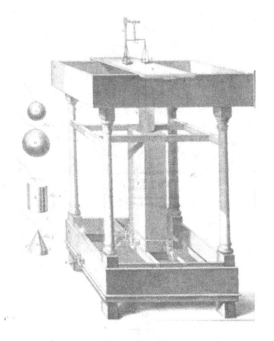

However, d'Alembert recalls his own work in the previous *Traité de l'équilibre*[12] where he assumed the impact theory obtaining the same values as Newton, $C_D = 4$ for elastic corpuscles and $C_D = 2$ for non-elastic ones [§.78]. Also he mentions Euler's work in the *Scientia navalis*, where he took $C_D = 1$ for a plate based on a discharge of a fluid stream against a plate. Really Euler worked theoretically with

---

[12]*Traité de l'équilibre et de mouvements des fluides*, Book. 3, Chap. 1.

two assumptions; one found in the transfer of momentum, basically similar to inelastic impacts, giving $C_D = 2$; the other on a transfer of live force, resulting in $C_D = 1$, and accepting the latter because it agrees better with the experiments.[13] There is another theory from Euler, based on the force that a lid with a hole undergoes before the hole is open, which D'Alembert also rejects.

The last author quoted is 's Gravesande, one of the most relevant experimenters of the age [§.79] and his experiments on fluid resistance.[14] Figure 19.2 shows a reproduction of the measuring apparatus.[15] It was quite large; the length of the four pillars being about one and a half meters.[16] The water fell from the reservoir on the top through a tube, which had the function of test chamber with 105 mm diameter and a length of 471 mm. The tube discharged to a channel of rectangular section. The flow regulation was made by four taps, two with the same opening width, the third double and the four triple. By opening and closing them adequately seven flow regimes could be achieved. The forces were measured by a scale placed on the top of the apparatus. The model was set in the middle of the tube fitted to the scale plate by an horsehair.

He uses four test models, shown on the left: two spheres, of 13.1 and 18.5 mm diameter, the second having a double cross section, a cylinder and a right-angled cone, both also of 13 mm diameter. For each one he made a test round registering the force for each of the seven velocities.[17] He assumed that the resistance had two addends, such as $k_1 v + k_2 v^2$, which he called *prima* and *secunda causa*, and that k_1 was equal for the three models of the same diameter. Therefore in order to obtain k_2, he used results of each round which are presented in a table separating both components. Instead of the true velocity he used the number of taps opened, that is 1 to 7 and 1 to 49, assuming the proportionality of taps and velocity. As an example, working with the data we have found that the velocity at level 6 was 0.36 m/s. The values for the globe were 20 and 26, 20 and 20 for the cone, and 20 and 39 for the cylinder.[18] Considering the *secunda causa* the resistance of a globe is 2/3 of the cylinder and for the cone $1/\sqrt{2}$, which is the ratio of half the diameter to the side.[19]

He found that the cylinder had a resistance equal to the kinetic height, and the sphere had 2/3 of this, that is $C_D = 1$ and $C_D = 0.67$.[20] The experimental data were presented in a table of his *Physices* [§.1945]. For the velocities 0.182 and 0.364 m/s,

---

[13] *Scientia navalis*, I.§.465–473.

[14] *Physices Elementa Mathematica Experimentis Confirmata*, we use the 3rd edition, 1749. Under *Qua Experimenta Resistantiis institumtur*, §.1908–1952.

[15] Op. cit. Tab. LIX, p. 542, descripted in §.1897–1907.

[16] The lengths are given in Rhenish foot, equal to 314 mm. The masses in Dutch pound, equal 0,494 kg.

[17] Cf. §.1908-ss.

[18] Ibid §.1911, 1923 and 1930.

[19] Ibid §.1918–1920.

[20] Ibidem §.1918, 1950.

the forces measured were 2.34 and 9.36 for the sphere and 3.51 and 14.04 for the cylinder, expressed in grains (1.48, 5.90, 2.21 and 8.85 mN); as the diameter was half inch (13.1 mm). The calculation turns out to be $C_D = 0.66$ and $C_D = 0.99$ in both cases. 's Gravesande attributed this difference between globe and cylinder is due to the fact that the component of the force in the body surface is proportional to $v^2 \sin \theta$, instead of $v^2 \sin^2 \theta$, which he admits contradicts a principle that was taken as true by all the authors of Hydraulics.

These ideas and experiments of Benjamin Robins are quoted in [§.114]. Robins made remarkable experiments in order to prove his theories.[21] He used a canon that fired spherical lead bullets of 3/4 in (19 mm) and a mass of 1/12 lb. (37.8 g) against a ballistic pendulum located at different distances from the muzzle. Then with a round of firing he obtained the velocity at each pendulum and consequently the deceleration, which was used to find the resistance. Obviously, he needed a quite large number of firings in order to know the tolerances involved in the process. Furthermore he employed several power charges to produce different muzzle velocities. Summarizing, the pendulum was placed at distances about from 7 to 76 m, and the velocities registered from 520 down to 400 m/s, which corresponded with Mach numbers 1.5 to 1.2. He found coefficients of $C_D = 1.2$ and $C_D = 0.78$ that are in good agreement with our present values. Robins completed these experiments with firings over a lake, measuring the flight time and reducing the power charge to obtain a muzzle velocity of 122 m/s, resulting in $C_D = 0.5$ coinciding with Newton's values. We add that Robins was also the inventor of the ballistic pendulum, and that his trials were the first made at supersonic velocities.

The experiment of George Wolffang Krafft are mentioned in [§.142, 145], whose apparatus is shown in Fig. 19.3.[22] It consists in a vessel with a hole near the bottom where several mouthpieces are inserted. The water jet hits a vertical plate fitted to an arm of a balance that rotated around an axis at H. A weight placed on the left side of the arm, not represented in the original figure, but which we imagine as W, compensated the force produced by the water upon the plate. The efflux water velocity was measured by the trajectory in a free exit until the point X. The diameter of the vessel was 15/16 English feet (286 mm) and the height 2 feet (610 mm). There were three mouthpieces with 13.6 mm diameter but with no details of the length, only that one was long, another short and flush the third. The aim was to correlate forces upon the plate with the velocities and the weight of the equivalent water column. Krafft made a large number of experiments, and he added to these the one that had been made before by Daniel Bernoulli using a similar apparatus, although with a small vessel.[23] Bernoulli obtained a correlation between force and column weight of 0.95 and Krafft from 0.81 to 0.92.

These are the more relevant experiments made before d'Alembert. However, the most significant experiments would have to wait until the next decade when Jean-

---

[21] See *Gunnery*, Chap. II, Prop. II and III.
[22] This Figure has been taken from the *Comm. Acad. Petrop.*, Vol. VIII.
[23] In the "De legibus...", *Comm. Acad. Petrop.*, Vol. VIII.

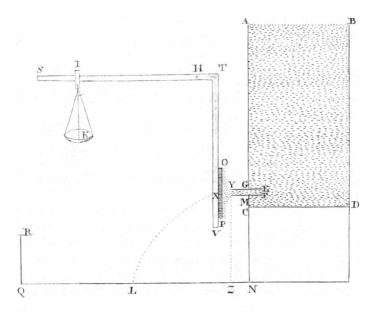

Fig. 19.3 Krafft apparatus

Charles de Borda, better known as the Chevalier Borda, tested single geometrical forms in water and air, whose results were published in 1763 and 1767.[24] Some years later in 1776, d'Alembert himself participated with Charles Bossut and Condorcet in another work focused on maritime matters. Probably his contribution was limited to the mathematical analysis of the data.[25] This work was continued by Bossut alone in 1778.[26]

19.2 Experiment with Pendulums

The pendulum as an instrument to measure the resistance had been used by other authors, such as Newton and Daniel Bernoulli. In the *Essay* there are several mentions of this device and an extensive study [§.95–99] for it use "when the velocity is very small". He presents the mathematical theory of a pendulum as an experimental apparatus, but he does not include any data and neither has he tried to apply the results obtained by other authors. His mathematical skilfulness shines here but adds no insight to the problem of resistance As far as we know, the

---

[24]Both with the title of "Expériences sur la Résistance des Fluides" in the *Memoirs of the Paris Adadémie*.
[25]*Nouvelles expériences sur la résistance des fluides*.
[26]"Nouvelles expériences sur la résistance des fluides" in the *Memoirs of the Paris Adadémie*.

conclusions obtained previously with pendulums were not clear, and probably d'Alembert wanted to help with a more detailed mathematical study, but in our opinion this should have been accompanied either by new experimental test data or by a re-elaboration of the known results. However, d'Alembert was in a theoretical realm here and quite remote from the practical reality.

We wish to highlight that the pendulum is not such a simple instrument as it looks. The damping effect is not only due to the fluid resistance, but also from friction in the fixing point and other string effects such as stiffness, vibration, etc. In addition, the motion is not stationary, which brings in the effect of the added mass; what is more, depending on the size of the basin, deposit or room where the experiment is carried out, the induced fluid motion in the successive oscillation could disturb the mass motion.

D'Alembert's idea was to measure the damping and from it to determine whether the resistance is only proportional to the square of the velocity or not, and in any case to obtain the numerical values of the regulating parameters.

The pendulum and its associated parameters are shown in Fig. 19.4. Firstly only the resistance for the square is considered [§.95]. The mass m starts from the point A, traverses the arc B until the lowest point D and it would ascend up to the A', which is symmetrical to A if there had not been any resistance, but how as it exists the mass will reach the point A'' below the former A'. The evaluation of the resistance of the bop is made relatively to a known condition, that is to say if at a velocity u_r the resistance is f, then at u it will be $f\frac{u^2}{u_r^2}$. The dynamic equation at a point M will give[27]:

$$gdx - \frac{f}{m}\frac{u^2 dy}{u_r^2} = udu \qquad (19.1)$$

This is a nonlinear differential equation that he resolves by an approximate method. His first step is to assume that there is no resistance, that is $f = 0$, which would give $2gx = u^2$. Now introducing the arc y instead of x by means of the approximation $\cos\theta = 1 - \frac{\theta^2}{2}$ and $s = a\theta$, the result for the velocity is $u_0^2 = \frac{g}{a}(2By - y^2)$. The final step is to substitute the u in the second term of the former Eq. 19.1 by u_0 and integrating this term, which leads to:

$$u^2 = \frac{g}{a}(2By - y^2) - \frac{2g}{ma}\frac{f}{u_r^2}\left(By^2 - \frac{y^3}{3}\right) \qquad (19.2)$$

This approximation is justified if the coefficient f/mu_r^2 is small. With $u = 0$ we will have the value of $y_{A''}$ as a second degree equation. However, a simplified solution is attained following a similar line of reasoning [§.96]. Taking $f = 0$, the

---

[27]He represents the reference velocity by a kinetic height and uses p for the acceleration of the gravity.

Fig. 19.4 Pendulum in fluid

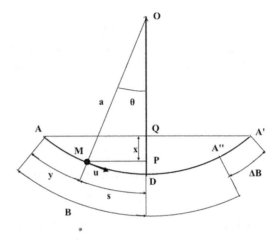

solution would be $y = 2B$ given by the first member of the above formula, which introduced in the second member leads to $\Delta B = 2B - y = \dfrac{4fB^2}{3mu_r^2} = \dfrac{2B^2}{3n_c}$. Where ΔB corresponds to the arc segment $A'A''$ and $n_c = \dfrac{mu_r^2}{2f}$. After k oscillations the formula can be written as $\Delta B_k = \dfrac{2B_k^2}{3n_c}$. Where ΔB_k means the arc lost in the vibration k-th. Now, supposing the period as almost constant, the former equation in finite differences can be transformed into a differential one like $-dB = \dfrac{2B^2}{3n_c}\dfrac{dt}{\tau}$, being $\dfrac{dt}{\tau}$ the differential of the number of vibrations [§.96]. The solution between any two vibrations, like v_1 and v_2, will be

$$\frac{1}{B_2} - \frac{1}{B_1} = \frac{2(t_2 - t_1)}{3n_c \tau} \tag{19.3}$$

He points out that if the resistance is proportional to the square of the velocity "$\dfrac{B'-B}{B'B}$ must be a constant quantity".

In the practical application [§.97] there are several intermediate constants and geometrical remarks, but at the end the constant n_c is found with two experimental determinations.

An additional point, that he calls a vibration, corresponds to a half period. It also implies that two successive vibrations finish at different heights of the vertical. Nonetheless it makes no effect on the theory.

In the next case the resistance will not only be proportional to the square, but also to the single velocity plus a constant value; that is to say, the friction and the viscosity. Obviously this is a rather more complicated case [§.98].

The resistance is now expressed by three terms $f = f_0 + f_1\dfrac{u}{u_r} + f_2\dfrac{u^2}{u_r^2}$, in which we maintain the former criterion in respect to a reference velocity u_r. Using a similar procedure to the previous case the value for the velocity will be

$$u^2 = \frac{g}{a}(2By - y^2) - \int_0^y \left[f_0 + f_1\frac{u_0}{u_r} + f_2\frac{u_0^2}{u_r^2} \right] dy \qquad (19.4)$$

Where u_0 is the velocity without resistance. After the pertinent calculations and taking $y = B$ in the upper limit of the integral, the value for the ΔB results in
$\Delta B = \dfrac{2a}{gm}f_0 + \dfrac{\pi B}{mu_r}\sqrt{\dfrac{a}{g}}f_1 + \dfrac{4B^2}{3mu_r^2}f_2.$

We agree with him that the determination will be very laborious but not so much as to consider it insuperable. His proposal to make it easier follows the same pattern as the introduction of an associated differential equation and a new parameter defined as $T = \dfrac{mu_r^2}{2f_2}$, which converts the former equation to a differential one

$$-dB = \left(\frac{af_0 u_r^2}{gf_2} + \frac{\pi B f_1}{2f_2}\sqrt{\frac{a}{g}} + \frac{2B^2}{3} \right)\frac{dt}{T} \qquad (19.5)$$

Now T takes the place of τn_c. The expression between brackets is a second degree equation in B that is written as $(B+G)(B+A)$, both G and A can be calculated as functions of $\frac{f_0}{f_2}$ and $\frac{f_1}{f_2}$. The solution, after some transformations, is:

$$(t_i - t_0)\frac{A - G}{T} = \ln \frac{B_0 + A}{B_i + A}\frac{B_i + G}{B_0 + G} \qquad (19.6)$$

With at least three observations all the parameters can be obtained, although "indeed by a very long calculation".

The section on pendulums closes with some remarks about two works of Daniel Bernoulli, both dedicated to the analysis of experimental data, and attempting to find the value of the viscosity in the resistance [§.99].

In the first of these works,[28] Bernoulli assumed that the effect of the viscosity was equivalent to reducing the weight of the body, just like d'Alembert does here. Bernoulli analysed the motion of a body falling in a tank, and next he took some of the experiments that Newton had made[29] in order to check his own formulas.[30] He thought that the viscosity was the cause of the difference between the descent times, both theoretical and measured. Therefore, working with the formulas and assuming

---

[28] *Comm. petrop.*, vol III. "Dissertationis de actione fluidorum in corpora solida et motu solidorum in fluidis", part VI.
[29] *Principia*, Book II, Prop. XL. He took the round three in the Newton's experiments.
[30] In *Comm. Acad. Petrop.*, vol II.

$C_D = 0.5$, he found that the viscosity was equivalent to 71/266, which is 0.27 of the total weight, a quite high percentage.[31] After that, he expressed some concerns about the accuracy of the experiments at very slow velocity in order to determine the viscosity. When the motion is faster the formulas agreed with the experiments, but at low speed it was not so easy.

In the second remark d'Alembert refers to two works about the motion of pendulums in fluids.[32] In the first work Daniel Bernoulli studied the motion of pendulum resisted by a force proportional to the square of the velocity, and in the second work the viscosity was added. In both, the theoretical analysis was complemented by an experimental correlation, whose data was taken from the experiments made by Newton with pendulums.[33]

Bernoulli applied Newton's data to these formulas in both cases, and at the end he concluded that when the motion is not too slow the resistance is almost proportional to the square of the velocity, that at slower velocities there are both the former resistance plus a constant one, and when the velocity is extremely slow it is very difficult to determine the law of resistance. However this does not invalidate the theory, because the experiments are very delicate and therefore it is difficult to extract any consequence. D'Alembert, who quotes these notes, suggests that probably it would be better to use the formula $fu^2 + ku + g$ that he had proposed.

Besides, he states that the formula given by Bernoulli in the second case does not agree with the equivalent one obtained by him.

---

[31] With the same formulas but taking another case of Newton's experiments, this value would be 0.45.

[32] *Comm. Acad. petrop.*, vol V. Dissertatio brevis de motibus corporum reciprocis seu oscillatoris, quae ubique resistentiam patiuntur quadrate velocitatis suae proportionalem. (106–125) and "Additamentum ad theoremata. De Motu Corporum curvilineo in mediis resistentibus, in quo resistentiae considerantur quae partim quadratis velocitatem partim momentis temporum proportionales sunt". (126–142).

[33] *Principia*, Prop. XXXI, Sch. Bernoulli took the first experiment.

Chapter 20
Other Motions

Under this title three type of motions are included, the flow in a vessel, the river currents and the jet against a plate; they all are somewhat related to the general theory. Furthermore we add the motion by plane sections, although we have found it difficult to classify.

20.1 Motion in a Vessel

D'Alembert had studied the discharge of a vessel in his *Traité de l'équilibre* with the assumption of motion by plane sections. Now, he wants to outline the application of the new principles to this problem, however, "as these researches here are not directly [related] to my subject, I will only state the principles" [§.148].

Figure 20.1 shows a vessel that has a part of it filled with the fluid *ABFE* held in place by the surface *EF* that acts as a plug. Suddenly, this plug is removed and the fluid starts to move due to the action of gravity. He notes that if the vessel was cylindrical, the fluid would fall down likes a heavy body, so that after a time t the velocity would be $u = gt$. However, the curved walls make both vertical and horizontal velocities to be a function of t, x and z; that is to say $u_x(t, x, z)$ and $u_z(t, x, z)$. He observes the tangency condition in the walls, so at any time $-\frac{dy}{dx} = \frac{u_z}{u_x}\big|_w$. Then he assumes $u_x = \theta(t)q(x,z)$ and $u_z = \theta(t)p(x,z)$, being q and p non-dimensional functions, as in the previous cases. In some way this problem is equivalent to the body in a fluid, but now the body has been converted into the vessel. In a similar way, we will have [§.149]:

$$du_x = d(\theta q) = q\frac{d\theta}{dt}dt + \theta\frac{\partial q}{\partial x}dx + \theta\frac{\partial q}{\partial z}dz \qquad (20.1a)$$

© Springer International Publishing AG 2018
J. Simón Calero (ed.), *Jean Le Rond D'Alembert: A New Theory of the Resistance of Fluids*, Studies in History and Philosophy of Science 47,
https://doi.org/10.1007/978-3-319-68000-2_20

Fig. 20.1 Discharge of a vessel

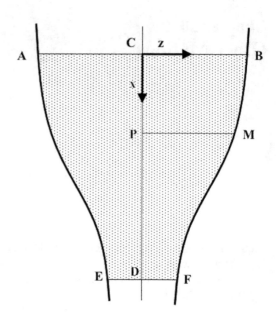

$$du_z = d(\theta p) = p\frac{d\theta}{dt}dt + \theta\frac{\partial p}{\partial x}dx + \theta\frac{\partial p}{\partial z}dz \qquad (20.1b)$$

These equations handled as before lead to $\frac{\partial p}{\partial z} = -\frac{\partial q}{\partial x}$, the well-known Eq. 17.8. Also, for the accelerative forces

$$\gamma_x = g - \theta\frac{\partial q}{\partial z}p - \theta\frac{\partial q}{\partial x}q - q\frac{d\theta}{dt} \qquad (20.2a)$$

$$\gamma_z = -\theta\frac{\partial p}{\partial z}p - \theta\frac{\partial p}{\partial x}q - p\frac{d\theta}{dt} \qquad (20.2b)$$

Making $\frac{\partial \gamma_x}{\partial z} = \frac{\partial \gamma_z}{\partial x}$ the resulting equation will be satisfied with $\frac{\partial p}{\partial x} = \frac{\partial q}{\partial z}$, the Eq. 17.11. Consequently, $dq = \frac{\partial q}{\partial x}dx + \frac{\partial q}{\partial z}dz$ and $dp = \frac{\partial q}{\partial z}dx - \frac{\partial q}{\partial x}dz$. "From these equations the general form of the quantities p and q will be determined".

When the motion starts, the upper and lower surfaces are horizontal and the "lost force" must be perpendicular to them, which implies $p=0$ both in AB and EF, because there are no lateral forces; even more, if the walls are not perpendicular, $q=0$ must be fulfilled by the tangency condition [§.150]. At $t=0$ the force in the column CD must be null, which gives $\int_{x_C}^{x_D}\left(g - \frac{d(q\theta)}{dt}\right)dx = 0$. He says that making $t=dt$ results in $\frac{d\theta}{dt} = g$, then $\theta = gt$.[1] We do not agree with this affirmation, because the integral in $t=0$ can be written as $g(x_D - x_C) = \dot{\theta}\int_{x_C}^{x_D}qdx$, but at dt

---
[1] Instead of $d\theta/dt = g$, it is $d\theta/dt = 1$, and $\theta = 1$.

Fig. 20.2 Motion with curve surfaces

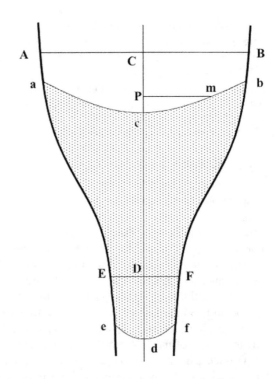

both x_D and x_C will change to $x_D + dx_D$ and $x_C + dx_C$, also $\dot{\theta}$ to $\dot{\theta} + d\dot{\theta}$. Then we will have $g(dx_D + dx_C) = \ddot{\theta} dt \int_{x_C}^{x_D} q dx + \dot{\theta}(q_D dx_D - q_C dx_C)$; which means that $\dot{\theta}$ is not constant.

The determination of q and p is made by the tangency condition on the walls. It seems that he thinks of the polynomial method [§.151].

After some time t the upper and lower surfaces lose their horizontality and become the curves acb and edf as shown in Fig. 20.2, which can be determined knowing the segments $a = Cc$ and $b = Dd$ [§.152]. His idea is based on the perpendicularity of the forces on the surfaces. Thus assuming the upper curve as $s(x)$, at any point its slope must be $-\frac{ds}{dx} = \frac{\gamma_x}{\gamma_z}$; introducing q and p, already known, in Eq. 20.2a, b as $q(x,s)$ and $p(x,s)$, also $\theta = gt$ will lead to the differential equation $-\frac{ds}{dx} = \frac{1-q-qt\frac{\partial q}{\partial x}-pt\frac{\partial q}{\partial z}}{-p+pt\frac{\partial q}{\partial x}-qt\frac{\partial q}{\partial z}}$. Now to relate the above a and b with time two more data are needed; one, that the volume of the fluid is given, and the second that the pressure on the channel CD must be null, that is $\int_{x_c}^{x_d} \left(g - q - qt\frac{\partial q}{\partial x} - pt\frac{\partial q}{\partial z} \right) dx = 0$.

"From which we will have the value of a and b in t, and the problem will be completely solved" [§.152].

The vessel, so far, has a continuous wall so the fluid mass is always the same. If there was an exit, for example at *EF*, the previous condition should be changed in the sense that the former curve *edf* should be *EdF* [§.154].

He recognizes that this method is more rigorous than the one used in the *Traité de l'équilibre*, "but the calculation is so difficult that we must almost give it up" [§.155]. This is true; nevertheless, we have to recognize his willingness in trying it.

20.2 Streams in Rivers

To speak plainly, the reason for the inclusion of this problem in the *Essay* is a claim against the manuscript *Recherches sur le Mouvement des Rivières* by Euler, presented in the Berlin Academy in May 1751,[2] which was known by d'Alembert shortly after. He affirms that "the method the author employs, though it seems to me less simple and less accurate than mine, has something in common with it" [§.160]. He continues to argue not only that he already had found his principles before that manuscript fell into his hands, but he also hints "it would not be impossible that the method outlined in my book was unknown to the author of the Memoir I speak of, and that it would not have helped him in his research on the flow of rivers". He gives only an outline of the problem and this in a rather careless way to our understanding. Grimberg has made a comparative analysis of both solutions and points out that they coincide in the hypothesis, but differ in the variables and methods.[3]

Figure 20.3 presents the riverbed *BN* and the flowing water *CMNB*. The fluid will meet the two equations $dq = \frac{\partial q}{\partial x}dx + \frac{\partial q}{\partial z}dz$ and $dp = \frac{\partial q}{\partial z}dx - \frac{\partial q}{\partial x}dz$, assuming q and p as a polynomial structure [§.156]. The condition for obtaining the equation of the surface *CM* is that the forces combined with the gravity must be perpendicular to it. Once this equation is known along with two points at the surface, *C* and *M*, and their points corresponding in the riverbed, *B* and *N*, the unknown coefficient q and p can be found.

He remarks that the solution would be easier if the riverbed was a shape taken at will, instead of being a given one [§.157]. We understand that he means that the riverbed was a mathematical expression able to support the use of complex functions. So the tangency condition on the bottom of the riverbed would be $\frac{dx}{du} = \frac{q}{p} = \frac{i[\Delta(x-iu)+\Delta(x+iu)]}{\Delta(x-iu)-\Delta(x+iu)}$ where the real and imaginary parts of q and p given are taken as in Eq. 17.23a, b.

When the river motion becomes steady, both the velocity tangency and the perpendicularity of the forces must be met at the surface [§.158]. Therefore $\frac{dx}{dy} = \frac{q}{p}$ and $\frac{dx}{dy} = \frac{\gamma_x}{\gamma_z}$; introducing the values of the acceleration from Eq. 17.3a, b and adding the gravity, we have $gp - p^2\frac{\partial q}{\partial z} = q^2\frac{\partial q}{\partial z}$.

---

[2]It was published in *Memoirs* of the Berlin Academy, vol. XVI, 1767. The date of presentation was given by Jacobi.

[3]Grimberg [1998], p. 73–75.

Fig. 20.3 Motion of a river

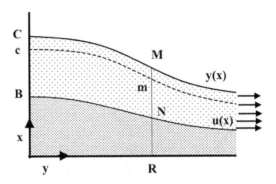

Finally, if the motion is not steady the above equations must be changed [§.160]. After a time dt the surface CM becomes cm and $q, p, \frac{\partial q}{\partial x}$ and $\frac{\partial q}{\partial z}$ become $q + \left(q\frac{\partial q}{\partial x} + p\frac{\partial q}{\partial z}\right)dt$; $p - \left(p\frac{\partial q}{\partial x} - q\frac{\partial q}{\partial z}\right)dt$; $\frac{\partial q}{\partial x} + \left(q\frac{\partial^2 q}{\partial x^2} + p\frac{\partial^2 q}{\partial x \partial z}\right)dt$; $\frac{\partial q}{\partial z} + \left(q\frac{\partial^2 q}{\partial x \partial z} + p\frac{\partial^2 q}{\partial z^2}\right)dt$. Also dx and dy become $dx + dq\,dt$ and $dy + dp\,dt$, also $d\left(\frac{dx}{dy}\right) = \frac{d^2x\,dy - d^2y\,dx}{dy^2}$. Substituting all these values we will have the rather complicated equation:

$$d\left(\frac{g - \frac{\partial q}{\partial x}q - \frac{\partial q}{\partial z}p}{\frac{\partial q}{\partial z}q - \frac{\partial q}{\partial x}p}\right) = 2\frac{\partial q}{\partial x}\left(\frac{g - \frac{\partial q}{\partial x}q - \frac{\partial q}{\partial z}p}{\frac{\partial q}{\partial z}q - \frac{\partial q}{\partial x}p}\right)dt + \frac{\partial q}{\partial z}dt$$
$$-\frac{\partial q}{\partial z}\left(\frac{g - \frac{\partial q}{\partial x}q - \frac{\partial q}{\partial z}p}{\frac{\partial q}{\partial z}q - \frac{\partial q}{\partial x}p}\right)^2 dt \qquad (20.3)$$

We agree with his last note "I have only indicated here the method, because the details would lead me too far" [§.160].

20.3 Jet Stream Against a Plate

For a quite long time the effect of a jet against a plate and the resistance of a submerged plate were considered as equivalent phenomena.[4] Furthermore, the first experiments concerning the resistance were carried out measuring the force of a jet from a hole in a vessel against a plate fitted to a scale.[5] It was Daniel Bernoulli who understood that this identification was faulty and backed the case experimentally.[6] D'Alembert recognized his work [§.137] and explained that this question had some relation to the resistances of fluids, and that "its solution is easily deduced from my

---

[4] Simón Calero [2008], p. 137-ss.
[5] Jean-Baptiste du Hamel in his *Regiæ Scientarum Academiæ Historia Parisiis*, quoted experiments in 1669.
[6] *Comm. Acad. Petrop.* Vol. VIII, "De legibus quibusdam mechanicis...", 1736 (1741).

Fig. 20.4 Jet against plate

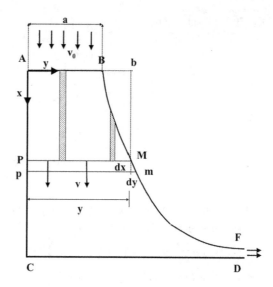

principles, and besides [this problem] will give me the "opportunity to make some new observations about this matter consistent with experience".

A general view of the flow is depicted in the Fig. 20.4. The stream is assumed with plane sections and flowing from an upper vessel through the exit *AB* with a velocity v_0. Due to the presence of the plate *CD* the jet will expand following the curve *BMF* and it will end parallel the plate. For this discharge he establishes two points: first, the velocity along the curve *BMF* is constant, and second, if the size of the hole is small the fluid moves by parallel sections [§.138]. The first one is justified for the same reasons given the analysis of the stagnation zones [§.36], where he said only normal forces act upon the fluid contour and no tangential ones. Therefore the velocity along *BMF* must be v_0. From the second one, it is clear that the vertical velocity in any slice has be constant, "which will not be far from the truth" [§.139]. Therefore, at a point of the contour, such as *M*, the vertical velocity, obtained by continuity $av_0 = vy$, must be equal to the vertical component of the tangential one, that is $v = v_0 \frac{dx}{ds}$. Combining both velocities $dx = \frac{ady}{\sqrt{y^2 - a^2}}$ is found [§.140]; whose solution is $x = a \ln \frac{y + \sqrt{y^2 - a^2}}{a}$ [§.144] which can be also expressed as $y = a \cosh \frac{x}{a}$. It easy to see that this equation does not meet the condition of tangency with *CD* at infinity, as will be pointed out later [§.142].

As an additional consequence the pressure must be constant as well upon the entire slice. To prove it, d'Alembert repeats [§.140] the same steps that he had made before [§.27]. Assuming the pressure at the exit as p_0, and using Eq. 16.8, it is:

$$p = p_0 + \frac{1}{2}\rho v_0^2 \left(\frac{1}{a^2} - \frac{1}{y^2} \right) \tag{20.4}$$

20 Other Motions

Next comes the calculation of the force upon the surface *PM*, which he names "pressure on *PM*". As we can see, *PM* is not a physical object but an imaginary one, therefore this force is an internal one that will turn out to be real when the surface coincides with the plate *CD*. But there is one thing more, the force upon *PM* will be $p_M y$, however the calculation that he makes gives the force on the fluid mass *ABMP*.

Effectively, "the pressure that would come from the part *BbM* must be subtracted from the previous quantity [$p_M y$]" [§.140]. This means that he is computing all the forces upon the fluid part, and obviously the one coming from *AB* is taken as null because p_0 acts upon the entire fluid, which is equivalent to taking it as zero. To calculate the pressure over a free surface like *BM* is not easy, so he imagines *BM* as a physical wall. Therefore the force obtained will be the thrust, or reaction, of the fluid upon the upper vessel.

According to this, and following his steps, the force at any point will be:

$$F = p - \int_a^y p\,dy = \rho v_0^2 a \left(1 - \frac{1}{y}\right) = 2\rho g a h_0 \left(1 - \frac{1}{y}\right) \quad (20.5)$$

Where $h_0 = 2gv^2$ is the height due to velocity or the kinetic height. It means that the force is less than double of the weight of a fluid column of h_v height and base *a*. He tries to explain all this in a rather confused article [§.141]. However, an easy way to arrive at this formula would be evaluating the momentum lost in the fluid dominion *ABMP*; the fluid mass rate is $\dot{m} = \rho a v_0$, therefore $F = \dot{m}(v_0 - v)$.

He points out that according to this equation the curve *BM* will cross the plane *CD* when $x = AC$ and $dx/dy \neq 0$. "It follows that the direction of the fluid when it reached the *CD* plane is not exactly parallel to that plane, but makes an angle with the plane *CD* which is more acute the farther it is from *AB*" [§.142]. Therefore, if the plane is large the force will be always a little less than $2\rho g a h_0$, which agrees with Krafft's experiments.[7]

In the case that the vessel is circular rather than rectangular, the motion is axisymmetric [§.143]; the continuity would be $a^2 v_0 = y^2 v$ and the contour equation would be $dx = \frac{a^2 dy}{\sqrt{y^4 - a^4}}$, whose solution leads to elliptic integrals, but the final force will be:

$$F = 2\rho g \pi a^2 h_0 \left(1 - \frac{a^2}{y^2}\right) \quad (20.6)$$

As D'Alembert says, it is "An expression that agrees again with *Krafft's* experiments". In fact these experiments and also Bernoulli's were made with this configuration.

So far, no forces acting upon the fluid have been assumed, but now the weight is added [§.144]. The first consequence is that the hypothesis of the constancy of the velocity along the contour cannot be maintained, because the gravity component

---

[7] *Comm. Acad. Petrop.*, "Vi venæ aquæ contra planum incurrentis experiment", Vol. VIII, 1736 (1741).

tries to lengthen the element Mm, written as ds in Fig. 20.4. Based on the well-known equation $s = \frac{1}{2}gt^2$ he takes $ds = gt dt \frac{dx}{ds}$ and $d^2s = g dt^2 \frac{dx}{ds}$, since $g\frac{dx}{ds}$ is the above mentioned gravity component. The value dt is substituted as before, arriving at $d^2s = \frac{gy^2 dx^2}{v^2 a^2} \frac{dx}{ds}$, which he integrates as $\frac{1}{2} ds^2 = \frac{y^2 dx^2}{v^2 a^2}(x+m)$, or:

$$dx^2 + dy^2 = \frac{2y^2 dx^2}{v^2 a^2}(x+m) \tag{20.7}$$

This equation is difficult to integrate. An approximate solution is to introduce in place of x in the right member its value without gravity, called $x_0(y)$. Then it will give a relation between dx and dy like $dy^2 = \left(\frac{2gy^2 x_0(y)}{v^2 a^2} - 1\right) dx^2$, " an equation that represents almost exactly the curve BMD, above all in the points that are not too close to D" [§.144].

For the force on the plate the weight of the fluid must be added, i.e. the formula $\Delta F = \rho g \int y_0 dx = \rho g a \sqrt{b^2 - a^2}$, calculated assuming the contour of the case without gravity. He notes that this does not agree with Krafft's experiments, although they all were made for a horizontal jet.

Next, d'Alembert engages in a lengthy commentary about Bernoulli's solution, quoting his experiments and also Krafft's ones. His comments about Bernoulli [§.145] are sometimes difficult to follow because in the explanations d'Alembert gives he translates the same symbols that Bernoulli had used which causes confusion with the symbols d'Alembert used formerly. Bernoulli thought that the motion was as if the fluid flowed in narrow channels through which a small body moved driven by a tangential force (Fig. 20.5).[8] That is how he envisaged what later we will be call streamlines.

Although the development was carried out using the kinetic height, we think that is easier for us to use the ordinary velocity and in order to simplify the presentation. At a point N the fluid moves with a tangent acceleration $\dot{u} = u du/ds$ and a centripetal u^2/R, being R the curvature radius equal to $R = ds dy/d^2 x$. The projection of both on the axis X will give the vertical acceleration $\dot{u}_v$ and consequently $du_v = \dot{u}_v dt$; after some operations we will obtain

$$du_v = du \frac{dx}{ds} + u \frac{d^2 x}{ds} = d\left(u \frac{dx}{ds}\right) \tag{20.8}$$

Integrating along the ENG curve, the change of the vertical velocity would be:

$$\Delta u_v = u \frac{dx}{ds}\bigg|_E^G = u_G \left(\frac{dx}{ds}\right)_G - u_e \tag{20.9}$$

He assumed that the slope in the point G was null, so $\Delta u_v = -u_e$ for any streamline. For the calculation of the force upon the plane, he evaluated the

---

[8]Ibid.

Fig. 20.5 Bernoulli solution

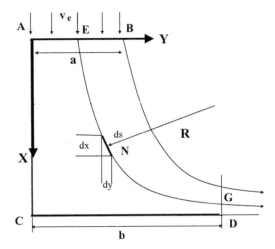

momentum change in a time dt; such as $Fdt = -\rho\Delta u_y \dot{m}_e dt = \rho a u_e^2 dt$. Taking into account the Eq. 20.7 it would be:

$$F = \rho a u_e^2 - \rho a u_G u_e \left(\frac{dx}{ds}\right)_G \qquad (20.10)$$

Which was reduced to $F = \rho a u_e^2$, which is equal to the weight of a column of base a and height twice the kinetic h_e.

D'Alembert makes a fine analysis of that theory, rightly pointing out that each streamline has a different velocity when arriving at the border of the plate. Therefore, the hole AB must be divided in elements da and the total force will be the integration of all them. He calculates the vertical component of force on the element ds following the same method as Bernoulli, which will be $df_y = -\rho \dot{u}_y ds$. Operating with the previous values this is:

$$df_y = -\rho u^2 \frac{d^2x}{ds} - \rho u du \frac{dx}{ds} = -\rho d\left(u^2 \frac{dx}{ds}\right) + \rho u du \frac{dx}{ds} \qquad (20.11)$$

Whose integral is:

$$f_y = -\rho u^2 \frac{dx}{ds}\bigg|_E^G + \rho \int_E^G u du \frac{dx}{ds} dx = \rho u_E^2 - \rho u_G \left(\frac{dx}{ds}\right)_G - P \qquad (20.12)$$

The term $P = -\rho \int_E^G u du \frac{dx}{ds} dx$ is always positive because u is decreasing. Therefore, considering the slope at G as zero, the force of any channel is $f_y = \rho u_E^2 - P = 2g\rho h_e - P$, and the total force some as $F = 2g\rho a h_e - \int P da$, always less than $2g\rho a h_e$.

He states that this formula coincides with Bernoulli's when $du = 0$, which means constant velocity in all the curves, "but this latter hypothesis, as well as the method itself, seems susceptible to some difficulties". In this line he builds a close channel formed by two streamlines and two horizontal lines and analyses the equilibrium conditions, assuming constant velocity. As he finds contradictions between Bernoulli's hypothesis and the consequences of the constant velocity, he concludes that the method is faulty, claiming that his own hypothesis of the constancy of the velocity in any slice comes nearer to the truth. We find this argument circular, because d'Alembert conclusion is based on $du = 0$, which was not an hypothesis in Bernoulli's equations. We think that Bernoulli's assumptions were more realistic, but had the added difficulty of obtaining the parameter for each streamline.

To finish with this article, d'Alembert expresses his preferences for Krafft's experiments, first because he made more, and they also gave a slightly lower value than the expected ones. The first is true, Bernoulli only made one experiment and Krafft seven. The second is not exactly true, the agreement was 0.95 for Bernoulli and from 0.81 to 0.92 in the seven of Krafft.

"Moreover, it could be applied to the research of the pressure of a fluid stream the method that I explained in this book [functions q and p]. But the calculation would be difficult" [§.146]. A stagnation zone would be produced in the centre, mFM, where $p = q = 0$ (Fig. 20.6), and on the plate $q = 0$ and $\frac{\partial q}{\partial z} = 0$. The pressure on the plate would be $\frac{1}{2}v^2 \int 2\pi y dy (1 - p^2 - q^2)$. He thinks of q only as a sum of powers of x and z, which means that q should contain x in all the terms. In this condition the problem is undetermined, so for this reason he tries "to look for another route to find the pressure of a fluid stream against a plane, though perhaps less rigorous and less direct".

He extends the problem to a plate moving inside a fluid [§.147]. "The values of p and q seem to me indeterminate in these cases, or rather indeterminable; in such a way that is as impossible to compare the theory with the experiment, even in this case that seems the simplest of all". That is somehow surprising, because they are two quite different phenomena, which Bernoulli clearly separated.

Summarizing, the entire issue is developed by means of the plane sections hypothesis, at least as an approximation in which he believes up to a certain extent. It is only in the two last articles [§.146–147], where he tries to explain how to apply his principles but with few words and less results. This somewhat different to a "solution... easily deduced from my principles" [§.137] as he had stated at the beginning.

20.4 Hypothesis of the Plane Sections

The hypothesis of the plane section was a common issue in the study of fluid motion in vessels. D'Alembert himself had considered it as a "truth of Nature" in the *Traité*,[9] and there the applications seemed to agree quite well with the experiments. He confesses to have been so captivated by this hypothesis that he thought to use it

---

[9]*Traité de l'équilibre*, §.10.

Fig. 20.6 Impact

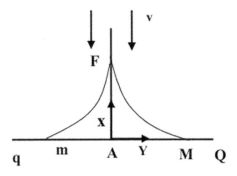

for deducing the fluid resistance theory. However, he notes that in many cases its application led to the result of null resistance and other consequences very contrary to experience. "Maybe it will not be useless to explain this more at length" [§.100].

According to his words, and just as a conjecture, probably in the wake of the *Traité de l'équilibre* he tried to find the resistance using this method but without success. However, in our opinion, once he established the new theory, the plane section hypothesis must be abandoned. Even when in some cases it could be acceptable, these would be exceptions, never a rule. Therefore, it is quite surprising that he pretends to prove that the consequences of an invalid theory are against the logic of Nature. We consider this reasoning to be flawed in its origin. Nevertheless, we will go through his explanations.

A body *AKB* [§.101], assumed to be two-dimensional and symmetrical, half of which is represented in Fig. 20.7, advances with velocity u in a vessel or channel limited by the wall QQ'. Any point N, moves to N' in the time dt and the fluid contained in the area $aANn$ is displaced to $TT'N'N$ and forced to move backwards with a velocity v. For the continuity condition it is clear that $(a-y)v = uy$. That means that the fluid between the body and the wall is moving in plane sections whose velocity increases from the point A to a maximum at K and decreases afterwards to zero at B. So this mass of fluid is subjected first to acceleration, later to deceleration, i.e. accelerative forces. The former velocity v is measured with respect to the channel, and its value relative to the body would be $v_r = u + v = \frac{au}{a-y}$. With this configuration d'Alembert analyses three different cases.

In the first one [§.102] the body is in repose and it is pushed suddenly with a velocity u_0. The body will respond with another velocity such as "u'_0 the actual velocity that it must have because of the resistance of the fluid". We can understand this as a pass to the limit with an acceleration u_0/τ when $\tau \to 0$, as we have done before. In any case the momentum lost by the body must be equal to that acquired by the particles in motion, so

$$S_B \rho_B (u_0 - u'_0) = \rho \oint_C v'_0 dy = \rho u'_0 \oint_C \frac{y dy}{a-y} = \rho u'_0 \int_A^B \frac{y^2 dx}{a-y} \qquad (20.13)$$

The identification of the two last integrals is founded on the hydrostatic theorem [§.23] and Eq. 16.5. Solving for u'_0:

Fig. 20.7 Body moving in channel

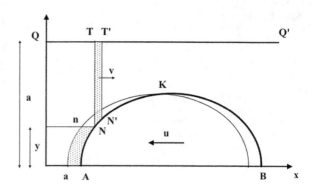

$$u'_0 = \frac{S_B \rho_B u_0}{S_B \rho_B + \rho \int_A^B \frac{y^2 dx}{a-y}} \tag{20.14}$$

Now, without giving any reason, the particular case when both densities are equal, $\rho = \rho_B$ is taken. Also, as $S_B = \int y dx$, the new velocity would be:

$$u'_0 = \frac{S_B u_0}{\int_A^B \frac{aydx}{a-y}} \tag{20.15}$$

D'Alembert makes three annotations to this result. One, that "in the first instant of the motion following the experiment $u'_0 = u_0$", while the formula shows $u'_0 < u_0$ except when $a = \infty$. This is quite surprising, because the model proposed is based on that the body after receiving u_0 responds with u'_0. The second is that u'_0 is smaller in the measure a is smaller; to which he argues: "I do not know any experiment that proves that the velocity lost at the first instant is greater in the manner as the vessel is narrower". This assertion makes no sense; moreover the fact that he does not know these possible experiments is not a valid argument. Third, the apparently most meaningful argument, i.e. that the vessel figure does not seem to influence the motion "because, as it has been proved above, the motion that the body communicates to the fluid particles extends up to a very short distance around it (*art.* 71 and 72)". What he calls proof is just an estimation based on very weak arguments that he accepts as true, and it would be contrary to the hypothesis by which the velocity is equal in the entire plane. Again he invokes experiment; we should point out that is a fictitious experiment, an experiment he invented to accommodate the desired solutions.

From our point of view, if the body is given the velocity u_0 and there had been a resistance, it would start moving slowly until it stopped and its momentum $S_B \rho_B u_0$ should have been transferred to the fluid. But in the end the entire fluid would be at rest, because its motion only occurs in the fluid dominion limited by two vertical planes at A and B; therefore the initial momentum $S_B \rho_B u_0$ would have vanished. Consequently, there cannot be any resistance. Another way to prove this is the calculation of $\oint_C \frac{ydy}{a-y} = [-a \ln(a-y) - y]_C$, in Eq. 20.12 which is zero, irrespective of the body density.

In the second case [§.103] the body moves with a variable velocity $u(t)$ and this will affect the velocity v of any parallel section and also its acceleration $\dot{v}$. In the calculations u and u', y and y' are used as values at t and $t+dt$. Here we have a discrepancy with the interpretation of y'. In the *Essay* it is taken as $y(x+udt+vdt)$, while it should be $y(x+udt)$, because the velocity v is relative to the channel and its acceleration is due to the change in the height $a-y$, which obliges the fluid particles at a point to move faster or slower according to the contour slope at this point. The physical displacement of the body udt cannot be added with the displacement of the fluid vdt. Then the value for $\dot{v}$ will be:

$$\frac{dv}{dt} = \frac{d}{dt}\left(\frac{uy}{a-y}\right) = \frac{y}{a-y}\frac{du}{dt} + u\frac{d}{dy}\left(\frac{y}{a-y}\right)\frac{dy}{dx}\frac{dx}{dt} = \frac{y}{a-y}\frac{du}{dt} + \frac{au^2}{(a-y)^2}\frac{dy}{dx} \quad (20.16)$$

We notice that dv/dt has two addends; the first only takes place in non-uniform motions, while the second always exists. Therefore:

$$S_B\rho_B\frac{du}{dt} = \rho\frac{du}{dt}\oint_C \frac{y}{a-y} + \rho a u^2 \oint_C \frac{dy}{(a-y)^2} \quad (20.17)$$

The second term is null, and applying t the hydrostatic theorem o the first one as before, we will have:

$$S_B\rho_B\frac{du}{dt} = \rho\frac{du}{dt}\int \frac{y^2 dx}{a-y} \quad (20.18)$$

He finds this equation absurd, since to meet it this $S_B\rho_B$ must be equal to $\rho\int\frac{y^2 dx}{a-y}$, which will not happen in an infinity of cases because one term depends on the body mass and the other on the vessel shape. Furthermore, if it is so, any value of du would be valid.

We have to add that in the derivation in the *Essay* the additional term $\frac{au^2 y}{(a-y)^3}\frac{dy}{dx}$ appears, which is due to the mentioned introduction of vdt, which leads to $\oint_C \frac{ydy}{(a-y)^3}$, which is zero. Therefore, it would have not altered the former conclusions.

One more comment. This case can be reduced to the first, assuming $u(t)$ as an integration of differential pulses, so $du = u_0 dt$. He has done the same in the body moving in a fluid [§.56, §.388].

In the third case [§.104] the particles contiguous to the surface will move tangentially, which seems more realistic. This velocity will be such that the velocity relative to the body, that we have called v_r, will be a component of the velocity along the surface v_t, that is

$$v_t = v_r\frac{ds}{dt} = \frac{ua}{a-y}\frac{ds}{dx} \quad (20.19)$$

Fig. 20.8 Velocities in a symmetric body

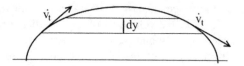

Also, for convenience the body will be symmetric with respect to the meridian section. The value for $\dot{v}_t$ is

$$\frac{dv_t}{dt} = \frac{a}{a-y}\frac{du}{dt}\frac{ds}{dx} + au^2\left(-\frac{1}{(a-y)^2}\frac{dy}{dx}\frac{ds}{dx} + \frac{1}{a-y}\frac{d^2s}{dx^2}\right) \qquad (20.20)$$

Now for the calculation of the total pressure he will integrate correctly along the contour and take the horizontal component that is $F_x = \rho \oint \frac{dv_t}{dt} dx$.

Just for the examination of the forces at both sides of any strip of dy the second term of dv_t/dt is disregarded, because its two addends change the sign while the first does not change[10] (Fig. 20.8). Therefore the final equation is:

$$-S_B(\rho_B - \rho)\frac{du}{dt} = \rho\frac{du}{dt}\oint \frac{a}{a-y} ds \qquad (20.21)$$

He introduces the term $S_B\rho$, which is equivalent to the buoyancy of the body. The factor $\dot{u}$ appears on both sides, which means that $\dot{u} = 0$, and as a consequence u constant, and the resistance null, "which is absurd".

Let us note that the solution is similar to the previous case, but his commentaries are different, even although in both they finish with an invocation to the absurd.

In the final scholium [§.105], he affirms the hypothesis of the plane section should be rejected because its results of null resistance are against experience. However, it is not a reason for rejecting it in the discharge of vessels, because in this case the hypothesis is fairly consistent with the experiments and "experience must be here our guide".

We insist that the hypothesis must be rejected in itself, because it is contrary to his main findings in fluid dynamics. We think that that the argument he uses are inconsistent and the entire section does not add any contribution to his *Essay*. It was unnecessary and misleading.

---

[10]In terms of xy, $\frac{ds}{dx} = \sqrt{1+y'^2}$ and $\frac{d^2s}{dx^2} = \frac{y'y''}{2\sqrt{1+y'^2}}$.

Chapter 21
The Oscillation of Floating Bodies

This chapter is dedicated to the oscillations of a body floating in a fluid, obviously a liquid. The procedure is to calculate the forces generated when the body is subjected to small perturbations, such as displacements and rotations, from the equilibrium position. The resulting forces and moments, jointly with the body mass and geometry, will determine the dynamics of the system, usually interpreted as pendulum oscillations.

The body displacements generate hydrostatic forces, but they also induce motions in the fluid and consequently hydrodynamics forces. D'Alembert does not ignore this fact, and considers this motion similar to a body in a fluid already studied by him; therefore the velocity induced will have the components up and uq leading to pressure terms in du/dt and u^2. However, on one hand he had assumed that the coefficient of the du/dt was null, and on the other he also considers that the effect of the u^2 can be neglected because both the velocity and the oscillation amplitude are very small [§.118]. In the $Mss.106$ he made a more detailed analysis in order to justify the nullity of the coefficient of du/dt as a consequence of the tangency of the velocity to the body surface, but he did not include these arguments in the *Essay*. In the end, only hydrostatic forces intervene, which means that this problem has little to do with the fluid resistance. The topic of the floating oscillations, apart from its theoretical interest, was very important in ships, the biggest machines of those days. It was linked to the static stability of a ship, a very old problem that after almost two millennia arrived at a solution with the "invention of the metacentre" more or less a decade before the writing of the *Manuscript*.[1] Geometricians such as César Marie de La Croix,[2] Pierre Bouguer,[3]

---

[1] See Chapter 4 "Inventing the Metacenter" of *Ships and Science* by Larrie D. Ferreiro.

[2] "Commentationes de statu aequilibrii corporum humido insidentium", *Comm. Acad. Petrop.*, Vol. X.

[3] For the metacenter see *Traité du navire*, Book II, Sect. II. The Sect. III is dedicated to the oscillations and in Ch. II the formula is given.

Leonhard Euler,[4] Johann[5] and Daniel[6] Bernoulli made contributions to this subject. They all only considered the static components, it was d'Alembert who envisaged the dynamic effects, even when he denied the *du/dt* component, from which the added mass is derived, and, as we have said, is quite significant in naval engineering.

Three cases are studied in the *Essay*: vertical displacement, vertical displacement coupled with rotation; and oscillations of an arbitrarily shaped body, that is, one, two and three degrees of freedom. The first two are flat or two-dimensional bodies, which are later transformed to axisymmetric ones. The methodology followed is to displace the body from its initial situation, and to calculate the forces and moments generated and next to apply the general equations of dynamics like $m\ddot{x} = F(x)$ and $I\ddot{\theta} = M(\theta)$. Once the differential equations are presented he makes comments and simplifications, but for their solution he refers to other works.

21.1 Rectilinear Oscillations

The first one, called "straight oscillations" [§.118], is shown in Fig. 21.1, where the body is subjected to the weight $g\rho_B V_B$ and the buoyancy $g\rho V_S$, so that:

$$\rho V_B \frac{du}{dt} = g\rho_B V_B - g\rho V_S \qquad (21.1)$$

The buoyancy depends on the vertical displacement, or better the difference with the equilibrium position, in which both forces are equal; leading to an equation type

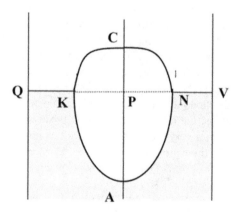

Fig. 21.1 Straight oscillations

---

[4]*Scientia navalis*, In the Chapter 4 "The oscillation of ships" and Vol. 2, §381–387.

[5]"De corporum aquae insidentium oscillationibus, et de invenienda longitudine penduli simplicis oscillationibus illis isochoni", *Opera omnia*, vol. 4, p. 286–296, 1742.

[6]"De motibus oscillatoriis corporum humido insidentium", *Comm. Acad. Petrop.*, Vol. XI.

$m\ddot{x} = kx$. He insists in that it is necessary to prove that the "fluid pressure... comes only from the gravity", i.e. that the components of du/dt and u^2 are null.

21.2 Curvilinear Oscillations

For the "curvilinear oscillations" [§.119] we present the body in Fig. 21.2, C and G being the centres of gravity and buoyancy respectively, separated horizontally by the distance β and vertically by ζ.

Initially the fluid level is at BD, and after a time t the body moves upwards x, and rotates counter clockwise the angle θ; therefore the rectangle BDD_1B_1 emerges, and the wedge BAB_2 also emerges while the opposite DAD_2 immerses. We have shown them separately in the Fig. 21.2 for clarity. D'Alembert makes a lot of effort for a detailed geometric calculation of these three geometrical forms [§.120–122] as a function of the variables x and θ; once these calculations are made they are converted in forces and moments with respect to the centre of gravity. It is easy to see that there will be a vertical force of $-(a+b)x - \frac{1}{2}b^2\theta + \frac{1}{2}a^2\theta$, a counter clockwise moment of $-\frac{1}{2}\rho g(b^2 - a^2)x - \frac{1}{3}\rho g b^3\theta - \frac{1}{3}\rho g a^3\theta$, and another additional moment of $\rho g V_S \zeta \theta$ due to the buoyancy center displacement to G_2.

According to the principles of dynamics, one equation for the forces and another for the moments will result, which are:

$$M_B \frac{d^2x}{dt^2} = \rho g V_S - \rho_B g V_B - \rho g(a+b)x - \frac{1}{2}\rho g b^2\theta + \frac{1}{2}\rho g a^2\theta \qquad (21.2a)$$

$$I \frac{d^2\theta}{dt^2} = \rho g V_S \beta - \frac{1}{2}\rho g(b^2 - a^2)x - \frac{1}{3}\rho g b^3\theta - \frac{1}{3}\rho g a^3\theta + \rho g V_S \zeta \theta \qquad (21.2b)$$

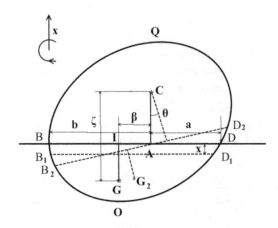

Fig. 21.2 Curvilinear oscillations

The terms $\rho g V_S - \rho_B g V_B$ for the force, and $\rho g V_S \beta$ for the moment are due to the initial outbalance of the body, since it is separated from the equilibrium position; they are both supposed to be small.

He compared this solution with Johann Bernoulli's one, who assumed that the centre of gravity C was fixed and the body was symmetrical. The motion is assimilated to an isochronous pendulum, measuring the frequency as its length, which, we have to say, was a common practice at the time [§.123].

For the integration of the two equations he refers to a previous work that he had sent to the Academies of Paris and Prussia [§.124]. These equations belong to the generic type

$$\frac{d^2x}{dt^2} + Ax + By + M(t) = 0 \qquad (21.3a)$$

$$\frac{d^2y}{dt^2} + Cy + Dx + P(t) = 0 \qquad (21.3b)$$

Two linear equations, whose variables can be separated by a linear transformation as $u = x + \nu' y$ and $w = x + \nu'' y$, obtaining the next two ones, that are easier to solve:

$$\frac{d^2u}{dt^2} + Eu + F(t) = 0 \qquad (21.4a)$$

$$\frac{d^2w}{dt^2} + Gw + H(t) = 0 \qquad (21.4b)$$

One particular case [§.125] is when $a = b$, that is the centre of gravity is the middle of BD, which gives two equations with separated variables:

$$M_B \frac{d^2x}{dt^2} = \rho g V_S - \rho_B g V_B - 2\rho g a x \qquad (21.5a)$$

$$I \frac{d^2\theta}{dt^2} = \rho g V_S \beta - \frac{2}{3} \rho g a^3 \theta + \rho g V_S \zeta \theta \qquad (21.5b)$$

Leaving out the constant terms, which would be very small or null, both equations are of the type $\ddot{x} \pm \omega_n^2 x = 0$. For the positive one the solution is a harmonic function such as $x = x_0 \cos \omega_n t$, but for the negative it is $x = x_0 \cosh \omega_n t$, which is divergent.[7] The first of these equations always gives a harmonic solution, but the second depends on the condition $\zeta \leq \frac{2a^3}{3V_S}$, which will give the maximum value of ζ required to keep the dynamic stability. Otherwise, "the value of θ will no longer contain circle arcs and the oscillation will not be infinitesimal"; we would say that the motion is unstable. We must remember that the metacentre in a ship was

---

[7] A physical example for the first can be a mass retained by a spring or a pendulum and for the second the same mass now rejected by the spring or the same pendulum at the upper position.

21 The Oscillation of Floating Bodies

the maximum height that the centre of gravity can reach above the buoyancy centre to maintain the ships stability. The classical definition for it is:

$$\overline{BM} = \frac{2}{3} \frac{\int y^3 dx}{V_S} \tag{21.6}$$

Where $y(x)$ is the hull section at the water level. In our case this section is a rectangle of constant width a. Therefore, as could not be otherwise, $\overline{BM} = \frac{2a^3}{3V_S}$.

If the fluid was not indefinite but contained in a limited vessel, the fluid level varies at the same time as the body moves [§.126] and some changes in the equations will be necessary. So, if the width of the vessel is w_F, and the emerged surface is $\Delta V_{S=}(a+b)x + \frac{1}{2}b^2\theta - \frac{1}{2}a^2\theta$, therefore the fluid level will vary $\Delta x_F = -\frac{\Delta V_S}{w_F - (a+b)}$, we include the minus sign to note that the value is contrary to the body fluid displacement. This value is introduced in the equations, which become a bit more complicated.

He presents the particular case when the body makes only rectilinear oscillations with $\theta = 0$, which will give:

$$M_B \frac{d^2 x}{dt^2} = \rho g V_S - \rho_B g V_B - \frac{\rho g (a+b) w_F}{w_F - (a+b)} x \tag{21.7}$$

Finally [§.127], if the body was not plane but a solid revolution whose section $QBOD$ was the meridian-section, the theory will still valid, although the geometric forms immersed and emerged wedges would be more complicated to express.

21.3 Irregular Bodies

"The problem becomes much more difficult when the body is of irregular shape" [§.128]. It is true, because now there are three rotations plus the vertical motion. We give only an outline of the method he used, which we think is correct; however, the complicated geometry, the variables used, and several misprints have made following the calculations difficult.

Figure 21.3 shows a body of this type whose flotation line is $BHDJ$. The centres of gravity and buoyancy are at C and G. This body is sectioned by a vertical plane $QBOD$ that contains both C and G. Now, let us assume a system of axis fixed to the body: one vertical (CX), another horizontal contained in the former plane (CZ) and the third perpendicular to them (CY).

First step will be to calculate the hydrostatic volumes produced when the body rotates around the former axis and when it moves vertically. The method is the same as used before with the emerging and immersing wedges, but more it is complicated due to the geometry. He assumes that the vertical rotation will only produce second degree effects that can be ignored [§.129]. For the other two rotations an additional

Fig. 21.3 Irregular body

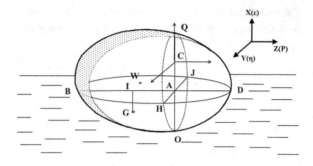

summered volume with a centre of gravity will be produced linearly depending on the angle. So, for η around the line HJ it would give $V_{W\eta} = a_\eta \eta$ for the volume and $y_{W\eta} = \beta_\eta \eta$ and $z_{W\eta} = \gamma_\eta \eta$ for its centre of gravity [§.129]. Similarly when it rotates around BD, the results would be $V_{WP} = a_P P$ and $y_{WP} = \beta_P P$ and $z_{WP} = \gamma_P P$ [§.130]. Additionally, a vertical displacement α would be $V_{Wx} = a_x \alpha$ and $y_{Wx} = \beta_x \alpha$ and $z_{Wx} = \gamma_x \alpha$ [§.131]. Taking the three volumes jointly the total submerged part will be $V_S - V_{W\eta} - V_{WP} - V_{Wx} = V_S - V_W$, the initial volume being V_S [§.133]. The centre of gravity will be at the point W, whose coordinates are a bit more complicated to find, but they can be expressed as $y_W = a_\eta \eta + a_P P + a_x \alpha = \omega_Y$ and $z_W = z_I + \beta_\eta \eta + \beta_P P + \beta_x \alpha = z_I - \omega_Z$; they all are linear functions [§.132]. From these formulas it is clear that in the case of no motions at all, the submerged volume would be V_S at z_I.

Now, the forces that act upon the body will be its weight gM_B applied at the gravity centre C, plus the buoyancy $\rho_F g(V_S - V_W)$ at the point W and upwards.

The dynamic analysis is twofold, one the rotation and other the vertical motion [§.134]. The latter is easier, because it is regulated by the sum $\pi'' = \rho_F g(V_S - V_W) - gM_B - gM_B \frac{d^2 x}{dt^2}$.

To solve the problem, d'Alembert notes that he will use the method previously taught in his work the *Precession of the Equinoxes*. The idea is to assume that the body is subjected to three forces F, G and π', one parallel to each axis and applied respectively in the points $P_F(\theta, 0, \zeta)$, $P_G(\xi, \chi, 0)$ and $P_{\pi'}(0, \nu', \mu')$, as shown in Fig. 21.4, "which must be destroyed"; that is, as we understand it, they have to be equal to the corresponding accelerative forces or d'Alembert forces. But additionally the previous π'' must be added to π', given $\pi = \pi' + \pi''$. Also the application point must change considering that the two first terms of π'' are acting at the former W and the third at C, as $\pi \mu = \pi' \mu' + g\rho_F V_S \omega_Y$ and $\pi \nu = \pi' \nu' + g\rho_F V_S (z_I - \omega_Z)$, where he has implicitly assumed that $V_W \ll V_S$.

The equilibrium will be expressed by these three equations:

$$F\zeta - \pi\mu = 0 \qquad (21.8a)$$
$$G\xi - \pi\nu = 0 \qquad (21.8b)$$
$$F\theta - G\chi = 0 \qquad (21.8c)$$

Fig. 21.4 Dynamic forces

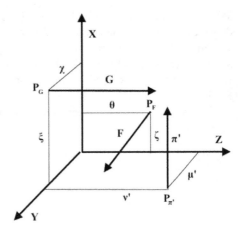

We could understand the problem considering that the body is impelled by an external moment $\vec{C}$ due to the hydrostatic forces, they are all vertical and are called π', which will induce a moment C_i around each of the three axis and which must be "destroyed" by another moment like $I_i \ddot{\theta}_i$. However, he takes $\vec{C} = \vec{r} \times \vec{F}$, which gives as result an arbitrary set of forces and application points, but π' being the only real one.

The motion of the body, defined by a trihedron fixed to it, with respect to the initial position, is made by three successive rotations. The first one is the angle P around the axis Z, with the velocity $\dot{P}$ and the acceleration $\ddot{P}$; next is the angle η over Y with $\dot{\eta}$ and $\ddot{\eta}$; and the last one is the angle ε around the body axis X. In the calculations d'Alembert takes $y = \cos\eta$ for the second rotation as a new variable instead of η. We can see here something quite similar to what would be called later Euler's angles. For the moments of inertia only two are considered, one with respect to Cp, and other to the plane ZX; we do not understand the reason, because the latter should be about the axis Cq.

In the development of the three components of $I_i \ddot{\theta}_i$, nonlinear terms appear containing $y, dy, d^2y, d\varepsilon, d\varepsilon^2, d^2\varepsilon$ and dP, dP^2, d^2P in rather complicated equations [§.134], with the lack of the appropriated moment of inertia. To obtain a solution he has to simplify them, neglecting the motion around X and taking the rest of the angles as small. This leads to three equations [§.135] which we present here slightly modified for a better understanding:

$$\rho_B V_B \frac{d^2x}{dt^2} = g\rho_F V_S - g\rho_F (V_B - V_W) \quad (21.9a)$$

$$K_\eta \frac{d^2\eta}{dt^2} = g\rho_F V_S (z_I - \omega_Z) \quad (21.9b)$$

$$K_P \frac{d^2P}{dt^2} = g\rho_F V_S \omega_Y \quad (21.9c)$$

Where V_B and V_S are the body and submerged part volumes, and ρ_B and ρ_F the densities, K_P and K_η the moment of inertia, some of which are given in a different way in the *Essay*. The three functions V_W, ω_Z and ω_Y have been defined previously, and they are linear polynomials of x, η and P. The above differential equations are a linear system whose solution had been made in his work about the Precession of the Equinoxes, to which it refers.

Finally [§.136], he makes some additional comments about the former hypotheses.

Chapter 22
Reflections on Fluid Equilibrium

D'Alembert finished with an Appendix in the *Essay* containing "some reflections on the laws of the Equilibrium of Fluid that I have not thought necessary to include in the body of the work in order not to interrupt the sequence of matters, but they seem to me worthy of being submitted to the judgment of the wise; and besides they have a fairly immediate relation with the subject of this book". Behind these words his main aim was the problem of the shape of the Earth.[1]

He comes back to the law of equilibrium equation $\frac{\partial(\rho Q)}{\partial x} = \frac{\partial(\rho R)}{\partial y}$, previously deduced [§.19], that now he will be obtain by another method. Let us recall that the former was based on the equilibrium of a close rectangular channel, now the condition is the same, but the channel is limited by curves upon which the forces are perpendicular [§.161]. In Fig. 22.1 this new channel is represented as $MNOm$, whose side Mm is perpendicular to the force A acting upon M. Besides, the two lateral MN and mO are also perpendicular at Mm and whose respective lengths are inversely proportional to the densities. This means that both have the same "weight" $d\zeta = \rho \cdot MN \cdot A = \rho' \cdot mO \cdot A'$, the apostrophe denotes the conditions at m, and $d\zeta$ is an auxiliary parameter. It is clear that there will not be any force acting on the channel Mm due to the perpendicularity, therefore the forces acting at N and O must be perpendicular to NO neglecting higher order terms. In this framework d'Alembert proposes to prove that in order for this to take place the following equation must be met $\frac{\partial(\rho Q)}{\partial x} = \frac{\partial(\rho R)}{\partial y}$.

First, it is clear that from the geometry and the force components there are several relations at M among the different parameters involved, such as $A = \sqrt{R^2 + Q^2}$, $\tan \alpha = \frac{Q}{R} = \frac{dy}{dx}$, $dx = \sin \alpha \, ds$, $dy = \cos \alpha \, ds$, etc.; also $dR = \frac{\partial R}{\partial x} dx + \frac{\partial R}{\partial y} dy$, $dQ = \frac{\partial Q}{\partial x} dx + \frac{\partial Q}{\partial y} dy$ and $d\rho = \frac{\partial \rho}{\partial x} dx + \frac{\partial \rho}{\partial y} dy$.

---

[1] In the Appendix the *Mss*.21–24 are included and complemented with six more articles.

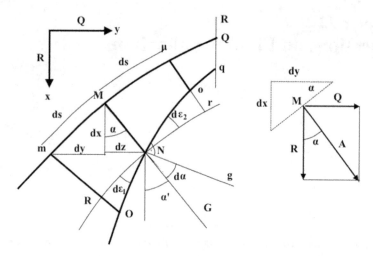

Fig. 22.1 Channels in equilibrium

The basic idea is to equate the angle $d\varepsilon_1$, obtained from the weight coming from M, to the angle $d\alpha$, from the external field of forces. D'Alembert starts with $d\varepsilon_1 = \frac{RO}{ds} = \frac{mO-MN}{ds}$ and after he makes a very fine and detailed analysis in which intervene the geometry and derivatives for both mO and MN. We can try to simplify the procedure, although following the same path as him. The quantity RO can be interpreted as $RO = d(MN)$ and as $MN = \frac{d\zeta}{\rho A}$, then $RO = d(MN) = d\left(\frac{1}{\rho A}\right) = \frac{\partial}{\partial x}\left(\frac{1}{\rho A}\right)dx + \frac{\partial}{\partial y}\left(\frac{1}{\rho A}\right)dy$. After operating, we will have:

$$d\varepsilon_1 = \frac{d\zeta}{\rho(R^2+Q^2)^2}\left(-RQ\frac{\partial R}{\partial x} + R^2\frac{\partial R}{\partial y} - Q^2\frac{\partial Q}{\partial x} + QR\frac{\partial Q}{\partial y}\right) \\ - \frac{d\zeta}{\rho^2(R^2+Q^2)}\left(Q\frac{\partial \rho}{\partial x} - R\frac{\partial \rho}{\partial y}\right) \qquad (22.1)$$

We have to note that dy is taken as negative because the slope of mM is. Now, enlarging mM with the segment $M\mu$, of equal length, and drawing the perpendicular μo with the same weight as MN and mO we will have $ro = -RO$ and $d\varepsilon_1 = d\varepsilon_2$.

The angle $d\alpha$ is formed by line Ng perpendicular to Ono and NG, which is an extension of MN. Now, for geometrical considerations $d\varepsilon_1 = d\varepsilon_2 = d\alpha$. As $\tan\alpha = \frac{Q}{R}$, it follows that $d\alpha = \frac{RdQ-QdR}{R^2+Q^2}$; and after another set of calculations we obtain for $d\alpha$:

$$d\alpha = \frac{d\zeta}{\rho(R^2+Q^2)^2}\left(R^2\frac{\partial Q}{\partial x} + RQ\frac{\partial Q}{\partial y} - QR\frac{\partial R}{\partial x} - Q^2\frac{\partial R}{\partial y}\right) \qquad (22.2)$$

As $d\varepsilon_1 = da$, equating the last two equations we find again $\frac{\partial(\rho Q)}{\partial x} = \frac{\partial(\rho R)}{\partial y}$. This lengthy demonstration could be shortened as Grimberg points out,[2] because the sides of the two mentioned angles are perpendicular tos each other.

At this point [§.162], d'Alembert recalls that when the density is constant at each layer the general equation is simplified to $\frac{\partial Q}{\partial x} = \frac{\partial R}{\partial y}$, and he raises the point of how both equations can occur at the same time. His answer is that at any of these layers the density will meet $\frac{\partial \rho}{\partial x} dx + \frac{\partial \rho}{\partial y} dy = 0$ and also $\frac{dy}{dx} = -\frac{R}{Q}$ for the perpendicularity; combining both expressions $Q \frac{\partial \rho}{\partial x} = R \frac{\partial \rho}{\partial y}$ is obtained. Now expanding $\frac{\partial(\rho Q)}{\partial x} = \frac{\partial(\rho R)}{\partial y}$ and applying the last equality the condition $\frac{\partial Q}{\partial x} = \frac{\partial R}{\partial y}$ is found. He remarks that this last one only occurs if the "weight" is perpendicular to the layer, while the former does not require this condition. Therefore he concludes "that the method of *art. 19* is the only really general one to determine the laws of the equilibrium of fluids". We quote here Truesdell's remark about the falseness of this assertion.[3]

With regards to the fluids of heterogeneous density, he thinks that the fluids will be in equilibrium if the general equation is met [§.163]. However he mentions the case of fluids of different densities that cannot mix together, which is confirmed by experiment. He advances a curious explanation for this: "But the reason which prevents this mixture is that gravity is *the same* for all these fluids, is that the equation could not take place when they are mixed".

The quantities R, Q and ρ have been considered as functions only of x and y. Now he introduces the possibility of another variable, and as an example he takes the particular case of a third variable z for R and Q, which would be constant for each layer, but different from one to other, and taking ρ as a constant [§.164]. This would imply a change in the angle $d\alpha$, given Eq. 22.2, so that this angle should be increased by the term $\frac{d\zeta}{R^2+Q^2}\left(R\frac{\partial Q}{\partial z} - Q\frac{\partial R}{\partial z}\right)$, so $\frac{\partial Q}{\partial x} = \frac{\partial R}{\partial y}$ will be now:

$$\frac{\partial Q}{\partial x} - \frac{\partial R}{\partial y} + R\frac{\partial Q}{\partial z} - Q\frac{\partial R}{\partial z} = 0 \qquad (22.3)$$

Leaving aside another simplification, at the end d'Alembert says that even when the former Eq. 22.3 is more general, only $\frac{\partial(\rho Q)}{\partial x} = \frac{\partial(\rho R)}{\partial y}$ and $\frac{\partial Q}{\partial x} = \frac{\partial R}{\partial y}$ will be applicable to the research of the fluid resistance.[4]

The former reasoning refers to a layer inside a fluid that can be assumed as unlimited. Now, he takes the case of a finite fluid, which is limited by a surface. In the unlimited fluid any inner element was pressed by the column above it; this does not occur when there is an external surface. The question that d'Alembert presents is what happens in this type of fluid [§.165]. One solution could be that the pressure is equal in the entire surface, since all its elements are equally pressed from the

---

[2]Grimberg [1998], p. 373.

[3]Truesdell [1954], p. LVI.

[4]In the *Mss.*23 he notes in hydrostatics or hydrodynamics.

inner side. Another possibility is to consider that this layer consists of globules, each one pressed by the next, but ignoring the layer below. This should be equivalent to an inflated elastic globe; it is easy to see that the lateral tension in the skin is related to the internal pressure so $pR_c = kT$, the curvature radius being R_c, which is in an inverse ratio with the force, that is to say the pressure. As consequence, the external surface must be either plane or spherical. He thinks that other external configurations can exist, so he assumes that one or several points are moving, which means that there will be a set of forces in order to maintain this condition. Therefore, any shape of the surface can be sustained by the corresponding forces. Other arguments are directly against the possibility of any globule layer and the equality of the forces. At the end he tries to follow the hypothesis given by MacLaurin, who considered the external surface as a level one.

A related problem arises when the fluid is a blend of various fluids of different densities, either they make a homogenous or heterogeneous mixture [§.166]; and since there are reasons to assume that each level layer need not necessarily be of the same density in all its extension. To prove that the isopotential and isodensity layers need not coincide [§.167], he presents a mass of fluid made up of layers of equal density (Fig. 22.2).

DAEF is a layer of density $\rho(r_0)$ whose geometry is defined by $r = r_0 + \alpha\eta Z(\theta)$, being α very small parameter common for all the layers, so that *DAEF* is very close to a circle. It is clear that the angle ε at the point *P* is $\tan\gamma = \frac{dr}{rd\theta} \approx \frac{\alpha\eta dZ}{r_0 d\theta}$. This point *P* is subjected to a central force along *PC* like $F_C = f_0 + \alpha f_C Z_C$ and a normal one along *PP'* as $F_T = \alpha f_T Z_T$. Let summarize the parameters involved; α is common for the entire fluid, r_0 jointly with $\eta(r_0)$ define the curve; f_0 and $f_C(r_0)$ are for the normal forces; and $f_T(r_0)$ the tangent ones and the functions Z, Z_T and Z_C. The goal of d'Alembert is to find the relation among all them in order for the equilibrium to exist.

The closed channel *PP'QQ'* must be in equilibrium, which means that [*QQ'*] + [*QP*] = [*Q'P'*] + [*P'P*] or [*QQ'*] − [*P'P*] = [*Q'P'*] − [*QP*]. The force upon *PP'* is the sum of the F_C and F_T components, that is $dF_{PP'} = (F_C \sin\gamma + F_T \cos\gamma)rd\theta = (F_C\gamma + F_T)rd\theta$, which, neglecting the high order terms, turns out to be $dF_{PP'} = \alpha\rho f_0\eta dZ - \alpha f_T Z_T d\theta$. For the force on *QP* we have $dF_{QP} = F_C dr = \rho(f_0 + \alpha f_C Z_C)(dr_0 + \alpha Z d\eta)$, which operating, and also neglecting terms, results in $dF_{QP} = \rho(f_0 dr_0 + \alpha f_0 Z d\eta + \alpha f_C Z_C dr_0)$. For the segments *QQ'* and *P'Q'* the forces are calculated as $dF_{QQ'} = dF_{PP'} + d^2F_{PP'}$ and $dF_{P'Q'} = dF_{PQ} + d^2F_{PQ}$, and for the equilibrium of the channel $d^2F_{PP'} = d^2F_{PQ}$. All this leads to:

$$\frac{dZ}{d\theta}d(\rho f_0\eta) - Z_T d(\rho f_T r_0) = \rho f_0 \frac{dZ}{d\theta}d\eta + \rho f_C \frac{dZ_C}{d\theta}dr_0 \qquad (22.4)$$

This will be the general equation of the equilibrium for that configuration. Now, if the layers are level it would be required that the forces along *PP'* were zero, which means $\eta f_0 dZ - Z_T d\theta = 0$

D'Alembert continues analysing several particular cases such as $\frac{dZ}{d\theta} = \pm Z_T$, which would produce $d(\rho f_0\eta) \mp Z_T d(\rho f_T r_0) = 0$, or $\frac{dZ}{d\theta} = \pm Z_T = \pm\frac{dZ_C}{d\theta}$, which

Fig. 22.2 Isodensity fluid mass

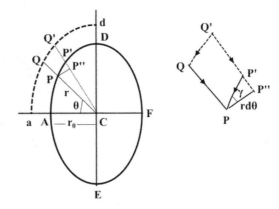

would give $\rho f_C dr_0 \pm \rho f_0 d\eta$. However, we do not see what the purpose of these transformations is.

All this has supposed that the density changes continuously. However, he affirms that the solution would be valid if the fluid had finite discontinuities, and that level surfaces crossing those discontinuities could exist [§.168].

The last three articles point directly to the shape of the Earth. "I will remark on this occasion, that it seems to me that the problem of the figure of the Earth has not yet been solved in a rather general way, with the hypothesis that the attraction is in inverse ratio to the square of the distance and that the Earth is made of a mass of fluids of different densities" [§.169]. He goes directly to this problem which had been addressed by Clairaut, from whom he declares he takes the main formulas.[5] Clairaut had studied the Earth as a solid rotating body, made up of solid spheroids of different densities and with an elliptical shape and surrounded by a finite layer of fluid. The problem was to find the shape of the fluid layer assuming both density and shape of the solid part to be known (Fig. 22.3).

Clairaut calculated the forces at any point at the surface of the external layer due to attraction of both the solid and fluid parts, according to Newton gravitation law. Assuming that the shape is very near to a circle, along CM he found a force $F_C = 4\pi \left[A + \frac{\rho_F}{3}(1 - a^3) \right]$, where $A = \int_0^a \rho r^2 dr$, $a = \frac{CE}{CF}$, ρ the density of the elliptical layer dr, ρ_F the fluid one and $CM = 1$. In the force along CV there are two components, one due the elliptical shape and the other from the Earth's rotation. The first one is $F_{VE} = \frac{8\pi}{5} q \left[D + \rho_F (\varepsilon_F - a^5 \varepsilon_a) \right]$, where $D = \int_0^a \rho d(\varepsilon r^5)$, ε is called "ellipticity" being $\varepsilon = \frac{EE' - NS}{EE'}$ and $q = MQ$. For the effect of the rotation, the ratio of the centrifugal force to the gravity is introduced as φ, and then $F_{VC} = 4\pi q \varphi \left[A + \frac{\rho_F}{3}(1 - a^3) \right]$. Finally, Clairaut said that the ellipticity of the external surface follows the ratio $2q\varepsilon_F = \frac{F_{VE} + F_{VC}}{F_C}$.

---

[5]*Théorie de la figure de la Terre*, Part II, Chap. II.

Fig. 22.3 Earth constitution by Clairaut

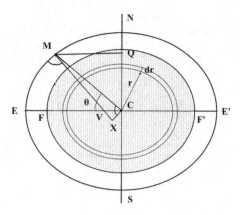

From his part, d'Alembert's considers the Earth to be made up entirely by a fluid arranged in spheroidal elliptical layers expressed as $r = r_0(1 + \varepsilon \sin^2\theta)$ (Fig. 22.2), each layer with a density $\rho(r_0)$ [§.169].[6] The rotation is also taken into account as a ratio of the centrifugal force to the weight in the equator. The problem is to find an equation that relates ε with r_0.

Using the same method as Clairaut, he calculates the forces and equilibrium in the channel $PP'Q'Q$, as he had done before (Fig. 22.2). The formulas given are rather complex, so we will skip them until the final general equation:

$$d^2\varepsilon + \frac{2\rho r_0^2}{\int \rho r_0^2 dr_0} dr_0 d\varepsilon - \varepsilon \left(\frac{6}{r_0^2} - \frac{2\rho r_0}{\int \rho r_0^2 dr_0} \right) dr_0^2$$
$$- \frac{r_0^2}{\int \rho r_0^2 dr_0} d\left(\frac{1}{r_0^4} d\left(\frac{M\rho dr_0}{r_0 d\rho} \right) \right) = 0 \qquad (22.5)$$

In this formula, $M = \dfrac{Kr_0^5}{\rho} \dfrac{d\rho}{dr_0}$, $K = \dfrac{\varepsilon \int \rho r_0^2 dr_0}{r_0^2} - \dfrac{\int \rho d(\varepsilon r_0^5)}{5r_0^4} - \dfrac{r_0 F}{5} + \dfrac{r_0 \int \rho d\varepsilon}{5}$ $-\dfrac{r_0 A\varphi}{2}$, and $F = \int \rho d\varepsilon$ are a set of linked constants up to φ and A. The first is the ratio of the centrifugal force and the second is related to the weight at the equator, although the value of A is not well defined.

Then, the former equation could be rewritten as:

$$\frac{d^2\varepsilon}{dr_0^2} + Q_1(r_0)\frac{d\varepsilon}{dr_0} + Q_2(r_0)\varepsilon + Q_3[r_0, \varepsilon, K, \varphi] = 0 \qquad (22.6)$$

Really, this is an integral-differential equation, due to the complexity of the last term.

Now, d'Alembert will try to analyse the conditions that makes $Q_3 = 0$, which would reduce the equation to:

---

[6]For coherency we use the variables r_0, ρ, ε and $\sin\theta$ instead r, R, ρ and z.

22 Reflections on Fluid Equilibrium

$$\frac{d^2\varepsilon}{dr_0^2} + Q_1(r_0)\frac{d\varepsilon}{dr_0} + Q_2(r_0)\varepsilon = 0 \qquad (22.7)$$

"Which is the only one that has been found so far, but that is not as general as the previous one" [§.170]. He explores the possibility of $M=0$, which could be either $d\rho=0$ or $K=0$. The first means constant density and the second that the spheroids are also level surfaces. Another possibility is that $\dfrac{\rho dr_0}{r_0\, d\rho} = Cte$, which implies $\rho = Ar_0^n$, or $d\left(\dfrac{\rho dr_0}{r_0\, d\rho}\right) = Br_0^4 dr_0$.

Finally, in the last article [§.171], he insists in the fact that the principle of the level surfaces ($M=0$) and the equality of the weight of the columns ($K=0$) give the same equation. Finishing: "I will discuss this subject later in depth".

Annexes

Annex I

Fluid Mechanics Equations

A particle of an ideal and non-viscous fluids moving along a trajectory is ruled by the following equations:

$$\frac{D\rho}{Dt} + \rho \nabla \cdot \vec{v} = 0 \tag{A1}$$

$$\rho \frac{D\vec{v}}{Dt} = -\nabla p + \rho \vec{f}_m \tag{A2}$$

$$\rho \frac{De}{Dt} = -\nabla \vec{q}_R - p \nabla \cdot \vec{v} \tag{A3}$$

Which are respectively the continuity, momentum and energy equations. The operator D/Dt, called substantial or material derivative, represents the variation of any physical property along the trajectory. For a fixed system of axis they former equations become:[1]

$$\frac{\partial \rho}{\partial t} + \nabla \cdot (\rho \vec{v}) = 0 \tag{A4}$$

$$\rho \frac{\partial \vec{v}}{\partial t} + \rho (\vec{v} \cdot \nabla) \vec{v} = -\nabla p + \rho \vec{f}_m \tag{A5}$$

$$\rho \frac{\partial e}{\partial t} + \rho (\vec{v} \cdot \nabla) e = -\nabla \vec{q}_R - p \nabla \cdot \vec{v} \tag{A6}$$

The first two are known as Euler's equations.

---

[1] We note that for a property like φ results $D\varphi/Dt = \partial \varphi/\partial t + \vec{v} \cdot \nabla \varphi$.

© Springer International Publishing AG 2018
J. Simón Calero (ed.), *Jean Le Rond D'Alembert: A New Theory of the Resistance of Fluids*, Studies in History and Philosophy of Science 47,
https://doi.org/10.1007/978-3-319-68000-2

If the forces derive from a potential as $\vec{f}_m = -\nabla U$, the second one is rewritten as:

$$\frac{\partial \vec{v}}{\partial t} + (\vec{v} \cdot \nabla)\vec{v} = -\frac{\nabla p}{\rho} - \nabla U \qquad (A7)$$

Introducing the vorticity $\vec{\omega} = \nabla \times \vec{v}$ it can be written as:[2]

$$\frac{\partial \vec{v}}{\partial t} + \frac{1}{2}\nabla\left(|\vec{v}|^2\right) - \vec{v} \times \vec{\omega} = -\frac{\nabla p}{\rho} - \nabla U \qquad (A8)$$

The motions in which $\vec{\omega} = 0$ are called irrotational.

Bernoulli Equation

In order to analyze the motion along the trajectory at any point, we express the velocity as $\vec{v} = v\vec{i}_s$, being $\vec{i}_s$ the unitary versor of the trajectory. Then the Eq. A2 will be:

$$\frac{Dv\vec{i}_s}{Dt} = -\frac{\nabla p}{\rho} - \nabla U \qquad (A9)$$

Which developed becomes:

$$\vec{i}_s \frac{Dv}{Dt} + v\frac{D\vec{i}_s}{Dt} = -\frac{\nabla p}{\rho} - \nabla U \qquad (A10)$$

Now projecting the forces upon the trajectory, that is multiplying be $\vec{i}_s$, it results:

$$\frac{Dv}{Dt} + v\vec{i}_s \cdot \frac{D\vec{i}_s}{Dt} = -\frac{\vec{i}_s \cdot \nabla p}{\rho} - \vec{i}_s \cdot \nabla U \qquad (A11)$$

As $\vec{i}_s \cdot \frac{D\vec{i}_s}{Dt} = 0$ and $\vec{i}_s \cdot \nabla(\square) = \frac{\partial()}{\partial s}$, the equation is:

$$\frac{\partial v}{\partial t} + v\frac{\partial v}{\partial s} = -\frac{1}{\rho}\frac{\partial p}{\partial s} - \frac{\partial U}{\partial s} \qquad (A12)$$

---

[2]We recall $(\vec{v} \cdot \nabla)\vec{v} = \frac{1}{2}\nabla\left(|\vec{v}|^2\right) - \vec{v} \times (\nabla \times \vec{v})$.

Rearranged:

$$\frac{\partial v}{\partial t} + \frac{\partial}{\partial s}\left(\frac{v^2}{2} + U\right) + \frac{1}{\rho}\frac{\partial p}{\partial s} = 0 \tag{A13}$$

When the density is constant we have the Bernoulli equation for a non-steady motion:

$$\frac{\partial v}{\partial t} + \frac{\partial}{\partial s}\left(\frac{v^2}{2} + \frac{p}{\rho} + U\right) = 0 \tag{A14}$$

If the motion is steady, the equation will take its more common expression:

$$\frac{v^2}{2} + \frac{p}{\rho} + U = C_S \tag{A15}$$

Where C_S is a constant for the particular trajectory, which in the steady motion is also a streamline.

If the motion is irrotational and the density constant, the velocity will have a potential φ, such as $\vec{v} = \nabla \varphi$. Therefore the Eq. A8 will become:

$$\nabla\left(\frac{\partial \varphi}{\partial t} + \frac{1}{2}|\nabla \varphi|^2 + \frac{p}{\rho} + U\right) = 0 \tag{A16}$$

Consequently:

$$\frac{\partial \varphi}{\partial t} + \frac{1}{2}|\nabla \varphi|^2 + \frac{p}{\rho} + U = C(t) \tag{A17}$$

Where $C(t)$ is a constant that does not depend on the particular trajectory. Once known φ, this formula can be used to find the pressures and the forces upon any surface.

Non Steady Motion of a Sphere

The potential flow a sphere placed in a moving is:

$$\varphi(r, \theta) = u_F\left(r + \frac{R^3}{2r^2}\right)\cos\theta \tag{A18}$$

Expressed in spherical coordinates, where R is the radius. This potential corresponds to a Rankine oval when the source and the sink coincide in a doublet. The pressure at any point of the sphere is given from A17.

$$p = -\rho\left(\frac{\partial \varphi}{\partial t} + \frac{1}{2}|\nabla \varphi|^2\right) \quad (A19)$$

This pressure is relative to the upstream conditions, which means that $C(t) = 0$. The first term in the bracket corresponds to the non-steady motion and the second to the steady one.

The force upon the sphere will be:

$$F = \int_\Sigma p\vec{n}d\vec{\sigma} = 2\int_{-\pi}^{\pi} p\pi R^2 \sin\theta \cos\theta d\theta \quad (A20)$$

The three components of the velocity along the versors $\vec{i}_r$, $\vec{i}_\theta$ and $\vec{i}_\psi$ are:

$$\varphi_r = u_F\left(1 - \frac{R^3}{r^3}\right)\cos\theta, \quad \varphi_\theta = -u_F\left(1 + \frac{R^3}{r^2}\right)\sin\theta, \quad \varphi_\psi = 0 \quad (A21)$$

Which results to be on the sphere surface $r = R$:

$$\varphi_r = 0, \quad \varphi_\theta = -\frac{3}{2}u_F\sin\theta, \quad \varphi_\psi = 0, \quad \frac{1}{2}|\nabla\varphi|^2 = \frac{9}{8}u_F\sin^2\theta \quad (A22)$$

It is clear that over the sphere there is only tangential velocity φ_θ. Therefore, the force due to the steady component turns out to be $F_s = 0$, which is according with the d'Alembert's paradox.

For the non-steady component,

$$\frac{\partial \varphi}{\partial t} = \frac{3R}{2}\dot{u}_F \cos\theta \quad (A23)$$

For the force we will have:

$$F_{nsF} = 2\pi R^3 \rho \dot{u}_F \quad (A24)$$

For the case of a sphere moving in a steady fluid, the potential velocity is:

$$\varphi(r,\theta) = -u_B\frac{R^3}{2r^2}\cos\theta \quad (A25)$$

Repeating the calculations, the force to maintain the sphere in an accelerated motion will be:

$$F_{nsB} = -\frac{2}{3}\pi R^3 \rho \dot{u}_B \quad (A26)$$

The result of A24 can be broken in two parts. One $\frac{4}{3}\pi R^3 \rho \dot{u}_F$ would correspond to the translation component of A18, that is $\dot{u}_F r \cos\theta$, and it is equivalent to the buoyancy of a sphere under the acceleration $\dot{u}_F$. The other would be $\frac{2}{3}\pi R^3 \rho \dot{u}_F$ comes from the doublet of the potential and turns to be obviously equal to accelerated motion whose result is given in A26. The value $\frac{2}{3}\pi R^3 \rho$ is known like virtual or added mass m_a. The reason of this name is because when a sphere of mass M is thrusted with a force F in a fluid, the dynamic equilibrium will be $F - m_a\dot{u} = M\dot{u}$, that is, equivalent to apply the force to a mass of $M + m_a$.

Annex II

Essay *Manuscript Correspondence*

Essay	Mss.		Essay	Mss.	
In-I			24	28	
In-II	1–2	(1)	25	29	
In-III	2–6	(1)	26	30	
In-IV			27	31	
In-V			28	32	
In-VI			29	33	(4)
In-VII			30	34	
1	7		31	33	(4)
2	8		32	35	
3	9		33	36	
4	10		34	37	
5	11		35	38	
6	12		36	39	(5)
7	13		37	40	
8	14		38	41	
9	15		39	42	
10	16		40	40	
11	17		41	44	
12	18		42	45	
13	19	(2)	43	46	
14			44		
15			45	47	
16			46	48	
17			47	49	
18			48	50	
19	20		49	51	(6)
20	21	(3)	50		
21	25		51	83b	(7)
22	26		52		
23	27		53		

(continued)

Essay	Mss.		Essay	Mss.	
54	80		96		
55	83c		97		
56			98		
57			99		
58	54		100	90	
59	55		101		
60	56–57		102		
61	58		103		
62	59		104	91	
63	60	(8)	105		
64	61		106	102	
65	63	(8)	107		
66			108	103	(14)
67			109	104	(15)
68			110	105	(16)
69			111		
70	62	(9)	112	114	
71	66		113		
72	67		114	115	
73			115	116	(17)
74			116	117	
75	73	(10)	117	116	
76			118	107	(18)
77	74		119	108	
78	75		120	109	+A
79			121		
80	76		122		
81			123	110	
82			124	111	+A
83			125		
84			126	112	(19)
85	77		127		
86	78		128		
87	79		129		
88	80, 83c	(11)	130		
89	82, 84		131		
90			132		
91	85		133		
92	86		134		
93	87 +A	(12)	135		
94	88 (13)		136		
95			137	92	

(continued)

Essay	Mss.		Essay	Mss.	
138	93	(20)	155		
139			156		
140			157		
141	93A		158		
142	94–95		159		
143	96		160		
144	97		161	21	(3)
145	98–101	(21)	162		
146			163	22	(22)
147			164	23	
148			165	24	(10)
149			166		
150			167		
151			168		
151			169		
153			170		
154			171		

Manuscript *Articles Not Included in the* Essay

50	
51	(6)
52	
64	
65	
68	
69	
70	
71	
72	
80	(11)
89	
81	
104	(15)
106	
113	
118	
119	

1. Enlarged and modified.
2. Some modifications. In *Mss.* Clairaut is mentioned, but not in the *Essay*.
3. Only the first lines of *Mss.* 21 pass to Essay. The rest goes to §.161–162, with some changes.
4. The §.31 only includes the last line of *Mss.*33.
5. The final paragraph of *Mss.*, related to *Mss.*52, is changed.
6. The §.49 consists only the last lines of *Mss.* 51.
7. The *Mss.*83 is divided in three parts as (a), (b) and (c). The (a) has not correspondence in the *Essay*, the (b) is the second part of §.51, and the (c) is the §.55.
8. Some minor changes.
9. Some changes.
10. Enlarged.
11. Only few lines from *Mss.*80 and a new argumentation substituting the *Mss.*83c.
12. Enlarged.
13. Enlarged with mathematical developments.
14. The points 5th and 6th of *Mss.*103 are not included in §.108.
15. The §.109 is almost new, only a formula comes from *Mss.*104.
16. The §.110 has a single statement that corresponds to the point 1st of *Mss.* 105.
17. The point 5th of *Mss.*115 is omitted in §.115.
18. The §.118 has many changes although with the same basic idea.
19. There are some differences.
20. The §.138 takes only the first few lines of *Mss.*93 and it is enlarged.
21. The last part of §.145 corresponds to *Mss.*101 but quite enlarged.
22. The references to Clairaut are omitted in the *Essay*.

Annex III

Notes About the Specimen Hydrodynamicum de Resistentia Corporum in Fluidis Motorum

Jakob Adami was a J. U. P. (*Juris Utriusque Doctor*, that is a Doctor in Canonical and Civil Law) as declared in the *Specimen*, but very little information about him is available. It seems that he was an amateur mathematician who lived in Aurich. He corresponded with Euler after 1746 and some of his letters have been preserved, in which Euler encouraged and appreciated his researches in hydrodynamics.[3]

This is not the place to make a review of this 66 page memoir written in Latin. We only will give a brief summary. Adami's theory is based in the live forces mechanics, using the potential ascent as a measure of the live force, in the sense expressed for Daniel Bernoulli in his *Hydrodynamica* [III.§.1]. Basically he

---

[3]Juškevič, p. 26.

Fig. 1 Body as a piston

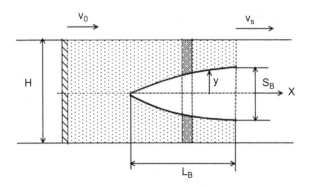

considers that the live force of the moving fluid is transferred to a body as a dead force, which is the resistance. Even more, the makes the different between the efficacious (*vis efficax*) and inefficacious (*vis inefficax*) dead force, the first one only is resisted by the inertia, and no effect is given by the second [§.III].

He starts assuming a body inside a channel whose fluid is pushed by a piston that moves accelerated (Fig. 1). The live force is transmitted to the layers and converted to efficacious force upon the body [§.VIII]. The resulting force turns out to be $H\frac{dw}{dx}(L_B - HE) + \frac{HS_B^2}{(H-S_B)^2}w$, where $E = \int \frac{dx}{H-y}$ and w is the potential ascent, $w = \frac{v_0^2}{2g}$ and $\frac{dw}{dx} = \frac{dv_0}{gdt}$. When the motion is steady the first term disappears. In this case if the piston is removed and the channel is full with moving fluid the results is the same [§.X]. He says that this force is due to the acceleration of the fluid in the narrowing channel between the body and walls, and that is necessary to add another force due to the single velocity, which is $\frac{HS_B}{2H-S_B}w$ [§.XIX]. Therefore, the total force will be $\frac{HS_B^2}{(H-S_B)^2}w + \frac{HS_B}{2H-S_B}w$ [§.XXII]. When the size of the body is infinitely small respect to the channel width, the former resistance is reduced to $\frac{1}{2}S_B w$ [§.XXIV]. According to this he proves that the resistance is equal to one-half of the weight of a column of fluid with the same base and height w [§.XXVIII], that is $\frac{1}{4}\rho S_B v_0^2$, equivalent to $C_D = 0.5$ in our present terminology. But if the velocity were not constant the amount $\rho V_B \frac{dv_0}{dt}$ must be subtracted of the former [§.XXX].

If the body has the shape as shown in Fig. 2, the fluid would be accelerated until the maximum body width and after decelerated to the initial velocity. In this case the resistance would be $H\frac{dw}{dx}(L_B + L_C - HE - HF) + \frac{HS_B}{2H-S_B}w$, where F is equivalent to E but applied to the afterbody [§.XXXIII].

Adam applies these formulas to the case of a globe [§.XXXI] in a proposition similar to the one presented by Newton in the *Principia*, Book 2, Prop. 39. He also

Fig. 2 Closed body

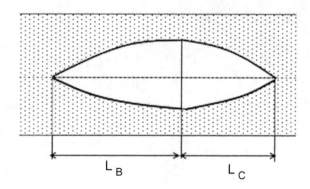

takes Newton's experiments [*Ibidem*, Prop. 40] about globes descending in a water tank and from the Saint Paul cathedral for applying the theory [§.XXXVIII–XLIX]. Something similar is done for a globe in horizontal motion, in order to compare with the Robins experiment as presented in the German translation of the *Gunnery* [§.LIV–LVIII]. Finally, he studies the problem of the efflux of fluid through a hole in a vessel, which is not a problem of resistance, but he mentions the Daniel Bernoulli experiment quoted in the *Hydrodynamica* [Sect. 3, Exp. Primum].

After this summary, we can conclude that Adam does not present a real theory but an application of the live force theories. Adam's *Specimen* can be considered as a sequel of the *Hydrodynamica*, while d'Alembert's *Essay* is a completely new approach to the resistance problem.

We finish quoting "Whoever Jakob Adami was, he has left no trace in the history of hydrodynamics".[4]

---

[4]Ibidem, p. 28. However, he presented another work to the Academy of Berlin in 1752: "De effluxu aquarum ex vasis, aperturam in latere habentibus: disquisitio mathematica".

Notes on the Translation and Manuscript

The original of the *Essay* used has been downloaded from the Bibliothèque national de France by the application *Gallica*. The copy has a stamp reading *Bibliothèque de l'Arsenal* and a handwritten letter from d'Alembert dedicated to the Marquis d'Argenson, which is reproduced below.

It seems plausible that this copy was one of the first ones and that he sent it to the Marquis as a personal sign of gratitude. D'Alembert's letter finishes with "on Friday 13". Looking at the 1752 calendar, the only Friday 13 was in October. This allows us to think that the *Essay* came to light in this month.

The *Essay* has been considered by some scholar as tortuous with careless wording,[5] for our part we only add that it is not easy to translate to English. We have tried our hardest to achieve clarity; however, we have kept a literary translation for some words, even when the text may look somewhat repetitive and awkward, in order to maintain the text closer to what could have been d'Alembert's thoughts and also his ambiguities, especially with reference to the physical concept of force and related matters. In this sense a basic glossary of those words is added here. Besides, sometimes for clarification we have made use of the Latin *Memoir* that d'Alembert had sent to the Berlin Academy 2 years before.[6]

In the original text of the *Essay* there are only two advertised errata, which we have corrected. However there are some more, and also several misprints which have all been corrected, although we have only underlined the major ones.

Another difficulty in understanding comes from the use of the mathematical symbols used. Sometimes the same symbol is applied to a physical variable as well as to a geometrical entity. Other times and quite often, the symbol for a variable

---

[5]We refer specially to Truesdell [1954], who apart of considering the work "tortuous and lengthy" (p. LI) also says that "the wording is even more careless than the average at the time" (p. LVII). Also Dugas, considers that "this memoir [seems] very arduous and very complex" (p. 12).

[6]We have followed the copy typed from the manuscript included in the Thesis of Gérard Grimberg. The translation from the Latin of the texts quoted has been done with the aid of Carlos Solís.

changes from one article to the next, and sometimes the same symbol is used for two different variables. We have noted these occurrences.[7]

Basic Glossary

French	English	Latin
choquer	to shock	ocurro
corps	body	corpus
corpuscule	corpuscle	corpusculum
élasticité	elasticity	elasticitas
équilibre	equilibrium	equilibrium
expérience	experiment	experimentum
finie	finite	finitus
force	force	vis
force accélératrice	accelerative force	vis acceleratrices
force centrifuge	centrifugal force	vis centrifuga
force suivant	force along	vis secundum
frapper	to strike	impingo
frottement	friction	frictio
globe	globe	globus
globule	globule	globulus
gravité	gravity	gravitas
impulsion	impulsion	impulsum
indéfinie	indefinite	indefinitus
mass	mass	massa
particule	particle	particula
pesant	heavy	gravis
poids	weight	pondor
pessanteur	weight	gravitas
pression	pressure	pressus
pression suivant	pressure along	pressus secundum
presser	to press	premo/conor
puissance	power	potentia
repos	at rest	quiescens
résistance	resistance	resistentia
ténacité	viscosity	tenacitas
tranche	slice	sectio
vase	vessel	vas
veine	stream	vena
vide	vacuum	vacuum

---

[7] We have to understand that the resources of the sub-index were not available yet.

Monsieur

Je vous dois sans doute des excuses d'oser vous dédier
cet ouvrage sans vous en avoir demandé la permission:
Mais, ou votre modestie n'auroit pas accepté mon hommage,
et je voulois me satisfaire; ou elle m'auroit interdit
tout éloge, et je voulois dire à mon aise la vérité. Je
vous prie d'être bien persuadé que de tout ce que j'ay écrit

ou que j'écrirai jamais, rien ne me sera plus cher et
plus précieux que les trois premieres pages de ce Livre
oserois je me flatter que vous voudrez bien les recevoir,
comme le present d'un Philosophe, & comme le seul
temoignage, mais le plus authentique que je puisse
vous donner du respect & de l'attachement inviolables
avec les quels je serai toute ma vie

     Monsieur

ce Vendredy 13   Votre très humble
       et très obeissant serviteur
         D'alembert

Monsieur

Je vous dois sans doute des excuses d'oser vous dédier cet ouvrage sans vous en avoir demandé la permission: Mais, ou votre modestie n'auroit pas accepté mon hommage, et je voulois me satisfaire; ou elle m'auroit interdit tout éloge, et je voulois dire à mon aire la verité, je vous prie d'etre bien persuadé que de tout ce que j'ay ecrit ou que j'ecrirai jamais, rien ne me sera plus cher et plus precieux que les trois premieres pages de ce livre oserois je flatter que vous voudrez bien les recevoit, comme le present d'un philosophe, et comme le seul temoignage, mais le plus authentique que je puisse vous donner du respect et de l'attachemens inviolable avec les quels je serai toute ma vie

Monsieur
Ce vendredy 13

Votre très humble et très obeissans serviteur, D'Alembert

Sir

Without doubt I owe you some excuses for daring to dedicate this work without having asked your permission. However, either your modesty would not have accepted my homage, and I wished to satisfy myself, or it would have forbidden me all praise , and I wished to speak the truth in my own way, I beg you to be persuaded that of all that I have written or that I will ever write, nothing will be more dear and more precious to me that the first three pages of this book I dare to pride myself that you will wish to receive them, as the present of a philosopher, and as the only testimony, but the most genuine that I can give you in this respect and of the inviolate attachment with which I will be all my life

Sir
This Friday 13
Your most humble and most obedient servant, D'Alembert

Bibliography

Primary Sources

Adami, Jakob. 1752. *Specimen Hydrodynamicum de Resistentia Corporum in Fluidis Motorum*, published as *Dissertation sur la Résistance des Fluides, qui a remporté le prix proposé par l'Académie Royal de Sciences et Beaux Lettres de Prusse, pour l'année MDCCL adjugé en MDCCLII*. Berlin: Libraires du Roi et de l'Académie.

Bernoulli, Daniel. 1729a. Disertatio de actione fluidorum in corpora solida et motu solidorum in fluidis. *Comm. Acad. Petrop* II: 1727.

———. 1729b. Theoria nova de motu aquarum per canales quoscunque fluentium. *Comm. Acad. Petrop* II: 1727.

———. 1732. Continuatio. *Comm. Acad. Petrop* III: 1728.

———. 1735. Experimenta coram societate instituta in confirmationem theoria pressionum quas latera canalis ab aqua transfluente sustinet. *Comm. Acad. Petrop* IV: 1729.

———. 1738a. Theorema de motu curvilineo corporum, quæ resistentiam patiuntur velocitatis sue quadrato proportionalem. *Comm. Acad. Petrop* V: 1730–1731.

———. 1738b. Additamentum. *Comm. Acad. Petrop* V: 1730–1731.

———. 1738c. *Hydrodinamica sive de viribus et motibus fluidorum commentarii*. Estrasburgo. English edition as *Hydrodynamics by Daniel Bernoulli & Hydraulics by Johann Bernoulli*. Nueva York: Dover Publications, Inc., 1968. German edition as *Des Daniel Bernoulli Hydrodynamik oder Kommentaire über die Kräfte und Bewegungen der Flüssigkeiten*. Munich: Veröffentlichungen des Forschungsinsttuts des Deutschen Museums für die Geschichte des Naturwissenschaften und der Technik. 1964.

———. 1741. De legibus quibusdam mechanicis, quas natura constanter affectat, nondum descriptis, earum usu hydrodinamico, pro determinanda vi venæ contra planum incurrentis. *Comm. Acad. Petrop* VIII: 1736.

———. 1747. Commentationes de statu æquilibrii corporum humido insidentium. *Comm. Acad. Petrop*. X, 1738. Traslation into French by Frans A. Cerulus, avaible in the Centre de Recherche en Histoire des Sciences, Université Catholique de Louvain, Belgium.

———. 1750. De motibus oscillatoriis corporum humido insidentium. *Comm. Acad. Petrop*. XI, 1739. Traslation to French by Frans A. Cerulus, avaible in the Centre de Recherche en Histoire des Sciences, Université Catholique de Louvain, Belgium.

Bernoulli, Jakob. 1693. De resistentia figurarum in fluidis motarum. *Acta Erud*. Leipzig.

———. 1705a†. Celeritates navis a quiete inchoatas usque ad maximam invenire. In *Opera*, Varia Posthuma.

———. 1705b†. Invenire veram legem, secundum quan aeris densitas decrecit. In *Opera*, Varia Posthuma.

———. 1716. Demostratio principii hydraulici de æqualitate velocitatis.... *Acta Erud.* Leipzig.

Bernoulli, Johann. 1692. Solution du Probleme de la Courbure que fait une voile enflée pour le vent. *Jour. Sav.*

———. 1700. De solido rotundo minimæ resistentia addenda iis quæ de eadem materia habentur in actiis an. super. Mens. Novemb. *Acta Erud.* Leipzig.

———. 1714. *Essay d'une nouvelle théorie de la manoeuvre des vaisseaux.* Basilea.

———. 1724. Discours sur le loix de la communication du mouvements. En *Opera Omnia.*

———. 1742a. *Opera omnia* (4 vol.). Genéve.

———. 1742b. *Hydraulica.* In Opera omnia, vol. 4th. The first part is also in Vol. IX de los Comm. Acad. Petrop., 1737 (1744) and the second in Vol. X, 1738 (1747). There is an English edition jointly with the Hydrodynamica by Daniel Bernoulli.

Borda, Jean Charles. 1763. Expériences sur la résistance des fluides. *Mém. Acad. Paris.*

———. 1766. Mémoire sur l'écoulement des fluides per orifices des vases. *Mém. Acad. Paris.*

———. 1767. Mémoire sur l'écoulement des fluides per orifices des vases. *Mém. Acad. Paris.*

———. 1768. Mémoire sur les pompes. *Mém. Acad. Paris.*

———. 1769. Sur la courbe décrite par les boulets et les bombes, en ayant égard à la résistance de l'air. *Mém. Acad. Paris.*

Bossut, Charles. 1777. *Nouvelles expériences sur la résistance des fluides.* Paris (with Condorcet and d'Alembert).

———. 1778. Nouvelles expériences sur la résistance des fluides. *Mém. Acad. Paris.*

Bouguer, Pierre. 1727. *De la mature des vaisseaux.* Paris.

Chapman, Frederik Henrik af. 1768. *Architectura navalis mercatoria.* Stockholm.

———. 1775. *Tractat om skepps-byggeriet.* Stockholm. There are two translations to French in 1779 and 1781, Paris and Brest.

Clairaut, Alexis Claude. 1743. *Théorie de la figure de la Terre, tirée des principes de l'hydrostatique.* Paris.

d'Alembert, Jean le Rond. 1744. *Traité de l'équilibre et du mouvement des fluides.* Paris.

———. 1747. *Réflexions sur la cause générale des vents.* Paris.

———. 1752. *Essai d'une nouvelle théorie de la résistance des fluides.* Paris.

———. 1758. *Traité de dynamique.* Paris. (1st edition 1743).

———. 1761. *Opuscules mathématiques,* vol. I.

———. 1768. *Opuscules mathématiques,* vol. V.

de La Hire, Philippe. 1702. Examen de la force necessaire pour mouvoir les bateaux tant dans l'eau dormante que courante.... *Mém. Acad.* Paris.

de Maupertuis, Pierre-Louis-Moreau. 1732. Of the Figures of Fluids, Turning Round an Axis. *Phil. Trans.* London.

Desaguliers, Jean Théofile. 1721. Experiment Relating to the Resistance of Fluids. *Phil. Trans.* London.

Euler, Leonhard. 1745. German translation and commentary on the *New principles of Gunnery* by B. Robins, Berlin (see Robins).

———. 1749. *Scientia navalis seu tractatus de construendis ac dirigendis navibus.* St Petersburg.

———. 1755a. Principes généraux de l'état d'équilibre des fluides. *Mém. Acad. Berlin* XI.

———. 1755b. Principes généraux du mouvement des fluides. *Mém. Acad. Berlin* XI.

———. 1755c. Continuation des recherches sur la théorie du mouvement des fluides. *Mém. Acad. Berlin* XI.

———. 1761. Principia motus fluidorum. *Novi Comm. Acad. Petrop* VI: 1756–1757.

———. 1773. *Théorie complète de la construction et de la manoeuvre des vaisseaux.* St Petersburg.

Huygens, Christiaan. 1669. Regles du mouvement dans la rencontre des corps. *Jour. sav.*

———. *Discours de la cause du Pesanteur.*

Juan y Santacilia, Jorge. 1771. *Examen Marítimo Teórico Práctico, ó Tratado de Mechánica aplicado á la Construcción, conocimiento y manejo de los navíos y demás embarcaciones.* Madrid.

Krafft, George Wolffang. 1741. Vi venæ aquæ contra planum incurrentis experimenta. *Comm. Acad. Petrop* VIII: 1741.
Lagrange, Joseph Louis. 1781. Mémoir sur la théorie de mouvement des fluides. *Nouv. Mém. Acad. Berlin*.
———. 1784/1785. Sur la percussion des fluides. *Mém. Acad. Sci. Turín* I.
———. 1788. *Mécanique Analitique*. Paris.
MacLaurin, Colin. 1742. *A treatise on fluxions*. Edimburg.
Mariotte, Edmé. 1686. *Traité du mouvement des eaux*. Paris.
Newton, Isaac. 1726. *Philosophia naturalis principia mathematica*. First ed. London, 1687; second ed. Cambridge, 1713; third ed. London.
Parent, Antoine. 1704. Sur la plus grande perfection possible des Machines. *Mém. Acad. Paris*.
Prandtl, Ludwig. 1928. Über Flüssigkeitsbewegung bei sehr kleiner Reibung, Third Mathematical Congress, Heidelberg, 1904. Translatated to English as *Motion of Fluids with very little Viscossity*, NACA TM 452.
Robins, Benjamin. 1783. *New Principles of Gunnery*. London, 1742. German translation commented by Euler: *Neue Grundsätze der Artillerie, aus dem Englischen des Herrn Benjamin Robins und mit vielen Anmerkungen versehen*, Berlin, 1745. Euler's comments were translated back to English by Hugh BROWN, London, 1777 as *The true principles of gunnery investigated and explained...*". There is also a translation of the German into French as *Nouveaux principes d'artillerie*. Paris.
Smeaton, John. 1759. An experimental enquiry concerning the natural powers of water and wind to turn mills, and other Machines depending on a circular motion. *Phil. Trans.* 51, London.
Torricelli, Evangelista. 1644. De motu gravium. In *Opera Geometrica*. Florence: Masse & de Landis.
van 's Gravesande, Willen Jacob. 1739. *Mathematical Elements of Natural Philosophy*. London.
———. 1749. *Physices Elementa Mathematica Experimentis Confirmata*. 3rd ed. Leiden.
van Musschenbroek, Pieter. 1739. *Essai de Physique*. Leyden.
———. 1741. *Elementa Physicæ conscripta in usus Academicos*. Leyden.

Secondary Sources

Bradley, Robert E. 2007. Euler, d'Alembert and the logarithm function. In *Leonhard Euler: Life, Work and Legacy*, ed. Robert E. Bradley and C. Edward Sanfifer. Amsterdam: Elsevier.
Crépel, Pierre. 2006. Les dernières perfidies de d'Alembert. *Math. Sci. hum.~ Mathematics and Social Sciences* (44e année, 176(4): 61–87).
Darrigol, Olivier. 2005. *Worlds of Flow. A History of Hydrodynamics from the Bernoullis to Prandtl*. Oxford: Oxford University Press.
Darrigol, Olivier, and Uriel Frisch. 2008. From Newton's mechanics to Euler's equations. *Physica D: NonlinearPhenomena* 237: 1855–1869.
D'Alembert. 1749. *Theoria resistentiae quam patitur corpus in fluido motum, ex principiis omnino novis et simplissimis deducta, habita ratione tum velocitatis, figurae, et massae corporis moti, tum densitatis & compresionis partium fluidi*. Manuscript presented at the Academy of Sciences of Berlin.
Demidov, Serghei S. 1982. Création et développment des équations différentielles aix dérivées partielles dans les travaux de J. d'Alembert. *Revue d'histoire des sciences*, vol. 35-1. Paris.
Dugas, René. 1950. *Histoire de la Mécanique*. Paris.
———. 1952. D'Alembert et l'essai d'une nouvelle théorie sur la résistance des fluides. *Revue française de mécanique* 6: 6–12.
———. 1954. *La mécanique au XVII^{e}siècle*. Neuchatel: Ed. du Griffon.
Eckert, Michael. 2006. *The Dawn of Fluid Dynamics*. Weinheim: Wiley-VCH Verlag.

Emery, Monique, and Pierre Monzani, eds. 1989. *Jean d'Alembert, savant et philosophe: portrait à plusieurs voix.* Paris: Édition des Archives Contemporaines.

Firode, Alain. 2001. *La Dynamique de d'Alembert.* Montreal/Paris: Bellarmin/VRIN.

Fraser, Craig. 1985. D'Alembert Principle: The Original Formulation and Application in Jean d'Alembert's *Traité de Dynamique. Centaurus* 28 (Part 1): 31–61. Part 2, pp. 145–159.

Grimberg, Gérard. 1995. D'Alembert et les équations d'Euler en Hydrodynamique. *Revista da SBHC* 14: 65–80.

———. 1998. *D'Alembert et les équations aux dérivées partielles en hydrodynamique.* Doctoral thesis, University of París 7, Denis Diderot.

Grimberg, G., W. Pauls, and U. Frisch. 2008. Genesis of d'Alembert paradox and analytical elaboration of the drag problem. *Physica D: Nonlinear Phenomena* 237 (14–17): 1878–1886.

Guilbaud, Alexandre M. 2007. *L'hydrodynamique dans l'oeuvre de D'Alembert 1766–1783 : histoire et analyse détaillée des concepts pour l'édition critique et commentée de ses OEuvres complètes et leur édition électronique.* Thèse de doctorat. Université Claude Bernard, Lyon-1.

Hankins, Thomas L. 1967. The reception of newton's second law of motion in the eighteenth century. In *Archives internationales d'histoire des sciences,* 78–79.

———. 1968. Prologue to the *Traité de Dynamique.* New York: Johnson Reprint Corporation.

———. 1970. *Jean d'Alembert. Science and the Enlightenment.* Oxford: Oxford University Press.

Juškevič, Adolf P, and René Taton. 1980. La Correspondance de Leonhard Euler, Introduction to the *Commercium Epistolicum* Vol. 4A/5 of the *Leonhardi Euleri Opera Omnia.* Birhäuser Springer.

Kline, Morris. 1972. *Mathematical Thought from Ancient to Modern Times.* New York: Oxford University Press.

Lafuente, Antonio, and Mazuecos, Antonio. 1987. *Los caballeros del punto fijo.* Barcelona: Ed. Serbal/CSIC.

Maugin, Gérad A. 2014. *Continuum Mechanics Thoough the Eighteenth and Nineteenth Centuries.* Berlin: Springer.

Neuser, Wolfang. 1993. The concept of force in eighteenth-century Newtonian mechanics. In *Hegel and Newtonianism,* ed. Michael John Petry, 383–397. Dordrecht: Springer.

Rouse, Hunter, and Simon Ince. 1963. *History of Hydraulics.* New York: Dover.

Ru, Véronique Le. 1994a. *Jean Le Rond d'Alembert philosophe.* Paris: Mathesis VRIN.

———. 1994b. La force accélératrice: un exemple de définition contextuelle dans le 'Traité de Dynamique' de d'Alembert. *Revue d'Histoire des Sciences* 47 (3): 475–494.

Sellés, Manuel A. 2006. Infinitesimals in the foundations of Newton's mechanics. *Historia Mathematica* 33: 210–223.

Simón Calero, Julián. 1996. La aparición histórica de la paradoja de d'Alembert, *Ingeniería Aeronáutica y Astronáutica,* n 343. Madrid.

———. 2001. La Mecánica de los Fluidos en Jorge Juan, *Asclepio,* vol. 53 (2), CSIC, Madrid.

———. 2008. *The Genesis of the Fluid Mechanics, 1640–1780.* Dordrecht: Springer.

Tokaty, Grigori Alexsandrovich. 1994. *A History and Philosophy of Fluid Mechanics.* New York: Dover. (First ed., 1971).

Truesdell, Clifford A. 1954. Rational Fluid Mechanics (1687–1765). Editorial introduction to vol. 12(2) of the *Leonhardi Euleri Opera Omnia.* "Societatis Scientarum Naturalium Helveticæ". Lausanne.

———. 1955. Rational Fluid Mechanics (1765–1788). Editorial introduction to vol. 13(2) of *the Leonhardi Euleri Opera Omnia.* "Societatis Scientarum Naturalium Helveticæ". Lausanne.

Westfall, Richard S. 1971. *Force in Newton's Physics. The Science of Dynamics in the Seventeenth Century.* New York: American Elsevier.

Index

A
Académie Royal des Sciences de Paris, 95, 136, 144, 176, 260
Académie Royale des Sciences et Belles Lettres de Prusse, 15, 76, 95, 101, 106, 260
Academy of Petersburg (Academia Sciemtarum Imperialis Petropolitanæ), 14, 63, 80, 107
Adami, Jakob, 136, 280, 282

B
Bernoulli, Daniel, 10, 11, 63–66, 79, 103, 107, 109, 138, 142, 148, 159, 163, 170, 224, 231, 233, 234, 236, 237, 240, 241, 247, 258, 280, 282
Bernoulli, Jakob, 141
Bernoulli, Johann, 91, 94, 137, 141–143, 150, 156, 258, 260
Borda, Jean-Charles, 145, 236–237
Bossut, Charles, 145, 237
Bouguer, Pierre, 143–145, 223, 225, 232, 257
Bradley, Robert, E., 136

C
Calero, Julián Simón, 139, 205, 247
Chapman, Frederik Henrik, 145
Clairaut, Alexis, 26, 130, 144, 162, 176, 178, 269, 270, 280

D
D'Alembert, Jean Le Rond, 135–139, 142, 144, 147–151, 153–159, 161–166, 169, 170, 173–181, 183–185, 187, 190–193, 197–200, 202–209, 211–213, 216–229, 231–238, 240, 241, 243, 246–254, 257–259, 262, 263, 265–268, 270, 276, 282, 283
Darrigol, Olivier, 147, 151, 155, 206
Descartes, René, 176
Dugas, René, 138, 220, 283

E
Euler, Leonhard, 64, 135–139, 143–145, 148, 200, 206, 220, 226, 234, 235, 246, 258, 263, 273, 280

F
Ferreiro, Larrie, D., 135, 257

G
Grimberg, Gérard, 135, 137, 144, 194, 198, 205, 246, 267, 283
Grischow, Augustin Nathanaël, 135, 136
Guilbaud, Alexandre M, 155, 186

H
Hamel, Jean-Baptiste du, 247
Hankins, Thomas L, 150
Hire, Philippe de la, 141
Huygens, Christiaan, 139, 140, 142, 144, 233

J
Juan y Santacilia, Jorge, 145

K
Kies, Johann, 135
Kline, Morris, 201
Krafft, George Wolffang, 105, 106, 109, 231, 236, 249, 250, 252

L
Lafuente, Antonio, 176
Lagrange, Joseph Louis, 145, 209
Lamb, Horace, 176

M
Mariotte, Edmé, 62, 63, 66, 139, 141–143, 159, 231–233
Maugin, Gérad A., 156

N
Newton, Isaac, 6–10, 12, 66, 138–144, 149–151, 157, 159, 170, 175, 176, 190, 219, 221, 223, 226, 227, 231–234, 236, 237, 240, 241, 269, 281

P
Parent, Antoine, 141
Prandtl, Ludwig, 206

R
Robins, Benjamin, 87, 137, 143, 145, 159, 206, 224, 226, 227, 231, 236, 282
Rouse, Hunter, 194
Ru, Veronique Le, 148

S
Sellés, Manuel, 135, 150
s'Gravesande, Willem Jacob, 64, 66, 75, 159, 222, 231, 234–236
Smeaton, John, 145
Solís, Carlos, 283

T
Torricelli, Evangelista, 141
Truesdell, Clifford, 136, 137, 154, 155, 166, 179, 198, 200, 209, 220, 267, 283

W
Westfall, Richard S., 150

CPSIA information can be obtained
at www.ICGtesting.com
Printed in the USA
LVHW081035090619
620619LV00010B/912/P

9 783319 885292